ENVIRONMENTAL IMPACTS OF ELECTRIC VEHICLES

ENVIRONMENTAL IMPACTS OF ELECTRIC VEHICLES

Prof. Dr. Mukesh Pandey

Delve
PUBLISHING

www.delvepublishing.com

Environmental Impacts of Electric Vehicles

Prof. Dr. Mukesh Pandey

Delve Publishing
4164 Lakeshore Road
Burlington, ON L7L 1A4
Canada
www.delvepublishing.com
Email: orders@arclereducation.com

ISBN: 978-1-77956-424-5 (Hardcover)

ABOUT THE AUTHOR

Prof. Dr. Mukesh Pandey is multi-skilled professional with around 30 years of rich exposure and versatile experience in Industries, Administration, Academics , Research domains and institution building. He received his B.E. from SATI Vidisha, M.Tech. from N.I.T. Bhopal and Ph.D. from RGPV, Bhopal. He played an instrumental role in the establishment of Rajiv Gandhi Technical University Bhopal, Madhya Pradesh since it's inception in 1999. He served as Rector(Pro Vice Chancellor), Director SoEEM & Dean of Faculty of Energy Technology. He has looked after as Coordinator for establishing IIIT, Bhopal (under PPP Mode by MHRD). He held various administrative positions as Director (R&D), Coordinator of RGPV- NBA Nodal Centre, Dy. Registrar, Member of Engineering Accreditation Evaluation Committee [EAEC] and Moderation Committee of NBA(National Board of Accreditation). Professor Pandey works in the area of Renewable Energy and Environment having focus on Solar Thermal Energy and he has developed an Energy Technology Park at RGPV, Bhopal by making use of the renewable energy resources; Solar Roof Top Plant on University Building, Energy Systems, Solar Thin Film systems, Solar wind Hybrid System, Bio-Diesel Reactor, CO2 Carbon Sequestration Unit, Dual Rotor Wind Turbine etc. Professor Pandey has an iconic track record of quality teaching, innovation and research he has been Principal Investigator of many Govt. Funded Projects by DST, MNRE, MPCST, AICTE etc including Principal Investigator of Innovative & Breakthrough Technology based 30KW CL-CSP SOLAR international R&D PROJECT (India-Japan Joint Venture Project) installed in RGPV campus. He has also facilitated as Reviewer for many national and international journals of repute and associated with many academic and research organizations at various levels. Over the past several years he has taught M.Tech. and Ph.D. Research Scholars in the areas of Energy, Environment, Direct Energy Conversion and Integrated Energy Systems and has been key-note speaker and resource person at several International and National Conferences and programmes.

TABLE OF CONTENTS

LIST OF FIGURES

LIST OF TABLES

PREFACE

Electric vehicles (EVs) offer a range of benefits that make them an attractive choice for many people. They are not only easier to drive but also have lower running costs compared to traditional vehicles. However, what truly sets EVs apart is their eco-friendly nature. They provide a more sustainable option for those looking to shift away from conventional fossil fuel-powered cars. In today's world, the transportation industry has a significant impact on the environment—contributing to resource depletion, air pollution, and climate change.

EVs present a promising solution to address these pressing environmental issues and help us move toward a more sustainable future. One of the key advantages of EVs is that they require fewer natural resources for their production compared to traditional vehicles. This shift supports the use of renewable materials and helps reduce the strain on our planet's resources. Additionally, when EVs are charged using electricity generated from renewable sources—like solar or wind power—they can significantly cut down on greenhouse gas emissions. This reduction in emissions helps to mitigate the harmful effects of climate change.

For EVs to fully realize their environmental benefits, a collaborative effort is needed. It's not just about producing more electric vehicles; we also need to invest in the infrastructure that supports them. This includes building more charging stations powered by clean energy sources and encouraging policies that support the adoption of EVs. Public awareness and understanding of how EVs impact the environment are also crucial. When people are informed about the benefits of EVs and the importance of using renewable energy, they are more likely to make choices that contribute to a greener planet. Given the urgent need for sustainable transportation solutions, understanding the environmental impact of EVs is more important than ever. By embracing electric vehicles and supporting the transition to clean energy, we can make a significant difference in reducing our environmental footprint and ensuring a healthier future for everyone.

This textbook, *Environmental Impacts of Electric Vehicles*, is structured into 10 chapters, each focusing on a critical aspect of EVs and their environmental implications.

Chapter 1 provides an introduction to electric vehicles, outlining their basic principles, technological advancements, and the environmental motivations behind their adoption. It sets the context for understanding the various environmental impacts discussed in subsequent chapters.

Chapter 2 examines the life cycle assessment (LCA) of electric vehicles. This chapter evaluates the environmental impacts associated with each stage of an EV's life, from production through use and disposal. It offers a comprehensive view of the overall ecological footprint of EVs.

Chapter 3 analyzes the manufacturing and supply chain impacts of electric vehicles, focusing on environmental challenges related to production processes and the logistics involved in manufacturing EVs, with emphasis on issues like resource extraction and emissions.

Chapter 4 addresses battery production and associated environmental concerns, discussing the implications of producing EV batteries, including raw material extraction, manufacturing emissions, and the challenges of managing battery waste.

Chapter 5 explores energy consumption and the carbon footprint of electric vehicles, assessing how EV use influences overall energy consumption and greenhouse gas emissions, and evaluating whether EVs fully realize their environmental benefits based on energy sources.

Chapter 6 covers the end-of-life and recycling of EV batteries. It discusses the processes involved in battery disposal and recycling, examining strategies to mitigate environmental impacts and improve recycling efficiency.

Chapter 7 evaluates the impacts of electric vehicles on air quality and human health. It compares the effects of EVs with traditional vehicles, assessing how the transition to EVs influences urban air quality and public health.

Chapter 8 addresses water and resource use in EV production, exploring water and other resource consumption during the manufacturing of electric vehicles, and highlighting the environmental pressures associated with these processes.

Chapter 9 reviews the environmental impacts of EV charging infrastructure, considering factors such as energy use, land requirements, and the development and operation of charging stations.

Chapter 10 examines policy, regulation, and environmental standards, discussing the role of government and regulatory frameworks in shaping the environmental performance of electric vehicles and offering policy recommendations to enhance their sustainability.

Offering valuable insights for researchers, policymakers, and industry professionals, this textbook provides a nuanced understanding of how EVs affect the environment and the role of regulatory frameworks in enhancing sustainability.

Chapter 1

Introduction to Electric Vehicles

Contents

INTRODUCTION

The internal combustion engine (ICE) vehicle is being replaced by an environmentally friendly option in the transportation industry by electric vehicles (EVs). Because EVs run entirely on battery power, they drastically lower air pollution and greenhouse gas emissions since they have no tailpipe emissions. Improving urban air quality and halting climate change depend on this change. While worries about range anxiety are decreasing due to the expansion of charging infrastructure, advancements in battery technology have increased the range of EVs, making them more feasible for daily use. Subsidies, tax breaks, and tighter emissions regulations are being used by governments across the globe to encourage the adoption of electric vehicles. Major automakers are also committing to electrify their fleets, indicating that the market may eventually be dominated by electric vehicles. Since EVs have fewer moving parts and don't require oil changes, they are more cost-effective in the long run even though they have higher initial costs. Moreover, advances in the integration of renewable energy are enabling EVs to be charged with clean energy sources, augmenting their environmental advantages. With ongoing technological advancements, electric vehicles are becoming more accessible, economical, and efficient for a larger range of people. There are still issues, like battery recycling and the requirement for a more extensive charging network, particularly in rural areas, but the popularity of EVs indicates that these problems will be solved. In addition to being a fundamental technological advance, the switch to electric cars marks a significant advancement toward a more resilient and sustainable future in which transportation promotes environmental sustainability rather than depletes it. It is anticipated that the use of electric vehicles will increase as awareness and technology develop, opening the door to a greener, cleaner transportation network.

1.1. BASICS OF ELECTRIC VEHICLE

The needs of different drivers can be accommodated in the plug-in electric vehicles (EVs or electric cars) currently available in the market like the conventional vehicles which have varied kinds of technology available to them. An electric power source that is off-board can be used by the drivers to plug in and charge their EVs, and this is one of the major features of these vehicles.

Herein they become different from the hybrid electric vehicles (HEVs) where the vehicle cannot be plugged in and a battery power supplements the internal combustion engine (ICE). EVs are basically of two types: Plug-in hybrid electric vehicles (PHEVs) and all-electric vehicles (AEVs). Further, fuel cell electric vehicles (FCEVs) and battery electric vehicles (BEVs) are included in the category of AEVs. Energy is lost during braking, and some of this energy is used to generate electricity by regenerative braking, which along with charging from the electric grid is used to generate both these kinds of vehicles.

Depending upon the driving habits and individual needs a specific kind of vehicle may suit the individual. As per one's requirements, the kind of PHEVs and BEVs best suited can be opted for. Only electricity is used to run AEVs.

Even though some of the luxury models have a range of up to 250 miles, most AEVs have a range of 80–100 miles. Depending upon the kind of battery and charger used, with Level 1 charging, it can take anything from 30 minutes to an entire day to charge up a depleted battery.

It may be a better choice to use a plug-in electric vehicle or a PHEV where this range does not seem to be sufficient. When the battery gets depleted for shorter ranges (6–40 miles) the PHEV switches over from running on electricity to a gasoline-run ICE.

PHEVs give the flexibility to the drivers to switch over to gasoline after fueling their vehicles as per requirement and as often as possible use electricity. Compared to conventional vehicle electricity-powered vehicles (from the grid) help to reduce tailpipe emissions, reduce fuel costs and cut the consumption of petroleum. PHEVs work like HEVs when the driving distances exceed the all-electric range so that fewer emissions are produced and at the same time, the fuel consumption too is lesser than that for similar conventional vehicles.

At other times like when air conditioning is being used or whilst rapid acceleration is being done, as per the model in use, the vehicle may also be powered by the internal combustion. Instead of gasoline, other alternative fuels too may be used in PHEVs like biofuels or hydrogen in a fuel cell (FC). Due to their lower dependency on oil and fewer emissions, EVs and more specifically the plug-in vehicles are gaining popularity. By 2022 over the globe the number of EVs is likely to be over 35 million, and by 2024, their sales per year would, in all probability reach 2.4 million in U.S. alone

All the same transformer and branch congestion is likely to result from the large impact on the power grid that shall be a consequence of the large amount of demand in charging these vehicles once they penetrate deeply. Associating the charging infrastructure with local power generation like renewable energy sources (RESs) can be an effective solution to bring about a mitigation of this impact. For instance, Nissan Leaf charging is provided with solar power by SunPower and solar power system is provided to Tesla EVs by SolarCity. There is a degree of fun, speed, and efficiency associated with plug-in EVs. At the same time, they are cheaper and easier to maintain. Simple plugging can help to 'refuel' the vehicle saving about $1.50 per gallon of gas in equivalent terms.

The carbon emissions reduce and whilst increasing our energy independence EVs help in contributing towards a healthier air. The best time to take advantage of the incentives being provided on the purchase of EVs is now as it reduces the price of these vehicles (International Energy Agency, 1999; Ayodele and Mustapa, 2020).

1.1.1. Types of Electric Vehicles

Electric vehicles that can be plugged into a series of sizes and shapes. Electricity powered options have a large variety, and these are very efficient, like neighborhood electric vehicles, bicycles assisted by electricity, and motorcycles. There are two basic designs of EVs for the Vermonters' trucks and passenger cars.

1.1.1.1. Battery Electric Vehicles (BEVs)

The typical architecture of BEVs consists of mainly three parts namely: rechargeable battery, controller, and electric motor. A rechargeable battery is utilized by the electric motor as a source of energy for the generation of propulsion.

The electric motor drives the vehicle so that it can move either forward or backward, and the power for this is supplied through the management by a two-quadrant controller. A four-quadrant controller is the one that can also support regenerative braking.

The inverter is yet another important part of a BEV. DC power or the power that the battery stores is converted into AC (alternating current) power by an inverter as AC motors that are cheaper and have simple schemes equip most of the electric motors (EMs).

1.1.1.2. All-Electric Vehicles

Instead of using internal combustion engines, all-electric cars are propelled only by one or more electric motors that draw energy from batteries. Regenerative braking is one method of charging the batteries in addition to plugging the car into an electric power source. While there are "life cycle" emissions related to the generation of electricity, all-electric cars do not emit any tailpipe emissions. Generally speaking, all-electric vehicles have a lower driving range per charge than conventional vehicles do per gas tank. Depending on the model, the majority of modern BEVs are built to go anywhere from 110 to more than 300 miles on a fully charged battery. To put this in perspective, in U. S. Less than 100 miles is covered on household travels. The range of an all-electric car varies depending on the road and the driver's habits. Because the battery powers the climate control systems in addition to the motor, extreme temperatures tend to reduce range. Range can also be lowered by speeding, aggressive driving, and carrying large loads.

1.1.1.3. Plug-In Hybrid Electric Vehicles (PHEVs)

In PHEVs, a conventional engine is powered by gasoline and an electric motor is powered by batteries. The PHEV's batteries can be charged in a few different ways, including by using regenerative braking and the conventional engine in addition to plugging the vehicle into an electric power source. Depending on the model, PHEVs have an all-electric driving range of between 15 and more miles. A PHEV can get the majority of its power from electricity for regular daily driving, provided the battery is charged. After that, the engine will start when the battery is almost completely dead, during hard acceleration, at high speeds, or when a lot of heating or cooling is needed. PHEVs emit no emissions from their tailpipes when they are solely powered by battery power. PHEVs often use less gasoline and emit fewer pollutants than comparable conventional cars, even when the conventional engine is operating.

1.1.2. Electric Vehicle Charging

Potential users have many questions regarding the charging of BEVs. We will go over the fundamentals of charging and its operation.

1.1.2.1. Types of Chargers

The plug itself and the location of the power conversion are two ways to categorize chargers. AC and DC charging are the two forms of power conversion. The power that originates from the grid itself is known as alternating current, or AC. AC chargers require the conversion of power from AC to DC (Direct Current), which is accomplished by the on-board converter within the vehicle. The battery receives the DC power after that, which powers the EV's fuel.

Because AC chargers have a lower required power range (between 16A and 63A), they are appropriate for charging sessions lasting at least 20 minutes. Since AC chargers are less expensive to produce, install, and charge, they are the more widely used type of charger. Since the converter for DC chargers is located inside the charger, the car will receive a DC charge without requiring internal conversion. Because DC chargers require more power, their installation and manufacturing costs are higher. For drivers who are always on the go, though, the charging process is substantially quicker. Regarding plug types, the kind needed for electric vehicles may vary depending on the manufacturer or the area. By examining the different kinds of chargers that are available in their area, users can make sure that their charging routines are efficient. This could have an impact on the kind of car they decide to buy. Although the number of chargers is rising yearly, the lack of standardization continues to irritate both the industry and drivers. Countries are making an effort to standardize charging methods, even though there isn't yet a single standard that applies to all nations and manufacturers.

1.1.2.2. Charging Time and Range

Different chargers are available for BEVs based on the manufacturer and location. Similar to this, the vehicle as well as the charger or charging technique being used can affect how long it takes to get a full charge. For instance, it may take up to 15 hours to fully charge a 63kWh battery in a stand-alone BEV model using a residential charger at 240V, depending on the kW. Nonetheless, the identical charge can be obtained in less than an hour using commercial chargers, which frequently have a higher kilowattage.

For those who drive ICE and are used to waiting only a few minutes at most for their gas tanks to fill up, this can be intimidating. Fortunately, EV owners have access to a similar technique: DC fast chargers have a charging capacity of 150–400kW, meaning an 80% charge can be reached in 8–20 minutes.

Vehicles that are driven long distances typically require charging two or three times per week from their drivers. However, the majority of EV users favor the top-up approach, in which they park in locations with easy access to charging infrastructure. Their car can charge as much as it can in the allotted time while they go about their daily business. This eliminates any possible range anxiety. However, as the popularity of BEVs has grown, producers and operators of charging stations have worked to expand the infrastructure available for charging, create more user-friendly battery swapping technology, and increase the range of the latest BEV models.

Because batteries and chargers vary depending on the manufacturer and location, it is challenging to determine the precise range a BEV can travel on a full charge. Nonetheless, EVs these days can travel 200–500 kilometers on a single charge, depending on the model of the car. For instance, the Tesla Model 3 LR has a 560-kilometer range before requiring a recharge. As EV battery research and development progress, this distance should rise.

1.1.2.3. Charging Costs

The cost savings that come with owning an EV is one advantage that many EV users highlight. Electric vehicles are generally more affordable in the long run due to their lower charging costs, even though the vehicle itself can cost a little more than a traditional internal combustion engine vehicle. According to the US. Department of Energy, assuming that electricity costs $0.13 per kilowatt-hour (the national average), charging an EV with a fully-depleted 66 kWh battery will cost about $9 to reach 100% charge. Costs can vary because the average domestic electricity rate in the UK is £0.14, or roughly $0.20, while the average rate in the European Union is €0.21, or roughly $0.26. Additionally, due to government support and subsidies, along with the lower cost of charging, EVs may ultimately prove to be a more cost-effective choice when buying a car than gas-guzzling cars (International Energy Agency, 1999; Ayodele and Mustapa, 2020).

1.1.3. Electric Vehicles as-a-Service

In mid-2020, McKinsey released a report stating that although EV sales increased by 65% in 2017 and 2018 over the course of 2018, they only grew by 9% in 2019. Following that, EV sales fell by 25% in the first quarter of 2020. Undoubtedly, the global COVID-19 pandemic is partly to blame for some of this, but overall, the rise in vehicle ownership is beginning to slow down.

This suggests that the current generation of drivers and passengers is increasingly choosing to use mobility services and subscriptions in place of buying their own cars. Although MaaS is not a novel idea, an increasing number of businesses are investigating the electrification of mobility services as transportation trends shift away from the conventional ICE vehicle. Another reason for the shift to electric MaaS business models is policy. Manufacturers are under pressure to produce more zero-emissions vehicles and services to comply with new regulations as more states, regions, and even countries are enforcing bans on ICEs.

One of the main selling points of the MaaS business model is that all services are managed from a single platform, which gives the service provider a smooth platform for management and the customer a simple platform for payment and user interface. An FMS, or fleet management system, is in charge of managing a large number of these mobility service platforms. While using a mobility service to find a ride may appear like a straightforward process to the average passenger, there are actually several parties involved and data exchanges taking place. A unified system is required for a seamless service experience when security precautions and payment/charges are added. This is where the FMS comes in. To help the service operator run their business more effectively, the FMS can gather vehicle data, analyze that data, and then provide the diagnostics. It can manage payment methods, guarantee verification of certificates from all parties, and optimize all features. The need for fleet management solutions has increased dramatically over the last several years, primarily due to the growth of the mobility ecosystem. With the use of alternative fuels, cars will continue to develop and may even become more autonomous, and mobility-as-a-service will undoubtedly keep up.

1.1.4. Challenges of Electric Vehicles

Even though EV trends are currently going in a positive direction, there are still difficulties with EVs. These are the top three problems facing the EV industry.

1.1.4.1. Range Anxiety

The feeling that EV users get when they worry that they might not have enough charge to reach their destination or the next charging station is known as "range anxiety," as was previously mentioned. Automotive manufacturers are aware of this significant obstacle and are taking steps to address it. The range will undoubtedly grow over time as battery technology and engine efficiency improve. The automotive sector will have to reassure prospective drivers about the expanding ranges to increase their willingness to switch from what they know to something new.

1.1.4.2. Charger Challenges

Many of the common problems experienced by EV drivers are related to the charger or the charging procedure itself.

1.1.4.2.1. Charging Times and Pricing

The cost of charging an electric car varies depending on the time of day since the vehicle draws power from the grid. For instance, the afternoon and early evening are regarded as peak hours, so charging a large vehicle during these times is not recommended. Charging during off-peak hours helps reduce this, but it takes far more deliberate action than just filling up the car when it's almost empty.

Vehicle-to-grid, or V2G, technology is being considered by many automotive experts as a potential solution to this problem. In this system, the vehicle would not only receive power from the grid but also contribute to it. The benefits of this kind of bidirectional charging system are twofold: the car would recoup some of its energy during peak hours and pay for the charge during off-peak hours. Although the technology hasn't made a significant impact on the market just yet, many automakers have been working to improve their cars' compatibility with bidirectional charging because they think it could change the game.

1.1.4.2.2. Charger Types

If you are one of the many of us who still drive an ICE, you probably don't think twice about where you get your fuel. Although cost may influence our choice, all gas stations will essentially sell the same fuel, and all gas pumps are compatible with all internal combustion engines.

With EVs, though, that's not always the case. As previously mentioned, the kind of fast DC charger varies depending on the manufacturer; consequently, even if there is a nearby charging station, the charger might not work with your car. It will take some time to resolve this issue because changing the infrastructure for the different ports will be more involved. However, experts believe that CCS and CHAdeMO may eventually converge into the current CCS standard, which is supported by the European Union. As the only company with a proprietary charger, Tesla may or may not find this to be the case. Tesla does, however, provide adapters for CHAdeMO and CCS plugs.

However, the problem with chargers is only one part of the bigger picture that consumers must deal with when it comes to the difficulties of standardization and interoperability in the electric vehicle sector. The software that oversees all the data from the cars, chargers, and charging services, as well as the interoperability of systems management and the communication that takes place between the vehicle and the charger, the charger and the charging network, are the bigger problems.

Many are looking to Plug and Charge (or PnC) as the next standard of vehicle-to-grid communications because infrastructure interoperability issues can be inconvenient for drivers, vehicle manufacturers, charge point operators, and service operators (often referred to as mobility operators).

The ISO 15118 standard was created in 2014 with the intention of improving driver convenience during the charging process. The PnC technology is managed by the standard, which makes it possible for drivers to plug in and charge their cars without the need for an RFID card or a mobile application. This makes the process extremely smooth. Users and service providers would benefit from both front-end and back-end interoperability by adhering to the ISO 15118 standard and related protocols such as the Open Charge Point Protocol. Since drivers are no longer restricted to a single charging network, this can be a major selling point for EV adoption, especially for those in the US or the EU who must travel across state or national borders.

1.1.5. Securing Electric Vehicles

The phrase "electric vehicle" conjures up images of plugging in and using chargers. But it's about much more than the charger alone. Information is transferred between the EV, the charging station, and the mobility operators while the vehicle is charging. Sensitive information such as payment or credit card details, personally identifiable information (PII), and other vehicle-related data are included in this data. The OCPP incorporates security features into the standard's specifications as well.

Having a security system that monitors the entire charging process is crucial, though, as the process involves more than just the charging station. The ideal security solution would confirm the identities of the car and the charging station and adhere to the previously mentioned ISO 15118 standards. It is imperative to support OCPP to guarantee the safe delivery of the contract between the mobility operators and the management systems of the charging station to the vehicle. For all parties involved, this enables a genuinely secure Plug and Charge procedure. Ensuring the identity of all parties involved can facilitate the secure sharing of information and data (International Energy Agency, 1999; Ayodele and Mustapa, 2020).

1.2. ELECTRIC VEHICLES OUTPERFORM GAS-POWERED CARS

Certain features that are seen mainly in the luxury segment of the vehicles like Bluetooth, heated steering wheel, navigation, and heated seats as well as electric panels can often be seen in the electric vehicles. As the weight of the batteries is distributed, in addition to providing all the other features as can be seen in most of the other vehicles, a tremendous amount of traction is provided by electric vehicles. As per several owners, in comparison to the AWD vehicles that they may have owned in the past, the suitability of EVs is far higher in the Vermont snow.

1.2.1. The Key Difference

Torque is generated from the start in electric motor vehicles, whereas fuel is burnt in the ICE so that heat can be generated to create propulsion. In a vehicle powered by gasoline, a delay is caused when the pedal is touched, whereas in an electric car when the accelerator is pressed, no delay is caused. The competition of the electricity-powered vehicles in the top levels of racing is quite evident in the Formula E racing series.

1.2.2. Battery Chargers for Electric Vehicles

Due to a life span that is longer, higher energy density, and lesser influence on the environment, Li-ion battery is used by most PEVs to store energy. All the same, charging Li-ion batteries is not a simple matter as it requires the current output of the charger and delicate control of voltage. If this is not done, the battery may get damaged due to current fluctuation and a large voltage.

The development of PEVs can be accelerated tremendously through a battery charger design that is advanced. As of now, off-board, and on-board are the categorizations used for PEVs battery chargers. Owing to its cost limits, space, and weight, the power level of the on-board charger is generally restricted.

On the other hand, the off-board charger is not bound by these limits. Furthermore, the distinction of the PEVs battery chargers can also be done based on whether the power flow directions supported by them are bidirectional where the power can be drawn from the grid as well as injected back into it or if it is unidirectional where the power is charged only to the battery from the charger.

Most of the time, in comparison to the bidirectional chargers, the hardware requirements are lower, and the interconnections are simpler in a unidirectional charger. Currently, single-phase chargers which are used commonly for charging at Level 1 and Level 2 equip most of the PEVs.

All the same, based on the assumption that the technique of vehicle-to-grid (V2G) has been realized and there is an availability of bidirectional chargers, a lot of research has been done. It is envisioned in the technique of V2G that power can be injected back into the grid by electric vehicles.

1.2.3. Efficiency-Aware Electric Vehicles

The fossil fuel consumption can go down tremendously through the efficient use of renewable energy. Many benefits can accrue from this, like a reduction in the contamination of air and for the customers, and enhanced cost performance. To a certain level, this objective can be achieved through the interaction of RESs and EVs. This section reviews the research efforts based on three objectives namely:

- Optimizing energy dispatch;
- Optimizing the management of energy; and
- Maximizing RESs utilization.

Problems like convex programming and linear programming can be formulated as optimization problems for the efficiency-aware objectives. Evolutionary algorithms (like genetic algorithms) and optimization tools (like IBM CPLEX) are some of the solutions that exist as of now for these problems that have been formulated.

1.3. ELECTRIC VEHICLE (EV) CONFIGURATIONS

The configuration of an EV is more flexible than that of a HEV. The reasons for this flexibility are:

- Flexible electrical wires, as opposed to rigid shafts or bolted flanges, are the primary means of energy transfer in electric vehicles. Therefore, distributed subsystems within the EV are indeed feasible.
- Different propulsion configurations, such as independent four wheels and in-wheel drives, are allowed by EVs.

The EV has three major subsystems:

- Electric propulsion;
- Energy source; and
- Auxiliary system.

The electric propulsion subsystem comprises of:

- The electronic controller;
- Power converter;
- Electric Motor (EM);
- Mechanical transmission; and
- Driving wheels.

The energy source subsystem consists of

- The energy source (battery, fuel cell, ultracapacitor);
- Energy management unit; and
- Energy refueling unit.

The auxiliary subsystem consists of:

- Power steering unit;
- Temperature control unit; and
- Auxiliary power supply.

The mechanical link is represented by the black line in Figure 1.1, the electrical link by the green line, and the control information communication by the blue line. The electronic controller provides the appropriate control signals to turn on or off the power converter, which in turn controls the power flow between the electric motor and the energy source, based on the inputs from the accelerator and brake pedals. The EV's regenerative braking is what causes the backward power flow, and if the energy source is responsive, this regenerative energy can be stored. The electronic controller and the energy management unit work together to regulate regenerative braking and its energy recovery. Additionally, it coordinates with the energy-refueling unit to manage refueling and track the energy source's usability. For all EV auxiliaries, particularly the temperature control and power steering units, the auxiliary power supply supplies the required power at various voltage levels.

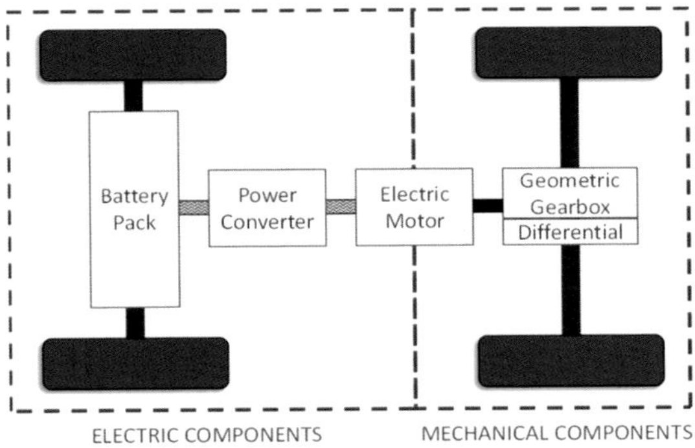

Figure 1.1. *General configuration of an electric vehicle.*

Source: Huertas JI, Mogro AE, Jiménez JP. Configuration of Electric Vehicles for Specific Applications from a Holistic Perspective. World Electric Vehicle Journal. *2022; 13(2):29. https://doi.org/10.3390/wevj13020029.*

In modern EV's configuration:

- Three-phase motors are generally used to provide the traction force;
- The power converter is a three-phase PWM inverter;
- Mechanical transmission is based on fixed gearing and a differential; and
- Li-ion battery is typically selected as the energy source.

1.3.1. Electric Vehicle (EV) Drivetrain Alternatives Based on Drivetrain Configuration

The differences in electric propulsion and energy sources allow for a wide range of possible EV configurations. These six alternatives are:

- A single EM configuration with a clutch and gearbox (GB) is depicted in Figure (1.2). It is made up of a differential (D), a gearbox, a clutch (C), and an EM. Power transfer from EM to the wheels can be connected or disconnected thanks to the clutch. The gear is made up of several gears with various gear ratios. The driver can alter the torque applied to the wheels by adjusting the gear ratios with the clutch and gearbox. In the lower gears, the wheels have high torque low speed, and in the higher gears, high speed low torque.

- A single EM configuration devoid of the clutch and gearbox is depicted in Figure (1.2). One benefit of this arrangement is that the transmission's weight is decreased. Nevertheless, to supply the required torque to the wheels, this configuration necessitates a more intricate control of the EM.

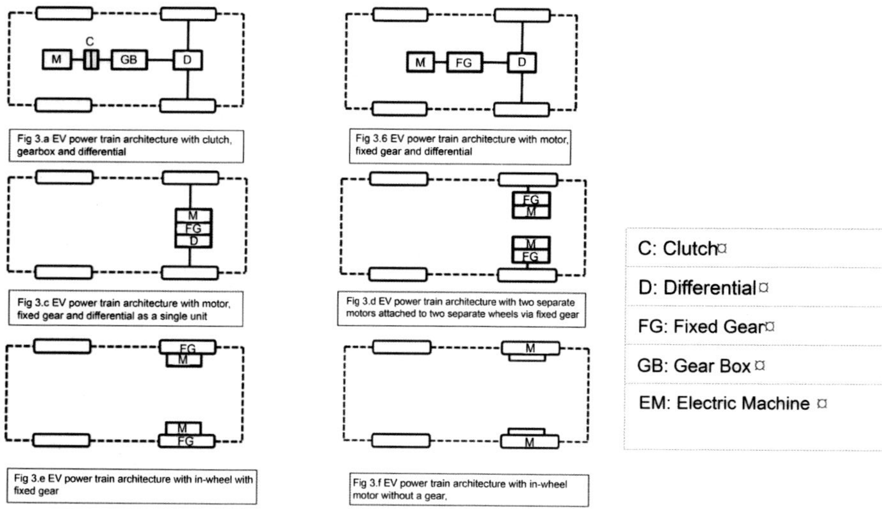

Figure 1.2. *EV configuration with clutch, gearbox and differential.*

Source: Karki A, Phuyal S, Tuladhar D, Basnet S, Shrestha BP. Status of Pure Electric Vehicle Power Train Technology and Future Prospects. Applied System Innovation. *2020; 3(3):35. https://doi.org/10.3390/asi3030035.*

- A single EM is used to configure an EV, as seen in Figure 1.2. The front wheel drive configuration is transverse EM. Its differential and fixed gearing are combined into a single assembly.
- Figure (1.2) illustrates a dual motor arrangement. In this setup, two electric motors can electronically provide the differential action of an EV during cornering.
- The EM can be mounted inside a wheel to reduce the length of the mechanical transmission path from the EM to the driving wheel. We refer to this setup as in-wheel drive. This arrangement uses fixed planetary gearing to lower the motor speed and raise the wheel speed to the required level.
- An electric vehicle setup devoid of mechanical gearing. The in-wheel drive can be achieved by installing a low speed outer-rotor electric

motor inside a wheel and completely forgoing any mechanical gearing.

1.3.2. Electric Vehicle (EV) Drivetrain Alternatives Based on Power Source Configuration

There are different EV configurations because of differences in energy sources in addition to variations in electric propulsion:

- **Configuration 1:** It is a basic battery-operated setup. The battery may be positioned under the vehicle's chassis, dispersed throughout the car, or bundled together at the back. In this scenario, the battery should be able to accept regenerative energy during braking and have a respectable specific energy and specific power. High specific power determines the driving range of an EV, while high power density determines the acceleration rate and ability to climb hills. Therefore, the battery in an EV should have both high specific energy and specific power.
- **Configuration 2:** This design makes use of two separate batteries in place of two. For high specific energy and high specific power, two different batteries have been optimized.
- **Configuration 3:** This setup makes use of a fuel cell. The fuel cell is a device that generates energy, while the battery stores it. The principle of operation for fuel cells involves the reversal of electrolysis. Hydrogen and oxygen gases combine to produce water and electricity through electrolysis and reverse processes. While oxygen gas is drawn from the air, the hydrogen gas used by the fuel cell can be kept in an on-board tank. Fuel cells work best when combined with batteries that have high specific power and high energy receptivity because they can provide high specific energy but cannot accept regenerative energy.
- **Configuration 4:** Using liquid fuels like methanol, hydrogen can be produced on-board and stored as a liquid, metal hydride, or compressed gas. In this instance, the EV has a mini reformer installed to provide the hydrogen gas required for the fuel cell.
- **Configuration 5:** Battery selection for fuel cell and battery combinations takes into account the battery's high specific power and high energy receptivity. The energy source in this setup is a battery and super capacitor combination. In this configuration, the super capacitor offers high specific power and energy receptivity,

while the battery is a high energy density device. Super capacitors typically have relatively low voltage levels, so a separate DC-DC power converter is required to connect the capacitor terminals to the battery.

1.3.3. Single and Multi-motor Drives

Conventional vehicles come standard with a differential. The outer wheel of a car must go on a wider radius than the inner wheel when it is navigating a curve. As a result, the differential modifies the wheel ratios. The wheels will slip if the relative speeds of the wheels are not adjusted, which will lead to tire wear, steering issues, and poor road holding. When it comes to EVs, two or even four EMs can be used to dispense the mechanical differential. Each wheel can be connected to an EM by using multiple EMs, allowing for independent speed control of each wheel and the electronic achievement of the differential action.

1.3.3.1. In Wheel Drives

The benefit of an in-wheel motor is that the mechanical transmission path between the electric motor and the wheel can be reduced by putting the motor inside the wheel. **Two possible configurations for in wheel drives are**:

- To achieve a reasonable wheel speed when using a high-speed inner-rotor motor, a fixed speed-reduction gear is required. A planetary gear set is typically used to reduce speed. The location of this planetary gear is between the wheel rim and the motor shaft. Typically, this motor is made to run at speeds of up to 1000 rpm in order to provide a high power density.
- If an outer rotor motor is used, the gearbox does not need to be removed because the outer rotor functions as the wheel rim and the motor speed is the same as the wheel speed.

The tradeoffs of the high-speed inner rotor motor are:

- It has the advantage of smaller size, lighter weight and lower cost; and
- Needs additional planetary gearset.

The tradeoffs of the outer-rotor motor are:

- Low speed and hence does not need additional gears; and
- The drawbacks are larger size, weight and cost because of the low-speed design.

The requirements of EMs used in EVs are:

- Frequent start/stop;
- High rate of acceleration and deceleration;
- High torque low-speed hill climbing;
- Low torque cruising; and
- Very wide speed range of operation.

The following are the main ways that the EMs for EVs differ from those of industrial motors in terms of load requirements, performance specifications, and operating environments:

- To accelerate and climb hills, EV motors must generate torque four to five times the rated torque, whereas industrial motors can typically produce torque twice the rated torque when operating in overload conditions.
- For highway cruising, EV motors must reach four or five times the base speed; for continuous power operation, industrial motors typically reach up to twice the base speed.
- Whereas industrial motors are typically optimized to provide high efficiency at a rated point, EV motors need to have both high power density and a good efficiency map, or high efficiency over a wide range of speed and torque.
- EV motors must be installed in mobile vehicles that are subjected to extreme operating conditions, such as high temperatures, inclement weather, and frequent vibrations, whereas industrial motors are typically fixed in locations (International Energy Agency, 1999; Ayodele and Mustapa, 2020).

1.4. APPLICATION IN MICROGRID

In a microgrid with respect to the efficiency-aware objective, efforts have been made to research the interaction in EVs with RESs. Optimal scheduling was developed by well-known researcher Tushar for joint control of electricity consumption of EVs and appliances. Minimizing electricity generation from the external grid and utilizing the electricity generated by RESs optimally is the objective.

Only wind power, a central controller, EVs, PV solar power and home appliances have been considered by the microgrid. Either a mode interacting with the grid or a self-sustained mode can be utilized to operate it. The formulation of the scheduling problem is done as a mixed-integer linear programming problem (MILP).

EVs can be considered as energy storage during the hours of renewable power generation; additional electricity shall be provided by the EVs during the period when the electricity demand is high. The prediction for the generation of solar power and wind power during the next 24 hours is done through the Markov chain state transition probability. As per the varying requirements, the differentiation of the home appliances is done for instance, whereas flexible scheduling may be required for some, others may require stricter scheduling.

A Poisson process is presumed to be followed by the arrival in microgrid of the EVs. There can be a random distribution of EVs with varying energy demands so that the duration of charging them can be randomly at home. The IBM CPLEX MLP solver is used to solve the formulated model that includes discrete and continuous decision variables.

In variation to the findings of the researcher named, the wind power has been considered as the single resource to interact with EVs by a well-known researcher named Wu and optimizing the wind power's utilization efficiency in the microgrid is the aim here. For the interaction of EVs with wind power, three energy dispatching methods that are coordinated have been proposed by them based on the flexibility in control of the EV's discharging and charging behaviors.

Individual svehicles' distribution of SOC and the volatility of power generation are the factors covered by their model. Interruptible dispatching, valley searching dispatching and variable-rate dispatching are the three approaches. The statistic model from the ICE vehicles of the families has been used to portray the driving patterns of EVs by arrival time, departure time, and the distance driven.

Normal distributions have been used to model the arrival and departure times. Algorithm normal distribution is used to model the driving distance. EV's residual energy that is available is represented by using SOC.

Results from the simulation show that the matching performance can be improved efficiently through the three methods that have been proposed. The microgrid model considered by Strnad et al. integrates the PV solar plants, thermal units, a diesel power generator, EVs charging station and BESS (battery energy storage system) (Figure 1.3).

The goal is to maximize utilization when energy is used to generate electricity and when BESS or EVs are charged using that electricity from PV solar plants and thermal units. In this manner, the impact that EV charging has on the microgrid can be reduced.

The production cost has an impact on the diesel power generator generation schedule. More specifically, the diesel generator is started when the microgrid experiences a power outage and only kicks in when the cost of the electricity it purchases from the grid is greater than the cost of producing the same amount of power from it. The power system of the Faculty of Electrical Engineering and Computing in Zagreb was used by the authors for their case study.

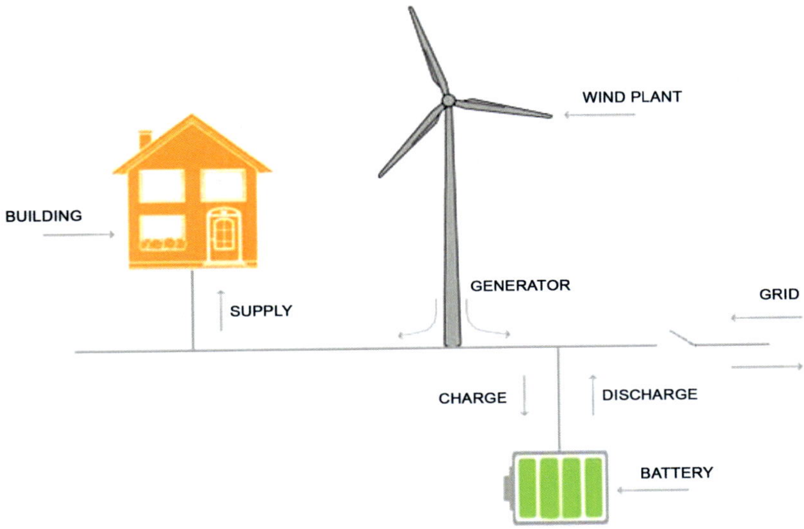

Figure 1.3. *Microgrid with RES BESS connected to each other.*

Source: https://commons.wikimedia.org/wiki/File:Microgrid_with_RES_BESS_GRIDconnected.png.

With an interval of 15 min, a simulation covering one summer and one winter workday has been provided by them. The amount of electricity required in this scenario, even when a slow charging rate is used 8 working hours suffice to charge the EVs and the RES can easily supply this amount of electricity.

For the maximum utilization of RESs in the microgrid charging policies have been proposed by well-known researcher Zhu. The semi-Markov decision process (SMDP) has been used for the formulation of the optimization process wherein RESs states, EVs numbers and battery states have been considered.

Only one PV solar panel was considered by them in the solar panel; however, they claimed that other sources of renewable energy like hydro and wind energy too could equally be applied to this model. To model the solar radiation, the continuous Markov chain has been used by the authors Tushar and to model the battery states, the linear method has been used, and the leftover store of energy in the battery is obtained by using this.

The transition probabilities are obtained by using the Poisson process decomposition property since they take EVs departure and arrival rates as Poisson distribution. The current PV solar panel system, the number of EVs charged by BESS, the number of EVs charged by the PV solar panel and the current remaining battery energy, all comprise the system. Further, the probabilities of state transition characterize the system dynamics. The SMDP model is used to solve the linear programming technology.

1.4.1. Maximizing RESS Utilization

Gottwalt, the researcher, claims that charging EVs—which are thought of as a sort of flexible load—can maximize the utilization of RESs. By employing an aggregator, the residential household demand is met concurrently with the EVs' coordinated charging behavior. The behavior of EVs owned by full-time employees is the only one modeled by the empirical driving profiles.

A German dataset is the basis for deriving their household demand. Conventional generator and renewable energy were the sources of electricity. Only when the energy demanded by the households and for charging the EVs is over the amount that is available from RESs does the conventional generator get initiated.

The generation from conventional means shall be minimized when the demand for EVs is coordinated well with the generation that is available from RESs. This has been formulated as a problem of mixed-integer programming by Gottwalt et al.

They have assumed that information about EV travel patterns, RESs in the future, and household load for a week is available. They first assessed fast operating uncoordinated charging in order to illustrate the advantages of coordinated charging. After that, the coordinated approach is meticulously simulated and assessed over a 52-week period.

IBM ILOG CPLEX 12.4 is used to solve the problem of the proposed mixed-integer programming. The electricity demand of the residents can be met through this work as one of the means to meet it. On the other hand, maximizing the RESs utilization through the demand from charging EVs has been the focus of the researcher named Mets.

To coordinate the supply of wind energy and the demand of EVs charging, a demand-side management (DSM) algorithm was adopted by them. As a result, there can be the maximum utilization of the renewable energy.

By using their approach, they guarantee that, for a predetermined amount of time, the supply and consumption of energy are equal. The owners of electric vehicles and a coordinator make up the power system's goal. The owners of

EVs can alter demand to preserve the order. If supply cannot keep up with demand, an imbalance in costs will inevitably result. Therefore, the initial problem of maximizing the RESs is transformed into a problem of minimizing the imbalance cost. This is expressed as a convex optimization problem. To solve this, the dual decomposition method is used. The objective function's constraints are moved using the Lagrange multipliers method.

A case study was carried out by the authors for four weeks so that the proposed method could be evaluated. The results of the simulation showed that in the most ideal scenario, 73% of the demand can be satisfied; however, in the real-world scenario, this is impractical. At the same time, demand for 68% of the energy can be supplied through the usage of their distributed algorithm. A demand of about 40% can be supplied in the business scenario of the usual kind (International Energy Agency, 1999; Ayodele and Mustapa, 2020).

1.5. ENERGY STORAGE IN ELECTRIC VEHICLES

A battery is made up of two or more connected electric cells. Electrical energy is produced from chemical energy by the cells. Positive and negative electrodes are connected by an electrolyte to form the cells. The chemical reaction that occurs between the electrolyte and the electrodes is what produces DC electricity. Reversing the current in secondary or rechargeable batteries can reverse the chemical reaction and return the battery to a charged state. **The 'lead acid' battery is the most well-known battery.**

Rechargeable batteries were first used in electric vehicles more than 25 years before rechargeable lead acid batteries were developed. A wide variety of materials and electrolytes can be mixed to create different kinds of batteries. Only a tiny number of combinations, nevertheless, have been produced as widely available rechargeable electric batteries that are appropriate for use in automobiles. These currently include lithium polymer and lithium iron, sodium sulfate and sodium metal chloride, lead acid, nickel-iron, nickel cadmium, and nickel metal hydride.

In this section, the different types of energy storage devices are presented. The following topics are covered in this lecture:

- Overview of Batteries;
- Battery Parameters;
- Lead acid batteries;
- Lithium ion batteries; and
- Metal air batteries.

1.5.1. Overview of Batteries

From the electric vehicle designer's 'which has a range of performance critters

- Specific energy;
- Energy density;
- Specific power;
- Typical voltages;
- Amp hour efficiency;
- Energy efficiency;
- Commercial availability;
- Cost, operating temperatures;
- Self-discharge rates;
- Number of life cycles; and
- Recharge rates.

The designer also needs to understand how energy availability varies with regard to:

- Ambient temperature;
- Charge and discharge rates;
- Battery geometry;
- Optimum temperature;
- Charging methods; and
- Cooling needs.

To avoid the performance and maintenance requirements of the various types of batteries as well as the majority of the drawbacks associated with their use—such as their limited lifespan, self-discharge, and decreased efficiency at higher currents—it is crucial to have at least a basic understanding of battery chemistry.

1.5.2. Battery Parameters

1.5.2.1. Cell and Battery Voltages

Every electric cell has a nominal voltage or the approximate voltage at which the cell produces electricity. The necessary total voltage can be obtained by connecting the cells in series. "Internal battery resistance" is the term. The voltage across the internal resistance R causes the voltage at the battery's terminals to differ from the voltage indicated by the fixed voltage E. Assuming that a current I is flowing out of the battery, then by basic circuit theory we can say that:

$$V = E - IR \qquad (1)$$

1.5.2.2. Charge (or Ahr) Capacity

One of the most important factors is undoubtedly the amount of electricity a battery can provide. For SI purposes, the charge that results from one Amp flowing for one second is called a Coulomb. A battery could have a 10 amp-hour capacity. It can thus supply 1 Amp for10 hours.

1.5.2.3. Energy Stored

A battery's capacity to store energy is determined by both its voltage and charge level. We use the Whr in place of the SI unit, the Joule because it is inconveniently small.

$$Energy\ in\ Whr = V \times Ahr \qquad (2)$$

1.5.2.4. Specific Energy

Specific energy is the amount of electrical energy stored for every kilogram of battery mass. It has units of $Wh.kg^{-1}$.

1.5.2.5. Energy Density

Energy density is the amount of electrical energy stored per cubic meter of the battery volume. It normally has units of $Wh.m^{-3}$.

1.5.2.6. Specific Power

A battery's specific power is its power output per kilogram. Given that the power the battery provides is largely dependent on the load attached to it rather than the battery itself, it is an extremely erratic and unusual quantity.

1.5.2.7. Ahr (or charge) Efficiency

A battery's amp hour efficiency would be 100% if it could, in an ideal world, return every ounce of charge that was placed in it. But no battery does; a battery's charging efficiency is not 100%. The exact value will change depending on the battery type, temperature, and charging rate. It will also change according to the charge's state.

1.5.2.8. Energy Efficiency

The ratio of the electrical energy supplied by a battery to the amount of electrical energy needed to restore it to its pre-discharge state is another crucial parameter.

1.5.2.9. Self-discharge Rates

Self-discharge is the term for when most batteries run out of power while being idle. This is significant because it implies that some batteries will need to be recharged after extended use. The type of battery and other variables, like temperature, affect the rate; higher temperatures significantly accelerate self-discharge.

1.5.2.10. Battery Temperature, Heating and Cooling Needs

While most batteries operate at room temperature, some require heating before use and cooling afterward due to their higher operating temperatures. In others, low temperatures cause a drop in battery performance, which is undesirable, but this issue could be fixed by heating the battery. The battery's temperature, heating requirements, and cooling requirements must be taken into account by the vehicle designer when selecting a battery.

1.5.2.11. Battery Life and Number of Deep Cycles

The majority of rechargeable batteries will only reach 20% of their full charge after a few hundred deep cycles. Nevertheless, the precise amount varies depending on the kind of battery, its usage, and specific design features. This is a crucial number in a battery specification since it affects the battery's lifespan, which in turn affects the operating costs of electric vehicles.

1.5.3. Lead Acid Batteries

The lead acid battery is the most well-known and frequently utilized battery for electric vehicles. Since lead acid batteries are frequently found in vehicles with internal combustion engines, they are well known. But stronger lead acid batteries that can tolerate deep cycling and that employ a gel electrolyte rather than a liquid are utilized in electric cars. Producing these batteries costs more. Lead dioxide is the active material in the positive lead acid cell plates, while spongy lead is present in the negative plates. The plates are submerged in a diluted sulfuric acid electrolyte. Lead sulfate and water are created when sulfuric acid reacts with lead and lead oxide, releasing electrical energy in the process. The overall reaction is:

$$Pb + PbO_2 + 2H_2SO_4 \leftrightarrow 2PbSO_4\ 2 + H_2O \quad (3)$$

The battery's individual electrode reactions. The diagram's upper region shows the battery draining. Lead sulfate is produced by both electrode reactions. Sulfuric acid is progressively lost by the electrolyte, which also gets more diluted. Lead and lead dioxide are used as the electrodes again. Additionally, the concentration increases as the electrolyte recovers its sulfuric acid. In all but the smallest of systems, the lead acid battery is the most widely used rechargeable battery. The primary causes of this are its low cost, dependable performance, relatively high voltage of roughly 2V per cell, and use of inexpensive materials (lead, sulfuric acid, and plastic containers). A good estimate of the internal resistance of a lead acid battery is thus:

$$R = No.\ of\ Cells \times \frac{0.022}{C_{10}}\ Ohms \quad (4)$$

1.5.4. Lithium Batteries

Rechargeable lithium cells have been available on the market since the late 1980s. Compared to other rechargeable batteries, they provide a significantly higher energy density, but at a significantly higher cost. Rather than the less expensive NiCad or NiHM cells that we have been discussing previously, lithium rechargeable batteries are a well-known feature of the priciest laptop computers and mobile phones.

The lithium batteries are of the following types:

- Lithium polymer batteries; and
- Lithium ion batteries.

1.5.4.1. The Lithium Polymer Battery

The positive electrode of the lithium polymer battery is made of transition metal intercalation oxide, while the negative electrode is made of lithium metal. Lithium and metal oxide combine to form lithium metal oxide and release energy in the ensuing chemical reaction. The chemical reaction is reversed when the battery is recharged. As a result, the lithium passes through the electrolyte as both a mobile ion and a reactant. The overall chemical reaction is:

$$xLi + M_yO_z \leftrightarrow Li_xM_yO_z \quad (5)$$

1.5.4.2. The Lithium Ion Battery

The positive electrode of the lithium ion battery is made of lithiated transition metal intercalation oxide, while the negative electrode is made of lithiated carbon. The battery was first commercialized in the early 1990s. Either a solid polymer or a liquid organic solution serves as the electrolyte. When lithium metal oxide and carbon combine to form carbon and lithium metal oxide, electrical energy is produced.

The overall chemical reaction for the battery is:

$$C_6 Li_x + M_y O_z \leftrightarrow 6C + Li_x M_y O_z \tag{6}$$

One crucial aspect of lithium ion batteries is that precise voltage control is required during the lithium cell's charging process. If it is too low, the battery won't be fully charged, and if it is slightly too high, the battery may be damaged. Alongside the battery, appropriate commercial chargers are being developed.

1.5.5. Metal Air Batteries

Since the metal air batteries cannot be recharged by simply reversing the current, they represent a completely different development. Rather, new electrodes need to be installed in place of the spent metal ones. Thus, one could think of the metal electrodes as a type of fuel.

1.5.5.1. The Aluminum Air Battery

The aluminum air battery's fundamental chemical reaction is essentially straightforward. Aluminum hydroxide is created when aluminum is mixed with oxygen from the air and water, releasing electrical energy in the process. The response cannot be reversed. The overall chemical reaction is:

$$4Al + 3O_2 + 6H_2O \rightarrow 4Al(OH)_3 \tag{7}$$

The negative electrode of the cell is made of aluminum, which is normally started as a plate that is one centimeter thick. The electrode gets smaller and smaller as the reaction goes on. The positive electrode is usually a porous material made of metal mesh onto which a layer of catalyzed carbon is pressed. It has the required porosity to allow oxygen to pass through while keeping the liquid electrolyte inside thanks to a thin coating of PTFE. An alkaline solution, typically potassium hydroxide, serves as the electrolyte. Replacing the used negative electrodes allows the battery to be recharged. Since the electrolyte will be tainted with aluminum hydroxide, it will typically also need to be replenished.

1.5.5.2. The Zinc Air Battery

Although the zinc air battery and the aluminum air battery are similar in many aspects, the zinc air battery performs significantly better overall, especially in terms of specific power, which is almost 10 times higher than the aluminum air battery. For this reason, it is appropriate for use in road vehicles. Similar in structure, it has a porous positive electrode where the electrolyte and oxygen interact. An alkaline liquid is the electrolyte. Solid zinc serves as the negative electrode.

Zinc oxide is created when zinc and airborne oxygen combine to release energy from the battery. Alternatively, zinc hydroxide may form, similar to the aluminum-air cell, depending on the condition of the electrodes and electrolyte. In most cases, the process cannot be reversed. Though the number of cycles is typically fairly low, a few manufacturers have claimed to produce electrically rechargeable zinc-air batteries. Replacing the negative electrodes is the more conventional method of recharging an aluminum air cell (International Energy Agency, 1999; Ayodele and Mustapa, 2020).

1.6. ELECTRIC VEHICLE MOTORS

The commonly used motors in EVs are:

- AC motors;
- Permanent magnet (PM) motors;
- Series wound DC motors; and
- Shunt wound DC motors.

Because they are so easy to control, DC series motors were used in several EV prototypes before that. However, the use of DC motors is becoming outdated due to their size and maintenance needs. Recently, AC, PM, and switched reluctance motors have been used in EVs and HEVs.

The technology for AC induction motors (IMs) is highly developed, and a lot of research and development has gone into induction motor drives. Although IM control is more complicated than that of DC motors, it is still easily manageable due to the availability of quick digital processors. The PM motor is the induction motor's rival. The rotor of permanent magnet motors is magnetized, and the stator structure is identical to that of an induction motor. The magnets in the PM motors can be inset within the rotor, or they can be surface mounted. In addition, the PM motors can be categorized as trapezoidal or sinusoidal based on the distribution of flux density in the air gap. Brushless DC (BLDC) motors are permanent magnet motors with a trapezoidal air gap flux distribution, and permanent magnet synchronous motors (PMSM) with a sinusoidal air gap flux distribution.

1.6.1. Permanent Magnet (PM) Machines

A PM machine drive with high power density, high speed, and high operation efficiency can be designed by utilizing high-energy magnets, such as rare earth-based magnets. These benefits make them appealing for use in EVs and HEVs. The major advantages of PM machines are:

- **High Efficiency:** Because PMs require no power for excitation, the PM machines have extremely high efficiency. Furthermore, there are fewer mechanical friction losses because there are no brushes or mechanical commutators.
- **High Power Density:** Very high flux densities in the PM machines have been made possible by the use of magnets with high energy densities. When compared to other motors of the same volume, high torque can be generated from a given motor volume due to high flux densities.
- **Ease of Control:** Because the control variables for PM motors are readily accessible and remain constant throughout the motor's operation, they can be controlled just as easily as DC motors.
- However, the PM machines also suffer from some disadvantages such as:
- **Cost:** Rare-earth magnets commonly used in PM machines are very expensive.
- **Magnet Demagnetization:** High temperatures and a strong opposing magneto motive force can demagnetize the magnets.
- **Inverter Failure:** PM motors carry a significant risk in the event that the inverters short circuit due to the magnets on the rotor. Every time the rotor is turned on, the short circuited windings experience an electromagnetic field (EMF). Those windings are carrying a very high current, and the rotor is in danger of being blocked by an equivalently high torque. The risks associated with obstructing one or more of a car's wheels are significant.

Based on the shape of the back e.m.f induced in the stator windings, the PM motors can be classified into two types:

- Permanent Magnet Synchronous Machine with sinusoidal back e.m.f; and

- Brushless Permanent Magnet DC Machines (BLDC) with trapezoidal back e.m.f.

Based on the construction of the rotor, the PM machines can be broadly classified into three categories:

- Inner rotor machine;
- Outer rotor machine; and
- Interior magnet rotor.

1.6.2. Principle of Operation of PM Machine

Rotor PMs in PM machines provide the required rotor flux. The stator mmf is produced by currents in the stator windings. If the stator mass flow (mmf) is rotating at the same speed as the rotor flux (i.e., rotor speed) and in the same direction, then the two will have zero relative speed. A set of polyphase currents phase-shifted from one another by the same amount of phase shift between the polyphase windings are injected, resulting in the revolving stator mmf. A rotating magnetic field that is constant in magnitude and travels at the angular frequency of the currents, for instance, is produced by a three-phase machine that has three windings that are electrically shifted by 120° into space (just as in the case of induction machines). Because the rotor is permanently magnetized, torque is generated when the flux generated by the rotor magnets begins to pursue the stator mmf. The rotor moves at the same speed as the stator mmf because there can be no relative speed difference between the stator mmf and rotor flux. The PM machines are therefore synchronous machines by nature. There is an induced emf in the windings as the moving magnets cause a change in flux linkages in the coils of the stator. The form of the flux linkage greatly influences the shape of the induced emf.

1.6.3. Permanent Magnet DC Motor

A PM motor uses PMs to create the magnetic field that the rotor field interacts with to produce torque, rather than having a field winding on the stator frame. For large motors, compensating windings in series with the armature may be used to enhance commutation under load. This field cannot be changed for speed control because it is fixed. In small motors, PM fields (stators) are useful for removing the field winding's power consumption. The majority of larger DC motors are stator winding-equipped dynamos. Historically, field windings were a more practical way to achieve the required amount of flux because PMs could not be made to retain high flux if they were disassembled. Large PMs, however, are expensive, hazardous, and challenging to assemble; this makes wound fields more suitable for large machines.

Miniature PM motors may employ high-energy magnets composed of neodymium or other strategic elements; the majority of these are neodymium-iron-boron alloys, to reduce overall weight and size. Electric machines with high-energy PMs can at least compete with all optimally designed singly-fed synchronous and induction electric machines due to their higher flux density. The structure of miniature motors is similar to the illustration, with the exception that they have three rotor poles or more (to guarantee starting regardless of rotor position) and their outer housing is a steel tube that connects the exteriors of the curved field magnets magnetically.

1.6.4. Brushless DC Motor

The BLDC design eliminates a few of the brushed DC motor's drawbacks. An external electronic switch synchronized to the position of the rotor replaces the mechanical rotating switch, also known as the commutator, in this motor. BLDC motors have an efficiency of at least 85%–90%. While brush gear DC motors are usually 75%–80% efficient, BLDC motors have been reported to have an efficiency of up to 96.5%.

The evenly distributed stator windings and the positioning of the rotor's PMs contribute to the distinctive trapezoidal back-emf waveform of BLDC motors. The stator windings of trapezoidal BLDC motors also referred to as electronically commutated DC or inside-out DC motors, can be single-phase, two-phase, or three-phase. Hall effect sensors are mounted on these windings to provide low-cost closed-loop control of the electronic commutator and to sense the position of the rotor. When exact speed control is required, such as in computer disk drives, VCRs, CD-ROMs, and spindles inside CDs, BLDC motors are frequently utilized. drives, and internal components of office equipment like fans, photocopiers, and laser printers. They have several advantages over conventional motors:

- They are significantly more efficient and run considerably cooler than AC fans that use shaded-pole motors. The fan's bearing life is significantly extended by this cool operation.
- A BLDC motor can have a much longer lifespan than a DC motor with brushes and a commutator since it doesn't have a commutator to wear out. A BLDC motor can be used in electrically sensitive devices like computers and audio equipment, but commutation also tends to produce a lot of electrical and radiofrequency noise.
- For applications requiring closed-loop control (servo control), the same Hall effect sensors that produce the commutation can also generate a useful tachometer signal. The tachometer signal in fans

can be used to generate a fan OK signal and give feedback on running speed.

- The motor can be easily synchronized to an internal or external clock, leading to precise speed control.
- Compared to brushed motors, BLDC motors are less likely to spark, which makes them more appropriate for settings containing flammable materials and fuels. Additionally, sparking produces ozone, which can build up in buildings with inadequate ventilation and endanger the health of building occupants.
- BLDC motors are usually used in small equipment such as computers and are generally used in fans to get rid of unwanted heat.
- They are also acoustically very quiet motors, which is an advantage if being used in equipment that is affected by vibrations.

The power of a modern BLDC motor can range from a few watts to several kilowatts. Electric vehicles employ larger BLDC motors, with a maximum rating of approximately 100 kW. They are also extensively used in electric model aircraft with high performance.

Electric currents do not flow through the rotor of the SRM, nor do brushes or PMs. Rather, torque is caused by a small misalignment between the rotor and stator poles. The stator field is rotated by the sequential energization of the stator field windings as the rotor aligns itself with the stator's magnetic field. The rotor's closest poles to the stator's energized poles will experience the magnetic flux produced by the field windings traveling along the path of least magnetic reluctance, which will magnetize those rotor poles and produce torque. Different windings will become energized as the rotor rotates, maintaining its rotation.

1.6.5. Synchronous Motor

Synchronous electric motors are identified by their rotor spinning coils that pass magnets at the same speed as the AC, creating a magnetic field that powers the motor. It has zero slip under typical operating conditions, to put it another way. In contrast, an induction motor needs to slip to generate torque. Similar to an induction motor, a synchronous motor's rotor is driven by a DC field. Brushes and slip rings are used to transfer current to the rotor. Synchronous motors get their name from the way the rotor poles are connected to one another and travel at the same speed. Another type uses discrete poles created by flats ground onto a traditional squirrel-cage rotor for low-load torque. Another has discrete poles and no rotor windings, like the ones used in older Hammond organs and clocks manufactured by Hammond before World War II. It doesn't start on its own. The older Hammond organs had an additional starting motor connected by a manually operated switch that was spring-loaded, but the clock needs to be manually started using a small knob on the back.

Finally, with a phase shifting capacitor for one phase, hysteresis synchronous motors are basically two-phase motors. They function initially like induction motors, but the rotor—a smooth cylinder—becomes momentarily magnetized when the slip rate drops enough. It behaves like a PMSM because of its dispersed poles. The rotor material can be easily demagnetized, but it will remain magnetized, much like a regular nail. The rotor poles do not drift once they are operating. Low-power synchronous timing motors, like those found in conventional electric clocks, can use shading coils to supply starting torque in addition to multi-pole PM external cup rotors. The two-spoke ring rotor in telechron clock motors functions similarly to a discrete two-pole rotor, and the motors have shaded poles for starting torque (International Energy Agency, 1999; Ayodele and Mustapa, 2020).

1.7. SUMMARY OF ELECTRIC VEHICLES

Electric vehicles have progressively been building the headline in current times. It has been observed that significant vehicle maker Volvo, for example, has declared that each vehicle they introduced after 2019 had an electric engine and that the recent UK Government needs to see their streets from petroleum and diesel-free vehicles by the end of 2040. Electric vehicles, sometimes called EVs, are set to be the norm. However, the question arises, what is the importance here for the normal purchaser? The majority of people do not have electric vehicles, but now is an ideal opportunity to begin thinking about the benefits and disservices of electric vehicles.

1.7.1. No Fuel, No Emissions

Many individuals get attracted to electric vehicles because of their no fuel, no emission quality. In case anyone wishes to lower their own effect on the climate through transport, at that point an EV is the way forward.

The electrical motor within an electrical vehicle, works on a close circuit, so an electric vehicle does not produce any of the gases regularly associated with global warming. No petroleum or diesel is required in a completely electric vehicle, which is an incredible carbon footprint.

1.7.2. Running Costs

Since the users are not paying for petroleum or diesel to keep their vehicles running, they can save a good amount of money on fuel. At the time of composing, it costs around £63.80 to fill the normal unleaded petroleum tank for drivers of medium-sized vehicles in the UK.

By examination, based on the electric vehicle people own and the tariff they are on, a full charge of their electric vehicle could cost just 96p. Anyone could even save £60 per year with the Good Energy Electric Vehicle Tariff, versus the Big Six; Standard Variable Tariffs, with the extra reward of the electricity going into your EV being 100% renewable.

1.7.3. Performance

The fun of owning a car usually comes from going out on the roads. In past times, the electric vehicles have not had the sleekest image. A lot many have had low expectations as to how good an electric vehicle can do as compared to the traditional engines.

As so many manufacturers have heaped into the market with their own take on the electric vehicles, the level of performance of those vehicles has super increased. Electric cars are lighter, as all of their power, is produced from a standing start-their acceleration capability can give a shocking surprise.

Few brands such as Tesla have done many things to enhance the perception of people regarding electric vehicles. The Tesla Model S is one of the fastest-accelerating cars available in the market, doing 0–60 mph in just 2.5 seconds. Well, this is not the only measure of performance of any car.

People would feel happy to know that electric vehicles are much more spacious as compared to conventional cars because of the absence of a large engine. Electric vehicles, in addition, provide a much smoother drive with a very low level of noise.

1.7.4. Popularity

These days, if people like to stand out in the crowd, they perhaps know that electric cars are the best option. They are becoming really positive. As more and more electric vehicles are finding their way onto roads, people can see the supporting infrastructure expand.

Already, there are more than 4800 charging points in the UK, for example. It offers almost 7500 individual charging points, and these numbers are gradually increasing. To compare, in the year 2016, there were 8,459 petrol pumps and after the year 2000, this number has been declining.

With tEVs being on the road, people can see this number gradually falling while the number of charging points will increase further. Along with the increasing number of charging points, the increased popularity of electric vehicles means more options to choose from for the car. Also, there are more affordable options for electric cars available now than earlier, for example, the Nissan Leaf and the Renault Zoe, with some of the most popular petrol and diesel models also available in an electric version, such as the Volkswagen e-Golf (International Energy Agency, 1999; Ayodele and Mustapa, 2020).

1.8. ADVANTAGES AND DISADVANTAGES OF ELECTRIC VEHICLES

1.8.1. Advantages of Electric Vehicles

Here are the advantages of electric vehicles.

1.8.1.1. Lower Running Costs

You save a ton of money on fuel because you don't need to buy gasoline or diesel to keep your electric vehicle running. When comparing the cost of charging an electric vehicle to that of gasoline or diesel, the difference is significant. By using renewable energy sources like solar power, you can further lower the cost of electricity.

1.8.1.2. Low Maintenance Costs

Vehicles running on gasoline or diesel need routine maintenance because they have a lot of moving parts. With electric vehicles, this isn't the case because they have fewer moving parts overall. This implies that over time, the maintenance costs of your electric car should be lower.

1.8.1.3. Tax and Financial Benefits

India has welcomed the use of EVs and the government has implemented various policies and incentives to promote the use of these cars. For example, buying an EV will cost less in road tax and registration fees than buying an ICE car.

1.8.1.4. Better Performance

EVs were once thought to be unfeasible. But as time has gone on, manufacturers have started to provide aesthetically pleasing and well-designed electric vehicles. Even electric vehicle performance has improved. Compared to fuel-powered vehicles, electric vehicles are lighter and accelerate flawlessly.

1.8.1.5. Zero Tailpipe Emissions

EVs have zero tailpipe emissions, lowering their carbon footprint. By charging your EV with renewable energy, you can further minimize your carbon footprint.

1.8.1.6. Easy to Drive and Quiet

EVs are simple to drive thanks to their simple controls and reduced number of moving parts. Additionally, when you need to charge a vehicle of this type, you can plug it into a public or residential charging station. Additionally, they are quiet, which lessens the noise that fuel-powered cars produce.

1.8.1.7. Convenience of Charging at Home

It is not necessary to locate the closest gas station in order to fill up. Get your EV charged at home and drive. Modern charging technology allows you to quickly charge an EV or even use battery swapping services so you can continue driving regardless of the availability of conventional fuel.

1.8.1.8. No Fuel, No Emissions

The way that EVs affect the environment is among their biggest benefits. Because pure EVs emit no tailpipe emissions, air pollution is decreased. The EV's electric motor doesn't release any hazardous gases because it runs on a closed circuit. Because they run entirely on electricity, pure electric vehicles are very environmentally friendly.

1.8.1.9. Spacious Cabin and More Storage

Electric vehicles provide greater cabin room and the possibility to turn these areas into storage because they have fewer moving parts. Since the conventional ICE is no longer present, you can also find storage spaces under the hood. In comparison to vehicles that run on gasoline, electric vehicle motors and batteries occupy less space.

1.8.2. Disadvantages of Electric Vehicles

1.8.2.1. Driving Range

People who have done their research on electric cars, are very much familiar with the term 'range anxiety.' The current units of electric vehicles are not capable of covering long distances that other fossil-fueled cars can, but the distance that can be traveled on a single charge has drastically improved recently and continues to do so. A lot many electric cars can now travel around 70–100 miles, and some even more than that, with just a single trip to the charging point. HEVs can do much to decrease range anxiety as the electric motor operates in tandem with a combustion engine.

Nevertheless, most of the trips via car are observed to be less than 30 miles, which many of the electric vehicles can easily do without any problem. Similarly, the way that people fuel their cars needs a different mindset to a fossil-fueled engine. Instead of just filling up in an infrequent manner, one simply needs to charge the car regularly, in a similar way in which a mobile phone is charged.

1.8.2.2. Recharge Time

Going to the motorway service station as the petrol gauge is flashing, refilling at the station, and getting back on the way takes almost five minutes. Thus, it can be said that it is a convenient process that everyone is familiar with. However, charging electric vehicles does not take much time. Certain estimates show that around 80% of the electric vehicle charging takes place on a slow charge at home overnight.

In addition, a lot many businesses currently have electric vehicle charging points in their parking lot. But charging could be a key issue when it gets down on highways. It would be hard to get to a place for charging and wait for a few hours and get back on the way.

Unluckily, there are no five-minute recharge available for electric cars. Gradually, rapid charging is becoming more common these days; one just needs to plan it into longer journeys as even a rapid charge takes 20–30 minutes.

1.8.2.3. Battery Life

A battery is an important part of an electric car, no one would be driving anywhere without the battery. For example, in the UK, the batteries in electric vehicles that are in use currently have a limited life expectancy, and thus will require to be replaced every 3–10 years based on the model. Replacement of a battery is a longer-term cost calculation that must be remembered when a person is considering purchasing an electric vehicle.

1.9. REGENERATIVE BRAKING

Regenerative braking, also referred to as regen, is the process of converting the kinetic energy of an electric vehicle into chemical energy that is stored in the battery so that it can be used to power the vehicle at a later time. The reason it's braking is that it slows down the car. Because the energy is recovered and put to new use in the battery, it is regenerative. The equation $E = \frac{1}{2}mv^2$ describes how the kinetic energy stored in a moving vehicle is related to its mass and speed. Your car has twice the kinetic energy if it is twice as heavy and has four times the kinetic energy if it is traveling twice as quickly, all other things being equal. The kinetic energy inside your car must go somewhere whenever it slows down. Let's examine the direction that this energy goes. Your car's rolling resistance, mechanical friction, and aerodynamics all use up some kinetic energy. These rays of energy heat the road, the air around your car, and the different spinning components. But when you stomp on the brakes, your brake pads convert the vast majority of the kinetic energy into heat. Regenerative braking in the Tesla Roadster recovers some energy that would have otherwise been lost when applying the brakes.

1.9.1. Working

Electric motors coupled to batteries power electric cars, trains, and other electric vehicles. Energy from the batteries powers the motors while you drive, rotating the wheels and giving you the kinetic energy you need to move. The entire process reverses when you apply the brakes and come to a stop: electrical circuits cut off the motors' power. The wheels are now turning the motors because of your kinetic energy and momentum, which causes the motors to function like generators and begin generating electricity rather than consuming it. These motor-generators return energy to the batteries, which charge them. Therefore, a significant amount of the energy lost during braking is recovered by the batteries and is usable for restarting. The majority of cars that use regenerative brakes also have regular (friction) brakes operating in tandem because, in reality, they take some time to slow down; this is also a good idea in case the regenerative brakes fail. Regenerative brakes don't save all of your braking energy for this reason, among others.

The term high voltage usually means electrical energy at voltages high enough to inflict harm on living organisms. Two factors considered in classifying a voltage as "high voltage" are the possibility of causing a spark in air, and the danger of electric shock by contact or proximity.

The International Electrotechnical Commission defines high voltage as above 1000 V for alternating current, and at least 1500 V for direct current—and distinguishes it from low voltage (50–1000 VAC or 120–1500 VDC) and extra low voltage (<50 VAC or <120 VDC) circuits.

When a person touches two points on a circuit, voltages higher than roughly 50 volts can typically result in dangerously high current flowing through them; hence, safety regulations are typically more stringent around such circuits. Applying voltages higher than 50 V to dry, intact human skin can result in heart fibrillation if they create currents in bodily tissues that happen to cross the chest region. The electrical conductivity of dry human skin determines the voltage at which electrocution is dangerous. The insulating properties of dry skin can prevent damage to living human tissue for up to 50 volts. Even voltage sources below 40 V can be deadly if the same skin gets wet, if there are wounds, or if the voltage is applied to electrodes that penetrate the skin.

Unintentional contact with high voltage that supplies enough energy has the potential to cause fatalities or serious injuries. This can happen because the body acts as a conduit for current, resulting in heart failure and tissue damage. Burns from the arc that is created by unintentional contact are another possible injury. If the victim's airways are impacted by the burns, the situation may become more dangerous. Injuries can also arise from the physical forces incurred by individuals who are thrown a significant distance or fall from an elevated height. Exposure of low energy to high voltage can be innocuous; an example of this would be the spark that appears in a dry climate when one touches a doorknob after stepping on a carpet. The current, or the rate at which charge transfers, is low, even though the voltage may reach several thousand volts. Electrical workers use insulated rubber gloves and mats as safety gear. These guard against electric shock to the user. To make sure it is still protecting the user, safety equipment is tested on a regular basis. Test laws differ from nation to nation. Testing businesses provide services ranging from glove testing to Elevated Working Platform (or EWP) testing, with the ability to test at up to 300,000 volts (International Energy Agency, 1999; Ayodele and Mustapa, 2020).

REFERENCES

1. Ayodele, B., & Mustapa, S. (2020). Life cycle cost assessment of electric vehicles: A review and bibliometric analysis. *Sustainability*, *12*(6), 2387. https://www.mdpi.com/2071-1050/12/6/2387 (accessed on 9 November 2024).
2. Berndt, D. (1997). *Maintenance-Free Batteries: Lead-Acid, Nickel/Cadmium, Nickel/Hydride: A Handbook of Battery Technology*. Research Studies Press; John Wiley & Sons.
3. Blomen, L. J. M. J., & Mugerwa, M. N. (1993). *Fuel Cell Systems*. Plenum Press.
4. Burke, A. (2000). Ultracapacitors: Why, how and where is the technology? *Journal of Power Sources*, *91*, 37–50.
5. Chan, C. C., & Wang, H. Q. (1996). New scheme of sliding-mode control for high-performance induction motor drives. *IEE Proceedings B*, *143*, 177–185.
6. Chan, C. C., Jiang, J. Z., Chen, G. H., & Chau, K. T. (1993). Computer simulation and analysis of a new polyphase multipole motor drive. *IEEE Transactions on Industrial Electronics*, *40*, 570–576.
7. Chan, C. C., Leung, W. S., & Ng, C. W. (1990). Adaptive decoupling control of induction-motor drives. *IEEE Transactions on Industrial Electronics*, *37*, 41–47.
8. Drive Electric Vermont. (n.d.). Introduction to electric vehicles. *Drive Electric Vermont*. https://www.driveelectricvt.com/why-go-electric/intro-to-electric-vehicles (accessed on 9 November 2024).
9. Good Energy. (2017). The advantages of electric vehicles. *Good Energy*. https://www.goodenergy.co.uk/blog/2017/09/28/driving-change-the-advantages-of-electric-vehicles/ (accessed on 9 November 2024).
10. Hannan, M., Azidin, F., & Mohamed, A. (2014). Hybrid electric vehicles and their challenges: A review. *Renewable and Sustainable Energy Reviews*, *29*, 135–150. https://www.sciencedirect.com/science/article/abs/pii/S1364032113006370 (accessed on 9 November 2024).
11. International Energy Agency. (1999). *Assessment of Electric Vehicle Impacts on Energy, Environment and Transportation Systems*.
12. Liu, L., Kong, F., Liu, X., Peng, Y., & Wang, Q. (2015). A review on electric vehicles interacting with renewable energy in smart grid. *Renewable and Sustainable Energy Reviews*, *51*, 648–661. https://www.sciencedirect.com/science/article/abs/pii/S1364032115006085 (accessed on 9 November 2024).

13. Miller, T. J. E. (1989). *Brushless Permanent-Magnet and Reluctance Motor Drives*. Oxford University Press.

14. Miller, T. J. E. (1993). *Switched Reluctance Motors and Their Control*. Magna Physics; Oxford University Press.

15. Rahman, M. A., & Qin, R. (1997). A permanent magnet hysteresis hybrid synchronous motor for electric vehicles. *IEEE Transactions on Industrial Electronics*, *44*, 46–53.

16. Staas, P. T., Grady, W. M., Aprapostathis, A., & Thallam, R. S. (1997). A statistical method for predicting the net harmonic current generated by a concentration of electric vehicle battery chargers. *IEEE Transactions on Power Delivery*, *12*, 1258–1266.

17. Van Mierlo, J. (2000). Views on hybrid drivetrain power management. In *Proceedings of the 17th International Electric Vehicle Symposium* (CD-ROM).

18. Wakefield, E. H. (1998). *History of the Electric Automobile: Hybrid Electric Vehicles*. Society of Automotive Engineers.

19. Zhan, Y. J., Chan, C. C., & Chau, K. T. (1999). A novel sliding-mode observer for indirect position sensing of switched reluctance motor drives. *IEEE Transactions on Industrial Electronics*, *46*, 390–397.

Chapter 2

Life Cycle Assessment (LCA) of Electric Vehicles

Contents

INTRODUCTION

Global greenhouse gas emissions are significantly impacted by motorized vehicles. Electric vehicles (EVs) are one way to lessen the impact on the environment. The deployment of EVs must be carefully considered, just like any reduction measure, to prevent problem shifting or rebound effects. The most common method for measuring the environmental impact over the full life cycle, from raw material extraction to end-of-life, is Life Cycle Assessment (LCA). It is difficult to determine the environmental impact of EVs using LCAs. The findings of various EV studies differ substantially. It is feasible to draw broad conclusions from these findings (such as the possibility that EVs will have less of an impact on the environment than conventional cars). However, answering more specialized questions is more difficult. It is challenging to ascertain which circumstances and circumstances these conclusions apply to.

The opaqueness of the factors influencing the LCA of EVs is one factor contributing to the calculation's difficulty and the results' variability. Important influencing factors of the LCA results of EVs include the energy mix, usage patterns, and material composition of the vehicles. Primary data from the EV's whole life cycle is typically not available to LCA practitioners. The production, use, and disposal of the vehicle involve several stakeholders, which distribute the necessary information throughout the whole supply chain and over time. Carrying out an LCA of an EV is made more difficult by factors like complex supply chains, a range of potential use patterns, unknown future developments, and confidentiality concerns.

The qualitative results are displayed in relevant research. The resultant environmental impact is indicated by the y-axis, while the x-axis indicates the distance traveled. A certain range is supported by the travel distance. The environmental effects of both the extraction of raw materials and the manufacturing of vehicles are represented by the y-intercepts of the lines. A range of values is possible, depending on the assumptions made for these initial stages of the life cycle. Here, two automobiles with low and high environmental impact during the extraction of raw materials and vehicle manufacturing are displayed. The line's slope indicates the environmental impact a vehicle has during its use in relation to the distance it travels.

The vehicle's energy consumption and the energy mix's environmental impact determine the slope. The energy consumption and energy mix values vary depending on the scenario, resulting in a minimum and maximum slope. Based on the presumptions made during the use phase, this produces a variety of final results. A range is defined with potential LCA outcomes when the raw material and manufacturing phase, use phase, and driving distance are combined. The perfect scenario, utilizing an energy mix with (nearly) no environmental impact, is depicted by the continuous line. In this instance, the car's consumption is unimportant. This qualitative chart does not take the end-of-life phase into account for clarity's sake. The end-of-life phase's integration broadens the range of potential outcomes and adds another set of parameters. A framework for EV life cycle assessment (LCA) elucidates various parameters and provides a detailed description of the EV system in relation to LCA (Figure 2.1) (**Nordelöf et al., 2014; Ayodele and Mustapa, 2020**).

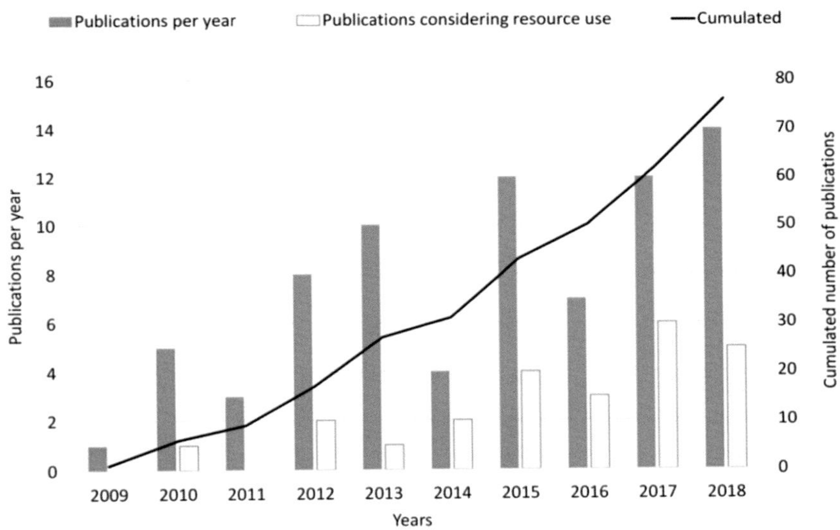

Figure 2.1. *Yearly and cumulated number of publications performing life cycle assessment (LCA) on electromobility as well as number of publications considering resource use.*

Source: Dolganova I, Rödl A, Bach V, Kaltschmitt M, Finkbeiner M. A Review of Life Cycle Assessment Studies of Electric Vehicles with a Focus on Resource Use. Resources. *2020; 9(3):32. https://doi.org/10.3390/resources9030032.*

2.1. ELECTRIC VEHICLES AND LIFE CYCLE ASSESSMENT

The term "electric vehicles (EVs)" refers to a variety of car models, including battery-electric and hybrid models. Both an electric and a combustion engine are found in hybrids. Depending on the kind of hybrid, such as plug-in hybrids, the car can be charged from the grid.

EVs are viewed as a way to lessen or do away with the drawbacks of driving modern fossil fuel vehicles. Their features provide benefits that address issues with traditional vehicles. These are the removal of tailpipe emissions, the decrease in noise, and the independence from fossil fuels. Mobility free of fossil fuels is made possible by their capacity to operate on a variety of energy sources thanks to batteries' capacity to store electricity. Using an EV produces very little noise and very few local emissions. These advantages are extremely valuable in megacities that have high noise levels and severe air pollution. The primary drawbacks of EVs are related to their driving range and cost. Another issue is the shift in resource consumption from traditional cars to electric engines with permanent magnets and lithium-ion batteries. This results in a greater demand for rare earth metals like neodymium as well as metals like cobalt, manganese, and lithium. At the moment, EVs' driving range is much less than that of conventional cars. Many consumers believe that the driving range of EVs is insufficient, Although this range is adequate for most daily travel requirements. Furthermore, because heating and cooling appliances consume a lot of energy, using them can drastically shorten the range. Purchasing an EV costs more than buying a comparable-sized conventional car. Nevertheless, despite these drawbacks, EVs are effective even in environments that might be deemed unfavorable. As an illustration, consider Norway, where the range is lowered by the need for intense heating due to its climate. However, EV adoption has been particularly strong in the Scandinavian nations as a result of provided incentives like tax breaks and a municipal charging infrastructure.

A significant portion of the electric vehicle's environmental impact happens during use and is closely related to the energy mix and energy consumption during use. Three categories of factors determine how much energy an electric vehicle uses: driving resistances, auxiliary use, and losses. In order to reach and sustain a specific speed, driving resistance must be overcome. Aerodynamic resistance, rolling, and acceleration are a few examples. The resistances are influenced by the frontal area and weight of the vehicle. Additionally, using auxiliary devices like air conditioning, heating, and ventilation results in higher energy usage. Because of the varying component efficiencies, losses also happen during the conversion of electric energy into mechanical energy.

2.1.1. Life Cycle Assessment

The mass and energy fluxes of the various product systems must be compared over the course of their life cycles using the LCA methodology in accordance with ISO 14040 to determine the environmental impact of EVs. LCA has shown to be a helpful technique for evaluating and measuring the effects that products have on the environment. In the past 20 years, it has gained widespread acceptance and been standardized and approved internationally.

The definition of the goal and scope, the inventory analysis, the impact assessment, and the interpretation of the findings are the four steps that make up the LCA procedure. The process must be viewed as an iterative process as opposed to a single exercise. For example, a change to the goal and scope definition might be necessary based on interim results from the inventory analysis, impact assessment, and interpretation.

The intended application, the reason for conducting the study, and the intended audience must all be specified during the goal and scope definition process. In addition, a detailed description of the product system under investigation is required, including information on its boundaries and functional unit (the system's measured performance). For reasons of transparency, several additional methodological choices and details must also be documented.

The goal of life cycle inventory analysis is to comprehend and account for the input and output flows that occur within the system under observation as well as the interactions it has with the outside world (elementary flows). The last step in calculating all incoming and outgoing flows that cross system boundaries can be assisted by material flow nets based on Petri-nets. A life cycle inventory is this balance of input and output that is referred to by the functional unit.

During the Life Cycle Impact Assessment, the possible environmental effects of extracted resources and released pollutants are assessed based on the Life Cycle Inventory. Emissions are first categorized based on how much of each impact category they contribute to. Second, the impact equivalents that could be contributed by them are stated, such as the CO_2-equivalents (CO_2-eq) for climate change.

Plotting the study's results to highlight the key issues is done during the interpretation phase. Furthermore, the study's dependability is examined through sensitivity and uncertainty analyses in addition to consistency and completeness checks (**Nordelöf et al., 2014; Ayodele and Mustapa, 2020**).

2.1.2. Life Cycle Assessment of Electric Vehicles

LCA on EVs is a difficult task since the complexity of a product greatly influences the complexity of the corresponding LCA study. Recently, several LCAs on EVs or parts specific to EVs, such as Li-ion batteries, have been released. Hawkins, among others additionally to Nordelöf et al. offered thorough analyses of EV LCAs. They located complete reports, journal articles, and conference papers, totaling 55 and 79 pertinent studies, respectively. They both present results that are wildly different, which is perhaps to be expected for a complicated product in a developing market. The lack of primary data and variations in methodological decisions can both be used as additional explanations for the divergence. The goal of the E-Mobility Life Cycle Assessment Recommendations (eLCAr) project guidelines is to improve methodological choices' transparency and harmonize the methodological approach (Nordelöf et al., 2014). The majority of LCAs concentrate on vehicle production and associated raw material acquisition because contemporary EVs have just entered the market and there are numerous ongoing research initiatives to further develop the essential core technologies (such as energy storage). Because of this and the dearth of long-term measurements, monitoring, and experiences, very few LCAs specifically address the use phase. The primary source of use profiles is standard driving cycles, and the corresponding energy consumption is computed in a general way as specified, for example, in the eLCAr guidelines. There are currently no published models of use phase scenarios that are based on real-world data, particularly measurements unique to each site (Hawkins et al., 2013).

2.2. FRAMEWORK

The characteristics of the vehicle, the area in which it is used, and user influences are some of the factors that determine how an EV affects the environment. The driving habits of various users and the energy consumption are influenced by the local climate and the use of heating and cooling systems. These factors, along with the local electricity mix, affect how environmentally friendly EVs are. Therefore, to respond to particular LCA questions, the vehicles must be understood as a component of the environment in which they operate. Neglecting these interdependencies may cause significant details to be omitted or overlooked. The LCA practitioner can assess the impact of parameters on the environment by tying external influences to the vehicle's use phase. By establishing a descriptive framework, an LCA practitioner can reduce the uncertainty of LCAs by translating external influencing factors into environmental impacts.

The energy consumption and external factors are related and shows the EV as a component of a larger system of influencing factors. The life cycle of an electric vehicle is defined by the flows of materials and energy required during its entire life cycle for both manufacturing and operation (mid-level). The life cycle and the LCA results are influenced by the top level external factors in which the EV is deployed. These outside variables can be categorized into three groups: the environment, the infrastructure, and the user. We concentrate on how these outside variables affect the use phase, or energy consumption. Energy consumption in the use phase (bottom level) is influenced by both external and internal factors (vehicle charact eristics such as weight and frontal area size affecting aerodynamic resistance) (Table 2.1).

Table 2.1. *Analysis of the Selected Battery LCA That Cover Resource Use Assessment*

Publication, Year	Assessed Battery Chemistries	Functional Unit	Impact Assessment Methods for Resources/Impact Categories	Conclusion of Resource Use Assessment	Results of Sensitivity Analysis with Regard to Resource Use
Majeau-Bettez et al. (2011)	Lithium-ion and nickel metal hydride (NiMH); NCM; iron phosphate lithium-ion (LFP)	50 MJ accumulated by the battery and delivered to the powertrain (roughly driving 100 km)	ReCiPe 2008/FDP and MDP	Highest MDP of NiMH batteries because of electrode materials nickel and lanthanum; NCM have higher impacts than LFP due to use of nickel, cobalt, and partly from copper; mining and metallurgy activities for nickel production are responsible for 80% of MDP. Electricity consumed (European electricity mix) during use phase contributes to more than 40% of FDP; highest burdens found for NiMH	Sensitivity analysis only carried out related to environmental impacts
McManus et al. (2012)	Lead acid battery, nickel cadmium; nickel metal hydride; lithium-ion; sodium sulfur battery	100 kg (of battery)	ReCiPe 2008/FDP and MDP	Highest burdens of lithium-ion batteries in FDP and MDP due to metal depletion from ferrite production; lead acid batteries have the lowest impacts	No sensitivity analysis was carried out

Faria et al. (2014)	Lithium manganese oxide battery (LMO)	200,000 vehicle km (service life of the vehicle)	CML-IA 2001/ADP elements	Cathode of lithium-ion battery contributes by 28% to ADP elements due to lithium or manganese use	No sensitivity analysis was carried out
Ahmadi et al. (2017)	Iron phosphate lithium-ion battery	Energy provided over the total battery life cycle in kWh	ReCiPe 2008/FDP and MDP	Trade-offs by extending the service life of battery pack: MDP increases due to higher demand for virgin materials but less fossil fuel use (FDP)	Sensitivity analysis considering battery degradation: only minor effect on metal depletion; greater influence on fossil depletion. The higher the degradation rate the lower the energy efficiency, which increases energy use
Messagie et al. (2015)	LMO; lithium iron phosphate (LFP)	1 kWh of energy stored in the battery	ReCiPe 2008/MDP	Higher MDP impacts for LMO batteries due to manganese and copper use; benefits due to material recycling	No sensitivity analysis was carried out
Sanfelix et al. (2015)	Lithium manganese oxide cells; hybrid systems from lithium iron phosphate cells prolong the lifetime	1 km driven under European average conditions	CML-IA 2002/ADP elements	The credit from recycling of a hybrid energy storage system offsets ADP impacts from manufacturing and use phase; metal use and the necessary mining operations for a hybrid energy storage system cause most of the resource depletion impacts	No sensitivity analysis was carried out

Peters et al. (2016)	Sodium ion; LFP; LFP-with lithium-titanate anode (LFP-LTO); LMO; NCA; NCM	1 kg and 1 kWh	ReCiPe 2008, Eco-Indicator 99/minerals, CML-IA 2002/ADP (different reserve bases); AADP, CexD, EcoSc	Worst results of NCM in CML-IA, AADP, EcoSC, and CexD but not in Eco-Indicator 99 and ReCiPe (for functional unit in kWh); best results of LFP with almost all methods besides CML-IA and sodium-ion batteries (for functional unit in kg) if CML-IA is applied Most of the methods show impacts from copper use; in CML-IA, AADP, EcoSC, CexD, highest impacts from use of tantalum, cobalt, nickel, cadmium, partly to lithium	No sensitivity analysis was carried out
Zackrison et al. (2016)	High-capacity lithium-air batteries	Vehicle kilometer	CML-IA 2002/ADP	Highest burdens due to production phase (89% copper, 5% lithium); recycling avoids burdens from depletion	Sensitivity analysis only carried out related to environmental impacts
Van Mierlo et al. (2017)	LMO, LFP; sodium-nickel chloride; lead acid nickel cadmium; nickel metal hybrid	1 kWh	ReCiPe/MDP	Much higher MDP for LMO batteries than for LFP batteries because of high manufacturing burdens. Which materials are responsible was not analyzed. Separate discussion of lithium availability. Materials recycling can have significant benefits. proper material recycling is needed to prevent lithium shortage in EV industry	No sensitivity analysis was carried out

Bobba et al. (2018)	LMO; nickel-manganese-cobalt (NMC)	Average yearly energy balance of the system in which the battery stores energy	CML-IA 2002/ADP	The use of repurposed batteries from mobility application reduce ADP impacts; no detailed analysis of the contribution of certain materials	Changing the allocation factors (>0) for the manufacturing/ end of life (EoL) of repurposed EV batteries, benefits from battery reuse largely diminish—especially for impact categories dominated by manufacturing/ EoL (i.e., ADP); the higher the residual capacity the higher the level of benefit especially for ADP

Source: Dolganova I, Rödl A, Bach V, Kaltschmitt M, Finkbeiner M. A Review of Life Cycle Assessment Studies of Electric Vehicles with a Focus on Resource Use. Resources. 2020; 9(3):32. https://doi.org/10.3390/resources9030032.

2.2.1. Influencing Factor: Vehicle

Particular features of the car affect how much energy it uses during use. Since these are built-in characteristics of the vehicle, they are regarded as internal factors in this framework.

2.2.2. Influencing Factor: User

The way an EV is driven and charged, and how much it uses its auxiliaries all have an impact on how the vehicle affects the environment. Driving with greater aggression consumes more energy, whereas driving with greater caution uses energy more effectively. In comparison to the use of grid energy in many countries, the share of renewable energy can be increased significantly depending on the charging behavior and the willingness to install renewable energy specifically for the EV (e.g., in the form of solar panels). Lastly, a major factor in energy consumption is the use of heating and cooling to reach the appropriate temperature. A reduced environmental impact is directly correlated with the willingness to tolerate a warmer vehicle in the summer and a colder vehicle in the winter.

2.2.3. Influencing Factor: Infrastructure

For the LCA calculation, one of the most important parameters is the electricity mix. An energy mix based solely on renewable energy sources produces a totally different outcome than one based on fossil fuels. It can be difficult to select the right combination that produces fair and trustworthy results while reflecting the realities of the real world. An energy mix that is solely based on renewable energy is used in many life cycle assessments.

But frequently, it's unclear whether this is an accurate depiction of the grid condition or whether the EV is simply receiving credit for renewable energy instead of being put to another use. In the latter instance, it is necessary to determine whether the crediting is warranted. In theory, regular household plugs can be used to charge electric vehicles. However, more advanced solutions are frequently needed in public spaces or workplaces to enable sufficient and secure charging. Major construction activities may be required for the installation of these charging stations, depending on the site's conditions. When one or a small number of cars use a single charging station, these actions may be important. The smart charging options are influenced by the infrastructure for charging that is available. Applications for smart charging can raise the percentage of renewable energy used to charge EVs.

2.2.4. Influencing Factor: Surrounding Conditions

The environmental impact of EVs is influenced by the surrounding conditions. The topography, climate, and kind of road are found to be important determinants of energy use. The need for the vehicle's heating and cooling systems is influenced by the weather. Variations in temperature occur both daily and seasonally, which causes variations in energy usage.

The windshield of the car may fog up and need ventilation, air conditioning, or heating, depending on how temperature and humidity interact. Resistance heating is primarily used in electric vehicles today. The amount of energy used for heating can be decreased with alternative technologies like a heat pump. A landscape with a flat topography requires less energy than one with hills. When determining the energy consumption, it is crucial to take the breaking recuperation into account as well. Certain parameters, like the speed limit and the frequency of stops, like at traffic lights, are determined by the type of road, such as city streets or highways. These variables affect how much is consumed.

2.2.5. Fields of Application of Framework

The framework gives LCA practitioners the technical background on EVs they need, which is helpful. It can be used in different Life Cycle Assessments. The goal and scope section of the study must address the influencing factors. Design choices and comparison analyses with other automotive technologies are a couple of examples of potential topics. EVs compete with other alternative fuel or propulsion system vehicles as well as conventional automobiles. Detailed analysis of the environmental impact aids in the identification of use cases and geographical areas where EVs are most beneficial.

This can help consumers and policymakers make informed decisions in increasingly diverse markets. An additional context for decision-making is an EV's design. Influential factors for environmental sensitivity can be important to consider when evaluating design alternatives. Issues may change from one phase to the next or between different parts of the vehicle. To guarantee reliable results, a thorough examination of the system's parameters and overall operation is required to assess the potential environmental benefits of various design options. For automakers and their suppliers, this is pertinent (Nordelöf et al., 2014; Ayodele and Mustapa, 2020).

2.3. LIFE CYCLE ASSESSMENT OF ELECTRIC VEHICLES IN COMPARISON TO COMBUSTION ENGINE VEHICLES

Researchers and businesspeople have been working to improve efficiency by identifying the best alternatives for transportation fuel, energy sources, vehicle systems, and other resources and using them more effectively. The world's need for sustainable resources and transportation systems is growing as a result of rising oil prices, energy security concerns, and climate change. According to the EPA report, up to 65% of all greenhouse gas emissions worldwide are attributed to industrial activities and CO_2, which is the main gas content in emissions from fossil fuels. Furthermore, the World Business Council for Sustainable Development report projects that the number of light-duty vehicles will rise to 1.3 billion by 2030 and to 2 billion by 2050, in addition to increased greenhouse gas emissions. As a result, the entire state of climate change, air pollution in cities, and the depletion of non-renewable energy sources would all drastically change. All of these worries have compelled researchers and manufacturers to introduce new technologies into the automotive sector.

With the combination of different technologies, including plug-in hybrid electric vehicles (PHEV), fuel-cell electric vehicles (FCEV), and hybrid electric vehicles (HEV), EVs have become the best option for vehicles. Compared to internal combustion engines (ICEVs), EVs have fewer tailpipe emissions, require less maintenance, have more efficient powertrains, and, lastly, produce less pollution in cities. Global greenhouse gas emissions have decreased as a result of the transportation sector using electricity rather than fossil fuels like gas, diesel, and gasoline (GHG). Potential reductions in greenhouse gas emissions are roughly 90% for EVs, 25% for HEVs, and 50–80% for PHEVs. Energy storage plays a crucial role in electric vehicle applications by regulating and controlling the flow of energy. The right energy storage device should be chosen based on several criteria, including life cycles, energy density, power density, cycle efficiency, and self-charge and discharge characteristics. The production of batteries using heavy metals, the electricity mix used for charging, battery disposal, and vehicle ownership are all crucial aspects of the electric vehicle life cycle that require investigation for possible environmental effects. EVs are becoming more and more popular in the global market, but their development is not without technical difficulties. As such, an analysis of the environmental impact related to the vehicle manufacturing, fuel cycle, use, and recycling phases of EVs is required.

2.3.1. Evaluation Technique

The general life cycle of a vehicle is determined by a number of factors, including technical specifications, energy supply and resources, the in-use phase, and production and recycling technologies. The propulsion system is the primary parameter in technical specifications, and the customer always considers this when making a purchase to balance resource and energy consumption. A conventional propulsion system uses an internal combustion engine, which uses fossil fuels as its primary resource during the in-use phase. With an EV, however, the majority of resources are used during the manufacturing stage because the battery, storage system, and drivetrain require a lot of work. Thus, the energy and resource supply constitute the second parameter, while the manufacturing technology and recycling strategy determined by the product's end-of-life phase define the third. Lastly, life cycle considerations are taken into account during the in-use phase. This includes the demand for transportation, the driving profile of each individual user, and the energy and resources (such as fuel and electricity) used to maintain overall balance.

Two assessment techniques are used to analyze the environmental and economic impact of waste generated over its entire life cycle: LCA and life cycle cost (LCC) analysis.

2.3.1.1. Stages of Life Cycle Assessment

The International Organization for Standards (ISO) states that ISO 14,040 and ISO 14,044 are commonly used in LCA to quantitatively examine the environmental impact of a system or product over its entire life, from the use of raw materials to system recycling or disposal (Table 2.2).

In general, LCA is described in major four phases:

- Goal and scope definition;
- Inventory analysis;
- Impact assessment; and
- Interpretation.

Table 2.2. *Summary of the Selected LCA that Cover Resource Use Assessment*

Author, Year	Title	Functional Unit	Impact Assessment Methods for Resources/ Impact Categories	Conclusion of Resource Use Assessment	Results of Sensitivity Analysis with regard to Resource Use
Notter et al. 2010	Contribution of Li-ion batteries to the environmental impact of electric vehicles	One average kilometer driven by a vehicle with electric drivetrain and Li-ion batteries on the European road network	CML-IA 2002/ ADP	EV have a 37% lower burden than ICEV	Sensitivity analysis only carried out related to environmental impacts
Bartolozzi et al. (2013)	Comparison between hydrogen and electric vehicles by life cycle assessment: A case study in Tuscany, Italy	200 km at nominal full load within an urban area	CML-IA 2002/ ADP	EV have an 80% lower burden than ICEV	No sensitivity analysis was carried out
Hawkins et al. (2013)	Comparative environmental life cycle assessment of conventional and electric vehicles	1 km driven under European average conditions	ReCiPe/MDP and FDP	MDP: EV has a roughly three times higher burden than ICEV FDP: EV perform between 25% and 30% better than ICEV (average EU mix)	MDP: increase of vehicle life reduces burdens by EV more significantly as for ICEV, but EV still has higher burdens even with highest vehicle lifetime FDP: decrease of energy use for EV and fuel use for ICEV reduces burdens by EV more significantly as for EV, therefore emphasizing the better result of EV

Messagie et al. (2014)	A range-based vehicle life cycle assessment incorporating variability in the environmental assessment of different vehicle technologies and fuels	1 km driven under European conditions	Eco-Indicator 99/mineral resource depletion (MRD)	EV have slightly lower (5%–10%) burden than ICEV	No sensitivity analysis was carried out
Girardi et al. (2015)	A comparative LCA of an electric vehicle and an internal combustion engine vehicle using the appropriate power mix: the Italian case study	Lifetime of the vehicle (150,000 km)	CML-IA 2002/ ADP	EV have 40% lower burden than ICEV	Sensitivity analysis only carried out related to environmental impacts
Tagliaferri et al. (2016)	Life cycle assessment of future electric and hybrid vehicles: A cradle-to-grave systems engineering approach	1 km driven by one vehicle	CML-IA 2002/ ADP_{fossil} and $ADP_{elements}$	ADP_{fossil}: EV has lower burden than ICEV (50% less in one scenario and almost two times in another) $ADP_{elements}$ (assumption of "high recycling rate"): EV has higher burden than ICEV (almost nine times more in one scenario and three times more than ICEV in another)	ADP_{fossil}: change of electricity in 2030 and 2050 with less fossil and more renewable and nuclear energy as well as more biodiesel fuel, does not change the lower EV burden $ADP_{elements}$ was not considered in the sensitivity analysis

Henßler et al. (2016)	Resource efficiency assessment—comparing a plug-in hybrid with a conventional combustion engine	Life cycle of one car	ESSENZ method/ADP_{fossil} and $ADP_{elements}$	ADP_{fossil}: EV (when using electricity from hydropower) has lower burden (40%) than ICEV $ADP_{elements:}$ EV has higher burden (170%) than ICEV	No sensitivity analysis was carried out
Cellura et al. (2016)	Electric mobility in Sicily: an application to a historical archaeological site	Transportation of one person for 1 km (1 pkm)	ADP applied as recommended by life cycle data system (ILCD) 2011	ADP is on average 500% higher for BEVs than for ICEVs; the highest impact for BEV when it is powered by solar energy (PV) (ca. 1100% higher impacts in comparison to ICEV average)	No sensitivity analysis was carried out
Choma et al. (2017)	Environmental impact assessment of increasing electric vehicles in the Brazilian fleet	Transportation of one person for 1 km (1 pkm)	CML-IA 2002/ ADP	EV have lower burden (between one-third and 80%) than ICEV	ADP: different electricity as well as fuel sources are considered, not changing the trend that EV has lower burdens than ICEV

Cimprich et al. (2017)	Extension of geopolitical supply risk methodology: characterization model applied to conventional and electric vehicles	Production of one vehicle	GeoPolRisk	EV has higher criticality compared to ICEV; EV has higher ADP potential than ICEV	No sensitivity analysis was carried out
Helmers et al. (2017)	Electric car life cycle assessment based on real-world mileage and the electric conversion scenario	100,000 km driven under European average conditions	ReCiPe/MRD, FDP	MRD: EV has three times higher burden than ICEV FDP: ICEV has three times higher burden than EV	Four different electricity and urban vs. mixed driving conditions were considered, emphasizing the results for MRD (ICEV has lower burdens as EV) and FDP (EV has lower burdens than ICEV)
Van Mierlo (2017)	Comparative environmental assessment of alternative fueled vehicles using a life cycle assessment	1 km driving distance	ReCiPe/MRD, Eco-Indicator 99—Metal depletion	Lower metal depletion scores for lithium iron phosphate-based batteries	No sensitivity analysis was carried out
Souza et al. (2018)	Comparative environmental life cycle assessment of conventional vehicles with different fuel options, plug-in hybrid and electric vehicles for a sustainable transportation system in Brazil	Vehicle with an occupation of 1.6 persons and a total life traveled distance 160,000 km	CML-IA 2002/ ADP_{fossil} and $ADP_{elements}$	$ADP_{elements}$: burdens of EV and ICEV are similar; ADP_{fossil}: EV has ca. 2.5 lower burden than ICEV	Changes in energy supply use for fuels and electricity ADP_{fossil}: emphasizes trend that EV has lower burdens than ICEV, except for ethanol based ICEV, which scores better than the EV $ADP_{elements}$: no changes, since the assumptions for manufacturing phase remained the same

Del Pero et al. (2018)	Life cycle assessment in the automotive sector: a comparative case study of internal combustion engine (ICE) and electric vehicle	Lifetime of the vehicle 150,000 km	ADP applied as recommended by life cycle data system (ILCD) 2011	EV has a higher burden (ca. 32%) than ICEV	No sensitivity analysis was carried out
Yu et al. (2018)	Life cycle environmental impacts and carbon emissions: a case study of electric and gasoline vehicles in China	Life cycle vehicle travelling (250,000 km)	CML-IA 2002/ ADP	ICEV has six times lower burden than both analyzed types of EV (with lithium-iron and nickel-based batteries)	Sensitivity analysis only carried out related to environmental impacts

Source: Dolganova I, Rödl A, Bach V, Kaltschmitt M, Finkbeiner M. A Review of Life Cycle Assessment Studies of Electric Vehicles with a Focus on Resource Use. Resources. 2020; 9(3):32. https://doi.org/10.3390/resources9030032.

2.3.1.1.1. Goal and Scope Definition

LCA is carried out for a functional unit within a predetermined system boundary and analysis period thanks to goal and scope definition phases. Depending on the kind of product or service that needs to be examined, the system boundaries may be wide or specific. Because the modern supply chain is so complex, it is helpful to analyze which process causes the most environmental concern because system boundaries are broad. System boundary determines which processes, based on the assumptions made at the beginning, the intended use, and the cut-off criteria, should be included in the LCA. The references to which input and output can be related are defined by the functional unit. This makes it possible to compare two very different systems. Because the process's energy consumption varies over time, the analysis period negatively impacts the LCA results. The credibility of LCA results is determined by the quality of the data used.

2.3.1.1.2. Inventory Analysis

The amount of energy and materials produced or consumed at each supply chain node is examined during the inventory analysis phase. There are three main steps involved in this process: first, a flow chart comprising the raw materials, manufacturing process, transport, uses, and waste management is constructed; second, data on material inputs, products, and by-products, solid waste, air and water emissions, are gathered; and third, each relationship to the functional unit is calculated.

2.3.1.1.3. Impact Assessment

The inventory analysis comes after the calculation of the overall environmental impact in this phase. The effects on the environment include freshwater, marine, and terrestrial eutrophication; human toxicity; ozone depletion; land transformation; and climate change. Since the equivalent mass of CO_2 is used as an indicator when reporting the impacts of climate change, indicators are chosen for each impact factor category. The US Environmental Protection Agency's Tool for Reduction and Assessment of Chemicals and Other Environmental Impacts (TRACI), ReCiPi, the Open LCA method of Available Water Remaining per Area (AWARE), and other models are available for characterization.

2.3.1.1.4. Interpretation

This completes the LCA phase, which follows a methodical process for interpreting the assessment results. Using the LCA analysis as a basis, the assessment results are evaluated, summarized, and concluded in this phase.

This phase includes a summary of the inventory analysis and impact assessment findings.

This phase's primary goal is to ascertain the degree of confidence in the finished product and convey it fairly and accurately.

2.3.1.2. LCA Types

As illustrated in Figure 2.1, the entire life cycle of an electric vehicle is separated into three primary stages: cradle to gate, cradle to grave, and cradle to cradle. There are also sub-stages that include well to tank, tank to wheel, and well to wheel.

- "Cradle to gate" LCA refers to the portion of a product life cycle that ends at the factory gate (before the vehicle is delivered to the customer) and ends with the raw materials. In this instance, the use and recycling phase is skipped.
- "From the cradle to the grave" LCA refers to an electric vehicle's entire journey from raw material processing to eventual disposal. Its entire life cycle comprises assembly, upkeep, and refueling; at the end, its components are disassembled and recycled.
- "Cradle to cradle" LCA refers to an electric vehicle's closed-loop life cycle in which leftover construction debris and recycled products are repurposed at the raw material processing stage (cradle). It reduces the negative effects of products on the environment, pushes manufacturers toward sustainable production, and integrates social responsibility into product development.

In particular, well-to-wheel LCA, which emphasizes the energy provider's life cycle—such as that of gasoline or electricity—is utilized in the transportation of fuel and automobiles. It can be further separated into two phases: the tank to wheel phase, which concentrates on the energy carrier as tank to the vehicle's propulsion system during operation, and the well to tank phase, which concentrates on the energy source to the vehicle's tank storage. Analysis of the vehicle's energy conversion efficiency, overall energy conversion, carbon footprint, and emission effects is done using well-to-wheel analysis.

2.3.1.3. LCA Tools

The need to forecast life cycles and assess the environmental impact of products and services to make them more affordable, viable for the market, and appealing to customers has led to the development of Life Cycle Assessment (LCA) tools and software in recent years by several institutions and organizations. The

LCA software calculates the various data that are accessible via databases, spreadsheets, and Excel files that are stored on different cloud servers by organizations, environmentalists, and industrial sectors.

GaBi, SimaPro, Umberto, and Open LCA were the most widely used and well-liked programs because they provided a range of features. GaBi and SimaPro are the most widely used tools available, having been around for more than 15 years. However, they both have closed codes and expensive license fees. Another preferred tool is Umberto, developed by Ifu Hamburg, as this software is more complicated than GaBi and SimaPro and the license price is almost the same, but it doesn't provide any additional features as previous software. Open LCA is user-friendly, easy-to-use software that makes it possible to calculate every step involved in the life cycle assessment. Green Delta developed this software with assistance from UNEP (United Nations Environment Programme), PRé-Consultants (developer of SimaPro), and Pe-International (developer of GaBi). This software's ability to work with various databases that GaBi and other programs use, as well as to change the source code, is another benefit. Open LCA was initially intended to analyze how processes and products would affect the environment, but it can now also analyze economic factors.

2.3.1.4. Life Cycle Cost Analysis

The LCC of the system is defined as the sum of total costs experienced during its life span.

LCC covers all expenses incurred from the moment a system design decision is made, throughout the product or system's operational life, and up until its disposal. Since the overall cost of production exceeds the initial costs of purchasing, manufacturing, and transportation by a large margin, LCC analysis assists the engineer in analyzing all of the costs associated with the system or product. LCC is composed of two main components: the second is the cost of operations, which includes operations, maintenance, downtime, obsolescence, operator training, and inventory costs; the first is the cost of ownership, which includes procurement, depreciation, interest, inflation, and disposal costs. The main cost components of an LCC analysis are explained below in Table 2.3.

Table 2.3. *Overview of Sustainability Assessment Methods in Transport*

SL. No.	Author	Aspects of Sustainable Development			Area of Transport
		Economic	**Social**	**Environmental**	
1	Yedla and Shrestha (2003)	Levelized service cost (LSC)	-	Emission reduction potential (ERP)	Selection of alternative options for transport system
2	Brey et al., (2007)	Purchase cost, Environmental cost, Fuel cost	-	Acoustic emissions	Alternative fuel-based vehicles (HEV, FCEV)
3	Tisita and Pailacachi (2012)	Implementation cost, Technology maturity cost, Cost of energy	Energy security, Employment, Social welfare	CO_2 Emissions	Alternative fuels for road transport (biofuels, hydrogen, electricity)
4	Shiau and Liu (2013)	Modal split of transit	Traffic accidents, Mobility and transport for older adults and disabled persons, Transport infrastructure in remote areas, Transit subsidy in remote areas.	Emission intensity of greenhouse gases (GHG), Emission intensity of air pollutants, Proximity of transport infrastructure to designated environmentally sensitive areas (ESAs), Recycling of end-of-life vehicles.	Urban passenger transport solutions
5	Onat et al. (2014)	LCC	SLCA	LCA	Alternative passenger electric vehicles (BEV, PHEV, HEV)
6	Maimoun et al. (2016)	Fueling station availability, Fuel price stability, Fuel price, Vehicle cost	-	Power Density, Water footprint, Tail-pipe emissions, Life cycle emissions	Alternative fuels for waste collection vehicles (natural gas, biodiesel, hydraulic-hybrid)

7	Kicin-ski and Solecka (2018)	Investment costs	Travel time, Standard of travel, Level of integration, Availability, Safety and security, Reliability	Profitability of the urban public transportation system; Environ-ment friendliness	Urban public transportation system
8	Ekener et al., (2018)	LCC	SLCA	LCA	Alternative fuels for road transport (bio-mass based and fossil fuels)
9	Ullah et al. (2018)	Project initial cost, Tariff, Fuel cost, Job creation	Social ac-ceptance, Loss of life expectance	External cost, Land require-ment	Alternative fuels for road transport (CNG, LPG, LNG)
10	Liang et al. (2019)	Fuel cost, Vehicle cost	Social ac-ceptability, Compliance with policy	GHG, PM10, NOx, CO and HCs	Alternative-fuel vehicles (based on LPG, CNG, biodiesel)
11	Balieu (2019)	-	-	LCA	Electrified road systems
12	Bronie-wicz and Ogrodnik (2020)	-	-	Occupied area, Length of invest-ment, Number of vascular plant species de-stroyed, Length of sections with high pollution risk, Number of demolitions of residential build-ings and others	Transport infrastructure projects
13	Gulci-men et al. (2021)	LCC	SLCA	LCA	Light rail transit system
14	Barke et al. (2021)	LCC	SLCA	LCA	Battery systems for electric aircraft

15	Du and Kommala-pati (2021)	LCC	-	LCA	Public trans-portation fleet (electric buses)
16	Nour et al. (2022)	LCC	SLCA	LCA	Selection of alternative fuel taxis (HEV, CNG, BEV)
17	Hasse et al. (2022)	LCC	SLCA	LCA	Technologies and alterna-tive fuels for road transport (synthetic bio-fuel, electricity, hydrogen)
18	Barke et al. (2022)	LCC	SLCA	LCA	Comparison of conventional and electric passenger aircraft
19	Gutierrez (2022)	Depreciation costs, Traction costs, Main-tenance costs, Operating cost	Social criterion of service by kilometers travelled by each vehicle	NOx emissions, Particulate mat-ter emissions, CO_2 emissions.	Management of the public transportation system
20	Popien et al. (2023)	LCC	SLCA	LCA	Electric vehicle traction battery

Source: Burchart D, Przytuła I. Sustainability Assessment Methods for the Transport Sector Considering the Life Cycle Concept—A Review. Sustainability. 2024; 16(18):8148. https://doi.org/10.3390/su16188148.

For automobile life cycle studies, LCC can be formulated by:

$$\begin{aligned} \textit{Total LCC} &= \textit{Purchasing cost} \\ &+ \sum_{i=1}{}^{n} (\textit{operationcostforascheduledyear}_i \\ &+ \textit{Scheduledmaintenancecost}_i + \textit{Costduetoobsolescence}_i) \\ &+ \textit{cardisposalcost} \end{aligned} \tag{1}$$

Most EV LCC analyses center on the cost of the battery as the primary driver behind the higher cost of contemporary EVs. Lithium-ion batteries retail for about $300 per KWh; the 24 KW battery for a Nissan Leaf EV will set you back $72,000, which adds significantly to the overall cost of the vehicle. The assumptions made when modeling the EV life cycle and the study's scope have a significant impact on the total LCC of an EV. For the whole consumer life cycle of EVs, Diao et al. created a model that takes into account the phases of acquisition, operation, and disposal. Retail price, insurance cost, charger cost, service tax, and third-party insurance are all related to the purchasing phase. On the other hand, running costs include things like electricity costs, upkeep, tire replacement, insurance, and battery replacement. Additionally, the cost of recycling the batteries and the EV's scrap value are included in the disposal cost (Nordelöf et al., 2014; Ayodele and Mustapa, 2020).

2.3.2. LCA results of EV in comparison of ICEV

Research summarizes the comparison of LCA results for different EV and ICEV types, as reported by different authors, for a functional unit of "1 km driven." The comparison of the works of different authors revealed that the LCA results differ for each researcher because they use different data for analysis, choose different software, or use different database sources. Technical aspects like the weight of the vehicle without passengers and cargo and the battery weight play a significant role in controlling the environmental impact during the LCA of both BEVs and ICEVs. Fuel type and consumption are crucial variables to examine in the use phase that follows.

Additionally, the vehicle's manufacturing date determines how much fuel it uses, with newer models using less fuel than older models. The environmental impact is also determined by the power source used to charge the BEV's battery. Large hydropower and thermal power plants have greater effects on the environment than solar and wind-generated electricity. Researchers use the region's electricity mix, as in a general comparison, to ascertain the overall impact of charging in that region. As the year goes on, the source of energy is shifting to renewable sources to produce electricity and produce cleaner energy. Life cycle sustainability assessment (LCSA) models in the transport sector are summarized in Table 2.4.

Table 2.4. *Overview of LCSA Models in the Transport Sector*

SL. No.	Author	Models	Indicators	Application in the Transport Sector
1	Lahaussois et al. (2017)	JEC WTW	Greenhouse gas emissions (GHG), Primary energy consumption, Fuel efficiency	Assessing the total impact of fuel on greenhouse gas emissions and energy consumption
2	Xu et al. (2015)	GREET	Greenhouse gas emissions (GHG), Primary energy consumption, Emissions of pollutants, Energy efficiency	Life cycle assessment of fuels and propulsion technologies in the context of greenhouse gas emissions, pollutant emissions, and energy consumption.
3	Song et al. (2017)	TLCAM	Greenhouse gas emissions (GHG), Air pollutant emissions, Primary energy consumption, Natural resource consumption, Public health impact, Environmental impact, Life cycle costs.	Assessing the environmental impact and life cycle costs of various technologies and fuels
4	Boulay et al. (2017)	AWARE	Water consumption, Water availability, Impact on water resources, Water stress	Assess the impact of water consumption on available water resources in different regions
5	Xu et al. (2015)	BEST BUS	Greenhouse gas emissions (GHG), Air pollutant emissions, Fuel consumption, Operating costs, Purchase and depreciation costs, Energy efficiency, Passenger comfort.	Evaluation of environmental and economic parameters of different types of buses

6	Stanciulescu and Fleming (2006)	GHGenius	Greenhouse gas emissions (GHG), Air pollutant emissions, Primary energy consumption, Fuel consumption, Life cycle analysis, Environmental costs.	Assessing the total greenhouse gas emissions and energy and environmental impacts of various transportation fuels and propulsion technologies
7	Ranhman (2015)	LEM	Greenhouse gas emissions (GHG), Air pollutant emissions, Primary energy consumption, Natural resource consumption, Environmental impact, Life cycle analysis (LCA), Environmental costs.	Assessment of greenhouse gas emissions, air pollutants, and energy and natural resource consumption over the full life cycle of fuels and vehicles
8	Kouloumpis and Azapagic (2018)	FELICITA	Ozone layer depletion potential, Acidification potential, Eutrophication potential, Photochemical ozone creation potential, Recyclability of materials, Capital costs, Operating and maintenance costs, Fuel costs, Worker injuries, Large accident fatalities, Direct employment	Assess the sustainability and environmental impact of transportation technologies by analyzing emissions, energy use, resource consumption, public health effects, and overall environmental impact.

Source: Burchart D, Przytuła I. Sustainability Assessment Methods for the Transport Sector Considering the Life Cycle Concept—A Review. Sustainability. 2024; 16(18):8148. https://doi.org/10.3390/su16188148.

2.4. LIFE-CYCLE ASSESSMENT CONSIDERATIONS FOR BATTERIES AND BATTERY MATERIALS

The quick decarbonization of the transportation sector and the electrical grid depends on energy storage. To meet the demand for on-demand energy storage on the grid and to allow electric vehicles (EVs) to store and use energy, batteries are probably going to be crucial. However, the widespread use of rechargeable batteries is often cited as a drawback due to the critical material use and upstream environmental impacts from manufacturing. A popular method for analyzing the possible effects of large-scale battery production, use, disposal, and/or recycling is life-cycle assessment, or LCA. Fundamentally, LCA is a process that measures the immediate and long-term environmental impacts connected to a good or service. Examining the environmental trade-offs between various technologies that offer a comparable service is another helpful use for this framework. However, there are a number of obstacles to overcome before applying LCA to batteries: methodological decisions, lack of primary data on battery manufacturing, etc.

There is currently no agreement in the field of LCA regarding the proper methodology for analyzing and reporting on the environmental impact of batteries. Research employs an extensive range of system boundaries, functional units, primary data sources (which report data at varying granularities), as well as impact, midpoint, and life-cycle inventory categories. This limits the potential of LCA to act as a feedback loop for initial scientific research and technology development, and makes cross-comparisons of various technologies difficult. Additionally, it may make it more difficult for us to identify and fix mistakes. For example, life-cycle inventory results frequently differ by one or more orders of magnitude, and reviews frequently fall short of providing an explanation for these variations.

Although work on developing non-LIB battery technologies has not stopped, LIBs are still expected to hold a dominant position in the market for the foreseeable future. These technologies include potassium- and sodium-ion batteries, solid-state batteries (such as Li-sulfur, Li-metal, and rechargeable zinc alkaline), flow batteries, and multivalent batteries. Lithium iron phosphate (LFP), lithium manganese oxide (LMO), lithium nickel manganese cobalt oxide (NMC), and lithium nickel cobalt aluminum oxide (NCA) are the four cathode materials that are commonly used to distinguish LIBs. The majority of batteries investigated in earlier life cycle assessments have an anode made of graphite carbon.

The LCA needs to be specifically designed to address a specific question to compare the environmental effects of competing battery technologies or to comprehend the entire impact of increased battery production and use. LCAs are commonly defined by four key phases, all of which are essential to completing a meaningful study:

- The goal and scope definition phase;
- The inventory analysis phase;

- The impact assessment phase; and
- The interpretation phase.

Researchers must choose the question they hope to answer during the goal and scope definition phase, and they should use that question to inform the definition of system boundaries, environmental metrics, and one or more functional units. When it comes to batteries, the outcomes of LCAs can be utilized to guide research and development (R&D) projects that try to minimize negative environmental effects, evaluate alternative battery technologies for a specific application, or calculate the environmental effects of widespread implementation in grid or vehicle applications.

Applying LCA to a clearly defined functional unit is the easiest way to normalize any environmental impact per unit of a good or service being offered (e.g., CO_2 emitted per kWh of electricity generated or liters of water withdrawn per bushel of corn produced).

However, because of how LCA has traditionally been used, batteries present a unique challenge. Batteries are essentially energy storage devices, and how they are cycled has an unpredictable effect on both the longevity of the batteries and the value of the service they offer. It is most logical to normalize results based on the service offered when a study compares various battery technologies applied to the same use case. This method's drawback is that it makes it extremely challenging to compare results from different studies because they all inevitably make different assumptions about how the battery will be used over the course of its lifetime. In contrast, dependable use-phase performance data might not be available for an evaluation of one or more early-stage battery technologies meant to guide future research. In this instance, drawing a system boundary that ends at the factory gate might be appropriate. Any potential differences in cycle life that might eventually affect the batteries' lifespan should just be noted.

Cradle-to-gate studies are those that define system boundaries that do not include the use phase or end-of-life (the term "gate" refers to a factory gate). Research encompassing the use phase and end-of-life is known as "cradle-to-grave"; this can involve recycling, reuse, and finally disposal. Even though battery life cycles involve recycling, the term "cradle-to-cradle" is not frequently used to describe systems that involve recycling because it usually refers to a zero-waste process.

2.4.1. The Life Cycle of Stationary and Vehicle Li-Ion Batteries

The typical life cycle for LIBs in EV and grid-scale storage applications starts with raw material extraction and going through materials processing, component manufacturing, cell manufacturing, and module assembly. Finished modules can be shipped as modules for off-site rack assembly at energy storage sites, assembled into packs and put in vehicles, or assembled into racks on-site for shipment to stationary storage facilities. All LCAs, regardless of scope, must start with raw material extraction. Next, researchers determine whether to extend the system boundaries through the use phase and end-of-life (EOL) and, if so, how to tie their analysis to a specific use case.

Researchers need to be careful to specify the form batteries take at the factory gate, whether it be a module, pack, or fully assembled rack (in the case of stationary storage), even in cradle-to-gate studies.

EVs, hybrid solar and battery systems, or standalone grid-connected storage are just a few examples of the use cases for which the majority of battery LCAs are specific. When it is possible, this is the best course of action since it allows for the comparison of various options by defining a common functional unit that takes into account differences in performance, such as cycle life and round-trip efficiency. Not all studies, though, take into account the battery use phase for a given application, and this isn't always possible for more sophisticated, pre-commercial battery technologies. Wang along with others compared the environmental effects of various cathode materials using a use-agnostic analysis, and Wang et al. carried out a cradle-to-gate examination of batteries made of lead acid, LMO, and LFP. Researchers still need to choose a pack or rack configuration associated with a stationary or electric vehicle application in order to perform a use-agnostic cradle-to-gate analysis of a LIB. The system boundary may need to be established during the module assembly stage of a truly use-agnostic life cycle assessment (LCA) because the pack or rack assembly, which includes parts like electrical control and thermal management, will vary significantly depending on the battery's intended use. Cradle-to-grave life cycle assessments (LCAs) take into account the battery life cycle stages of collection, recycling, and/or disposal. There are several ways to recycle; these include hydrometallurgical and pyrometallurgical recycling, which are more adaptable and concentrate on removing only valuable metals from batteries, and specialized, chemistry-specific direct recycling.

2.4.1.1. Raw Material Extraction and Delivery

Concern over material constraints has arisen due to LIBs' reliance on limited resources, sharp production growth (roughly doubling every five years), and unclear future recycling practices. Olivetti and associates investigated any potential bottlenecks in the vital supply of materials needed to produce LIBs. The literature provides extensive documentation on the disintegration of battery components, starting from active material and progressing to individual cells, modules, and packs. Even though the use of critical materials is frequently addressed as a single problem, there are actually three distinct problems that merit discussion. First, there is the issue of resource availability in relation to consumption and whether increasing the production of batteries will result in the depletion of essential material reserves or a price increase. Second, highly concentrated production carries a geopolitical risk that can result in conflict, unstable prices, and fictitious shortages. Here, we define this issue as supply chain risk and contend that resource depletion and supply chain risk are too frequently confused by researchers. Supply chain risk is not well captured by traditional LCA techniques. Third, mining operations have an impact on the environment and society, which falls well within the scope of LCA (Nordelöf et al., 2014; Ayodele and Mustapa, 2020).

The applications of multi-criteria decision analysis in transportation are summarized in Table 2.5.

Table 2.5. *Overview of Applications of Multi-Criteria Decision Analysis in Transportation*

References	MCDA Methods	The Objectives of the Study
Yavuz et al. (2015)	WSM (weighted sum model)	Home health care service provider
Hayashi et al. (2014)		Biodiesel fuel production
Pilavachi et al. (2009)		Hydrogen production
Quintero et al. (2008)		Ethanol production
Zhou et al. (2007)		Light vehicles
Brey et al. (2007)	DEA (data envelopment analysis)	Alternative-fuel based vehicles (HEV, FCEV)
Maciol and Rebiasz (2018)	TOPSIS (technique for order preference by similarity to ideal solution)	Light vehicles
Büyükozkan et al. (2018)		Buses
Oztaysi et al. (2017)		Light trucks and vans for utility company
Mukherjee (2017)		Road sector in general
Maimoun et al. (2016)		Alternative fuels for waste collection vehicles (natural gas, biodiesel, hydraulic-hybrid)
Onat et al. (2016)		Light vehicles

Vahdani et al. (2011)	Fuzzy TOPSIS, PSI (preference selection index)	Buses in urban areas
Yedla and Shrestha (2003)	AHP (analytical hierarchy process)	Conventional fuel vs. CNG cars
Dinh et al. (2009)		Biodiesel production
Shiau and Liu (2013)		Urban passenger transport solutions
Aydın and Kahraman (2014)		Buses
Osorio-Tejada et al. (2017)		Road freight transport
Tsita and Pilavachi (2012)		Various fuel types
Tsita and Pilavachi (2013)		The road sector in general
Mardani et al. (2015)		Transportation systems
Ullah et al. (2018)		Alternative fuels for road transport (CNG, LPG, LNG)
Sehatpour et al. (2017)	PROMETHEE (preference ranking organization method for enrichment evaluation)	Light vehicles
Ziolkowska (2013)	Fuzzy PROMETHEE	Biodiesel, ethanol
Stević et al. (2020)	MARCOS (measurement of alternatives and ranking according to compromise solution)	Supplier selection
Ghose et al. (2019)	COPRAS (complex proportional assessment)	Material for electric vehicle
Kicinski and Solecka (2018)	ELECTRE III	Urban public transportation system
Ekener et al. (2018)	MAVT (multi-attribute value theory)	Biomass based and fossil transportation fuels

Source: Burchart D, Przytuła I. Sustainability Assessment Methods for the Transport Sector Considering the Life Cycle Concept—A Review. Sustainability. 2024; 16(18):8148. https://doi.org/10.3390/su16188148.

2.4.1.1.1. Resource Depletion and Conflation with Supply Chain Risk

Olivetti and associates compiled the information on the rates of consumption of nickel (Ni), manganese (Mn), cobalt (Co), lithium (Li), and natural graphite in relation to their reserves. They discovered that for Co, Li, and natural graphite, the static depletion index—a measure of known reserves to primary mine production—has increased, indicating that sustained demand has prompted more exploration and extraction. The fact that Mn and Ni did not exhibit an upward or downward trend suggests that the production to known reserve ratio has stayed mostly stable. This emphasizes how difficult it is to quantify finite resource depletion as part of an LCA because society's perception of available resources is dynamic.

Technological developments in exploration and recovery are driven by growing demand. The possibility of material recovery through recycling introduces yet another level of intricacy. According to Olivetti et al., as you correctly point out, there is a lag between the time batteries are made and when they run out of life, so recycling is unlikely to solve any short-term (10–20 years) material supply constraints, even for what can be recovered. Recycling may be a crucial long-term tactic if battery technologies do not advance away from dependence on these essential materials in the ensuing decades. Hydrometallurgical recycling facilities will recover all of the aforementioned metals along with Li and aluminum (Al), whereas pyrometallurgical recycling facilities will recover Ni, Co, Mn, and Copper (Cu). Even more materials can be recovered through direct recycling, many of which can be put to new uses right away. In light of this, it is worthwhile to reexamine Peters et al.'s claim. that on a normalized basis, abiotic depletion—an impact category that represents the depletion of non-renewable resources like minerals and fossil fuels—is the most significant effect of batteries, outweighing the significance of GWP. Typical LCA practices tend to miss out on many of the subtleties in critical material supply and use outlooks. Condensing these impacts into a single score is likely to lead to more confusion than helpful insights, especially when critical material depletion and fossil fuel use are combined. This is especially true for materials with rapidly increasing demand, like Co and Li. The geographic diversification—or lack thereof—of the reserves and supply of some of these materials actually presents a bigger worry.

The countries of Australia, Chile, and Argentina produce the majority of Li. Australia, China, Argentina, and Chile together account for 90% of the world's supply of lithium, and demand is still rising. While Chile and Argentina produce Li2CO3 from brine, Australia makes lithium concentrate from spodumene, a lithium aluminum inosilicate. According to Olivetti et al., although there are numerous ways to increase Li supply, it is a contentious topic that has been extensively covered in the literature concerning battery manufacturing. Although various parts of Europe, the US, Bolivia, the DRC, and other countries have potential lithium resources, these locations have not yet been developed for the commercial production of Li. It is feasible to extract lithium from two types of deposits: brine lake deposits and hard rock deposits. Most of the reserves are found in South American brine lakes, and the remaining deposits are mostly found in hard rock, mostly in Australia. In 2016, 34% of the world's lithium demand came from Li-ion batteries. The ability to ramp up the production of Li quickly enough to prevent short-term supply constraints and price spikes is likely to be the bottleneck for Li supply.

2.4.1.1.2. Environmental Impacts of Raw Material Extraction and Processing

To put things in perspective, the primary energy required to produce an LMO-graphite battery pack is estimated to be accounted for by battery materials to the tune of about one-third, with battery manufacturing accounting for the majority of the remaining energy demand. The specifics of the manufacturing process and the size of the facility, however, can cause this ratio to vary significantly. Al is estimated to contribute the greatest amount to the materials' energy footprint in an LMO battery by a considerable margin, followed by the cathode, battery management system (BMS), Cu, and graphite. Al, Cu, and graphite make up the largest shares of LIB packs (using graphite anodes) by mass. Al, for example, is used in the cathode current collector, positive terminal assembly, cell container, module heat conductors, and module enclosure. This presents a challenge for conducting the life-cycle inventory for material inputs to LIB manufacturing because the impacts are spread across a wide array of components. Battery designs are constantly changing, and comprehensive material breakdowns are frequently confidential.

Typically, a mix of diesel fuel for powering mining machinery and moving materials, electricity for mechanical processes, and natural gas for processing heat energy is used in the mining and processing of raw materials. Materials are typically classified as more or less energy-intensive based on their processing and refining stage. For instance, the energy footprint and consequent effects of producing aluminum are primarily caused by the process of smelting and refining, which uses a lot of electricity. China is where most of the world's aluminum is produced and refined, which has an even greater impact because coal still makes up the majority of China's average grid mix. Solvent extraction and electrowinning, commonly referred to as electroextraction, account for the majority of Cu's energy footprint. Because Co is a co-product of the production of Ni or Cu, which means that some sort of allocation is necessary and that the production ratios will vary depending on the location, researchers should use caution when obtaining data from previous studies. To put it another way, the effects of extraction and processing must be fairly allocated to various outputs. This can be achieved through either mass- or market-value-based allocation, or system expansion (which avoids formal allocation).

2.4.1.2. Use Phase and End-of-Life

2.4.1.2.1. Use Phase

Numerous studies, frequently citing the complexity and uncertainty surrounding battery performance and lifetime, fail to take the use phase of a LIB into account when conducting an LCA. When comparing various battery technologies, it is imperative to take into consideration their disparate roundtrip efficiencies and lifetimes. Other attributes might matter more or less, depending on the application. For instance, in an electric car, truck, or airplane, pack weight will affect vehicle efficiency; however, in stationary applications, weight is much less important.

The number of charge/discharge cycles that a battery can withstand under specific circumstances before its storage capacity deteriorates to a predetermined state is known as cycle life; for electric vehicles, this is usually 80% of the battery's initial capacity, and for stationary storage, it is 60%. Operating conditions will affect a battery's actual cycle life; when data is available, it is best to adjust for the expected use case before calculating lifetime energy throughput. Extreme temperature usage of batteries accelerates battery aging. Not only do higher temperatures speed up and increase the efficiency of reactions, but they also accelerate aging processes. Reduced space for Li-ion insertion on the anode due to low operating temperatures may result in electrode materials contracting. This could also lead to lithium plating, which is a significant factor in capacity fade. Accelerated battery aging can also result from high or low state of charge (SOC) battery operations. Increased battery aging is primarily caused by active material loss in the positive electrode at the extremes of the SOC. The number of years before a battery deteriorates to a certain state while it is inactive (for example, in severely underutilized batteries) is known as the shelf life metric. These high-level metrics can help estimate the total energy discharged over a battery's lifetime as a useful functional unit for LCAs, as they capture the combined effects of multiple battery degradation mechanisms on capacity fade.

When throughput (kWh discharged over a battery's lifetime) is the only functional unit considered in LCA, battery efficiency—another important aspect of the use phase—is overlooked. Irreversible side reactions prevent batteries from recovering a tiny portion of their energy. In practical terms, the ratio of total charge extracted from the battery to total energy put into the battery during charging depends on the charge and discharge C-rate and temperature. It is typically closer to 90%. Coulombic efficiency, which represents idealized slow charging and discharging, is approximately 99% for LIBs. For example, frequent use of ultra-fast charging and rapid discharging will reduce battery efficiency. This implies that a battery's lifetime energy losses and total throughput will be influenced by the way it is cycled. Although the term "round-trip efficiency" is not always used consistently, it can also refer to the efficiency of a pack or system rather than individual cells. In addition to losses at inverters or in parasitic loads, like thermal management systems, researchers must take care not to overlook or count energy losses in the battery itself twice when accounting for energy losses during charging and discharging. Battery efficiency declines with time, adding to the complexity, even though capacity fade is typically the deciding factor when it comes to decommissioning a battery. Since it can capture the effects of many of the mechanisms that lead to capacity fade, power fade, and reduced efficiency, impedance is frequently used as a measure of LIB health. However, this metric is not easily converted to any of the useful efficiency and lifetime throughput metrics required to account for use-phase performance in an LCA.

It is not surprising that the majority of battery LCAs overlook battery aging and the use phase due to the intricacy of simulating battery behavior and the scarcity of performance data from practical applications. Sathre and associates carried out a sensitivity analysis to determine the battery parameters and their influence on cumulative energy balance and cumulative greenhouse gas emissions. The study focused on second life applications for LIBs from EVs. They observed a significant degree of uncertainty in the timing of the point at which capacity fade accelerates and LIBs reach an inflection point. Hiremath and associates illustrated the various life-cycle effects of LIBs used in various stationary storage applications. They found that the use phase greenhouse gas emissions per MWh delivered can vary by almost two orders of magnitude between applications. Their analysis provided the power rating, discharge duration, energy rating, and cycle frequency for multiple stationary storage applications. Longo and associates examined two hypothetical EV batteries, one with a cycle life of 3000 cycles and a frequency of 2 cycles per day, and the other with a cycle life of 3500 cycles and a frequency of 1.6 cycles monthly. This led to a 5% variation in global warming potential between these two battery scenarios during their use. It would be great to have a more complete accounting of use-phase cycling and how it affects lifetime and efficiency, but trustworthy data for LCAs are hard to come by. A set of standardized scenarios that capture variations in C-rate, operating temperatures, SOC, and the anticipated effects on capacity fade, battery lifetime, and efficiency would be extremely beneficial for future research.

2.4.1.2.2. Battery End-of-Life and Recycling

When a battery reaches its end of life, it needs to be recycled or disposed of properly. Reuse and recycling have long presented a methodological challenge in LCA, posing issues regarding the appropriate allocation of credits for recovered materials and the consequently avoided impacts of virgin material production. Recycling can be classified as either open-loop, where materials are recovered for use in other production systems, or closed-loop, where materials are recycled within the same production system (e.g., cathode materials recovered for use in new cathodes). The majority of research on batteries uses a closed-loop strategy and investigates one or more of the three primary recycling modalities: direct, hydrometallurgical, and pyrometallurgical.

Pyrometallurgical recycling is a smelting technique used to recycle LIBs and nickel metal-hydride (NiMH) batteries. It is capable of recovering transition metals such as Co, Ni, and Cu. Li and Al are among the other materials that are oxidized during the process to produce process heat and are typically not recovered.

A higher percentage of battery materials are recovered by mass through hydrometallurgical and direct recycling, which are based on leaching and physical separation processes, respectively. While only direct recycling can recover the electrolyte (through cell flushing), both processes are intended to recover the cathode (including Li in addition to metals like Co or Mn), Al, and the anode. Compared to pyrometallurgical recycling, hydrometallurgical recycling achieves greater savings across a wide range of life-cycle inventory metrics, except for water use. Since direct recycling is less popular and the materials recovered and process configurations differ, it is more difficult to compare. Gaines notes that there is more of a continuum than a sharp divide between direct and hydrometallurgical recycling; a hybrid direct/hydrometallurgical approach may end up being more advantageous than a pyrometallurgical process as the Co content of LIBs decreases. While not the subject of this study, EV batteries may be examined, repaired as necessary, and used in stationary storage applications to increase their lifespan. Installing and using cell cooling systems is one of the additional effects of setting vehicle batteries for stationary use. Additionally, once the battery reaches its inflection point, capacity can drop off quickly.

Good use-phase performance of the recovered materials is ultimately what makes these recycling processes attractive, and direct recycling processes may have the least amount of certainty in this regard. Negative effects on cycle life or round-trip efficiency from recovered cathode materials could easily outweigh the advantages of offsetting raw material production. Up until now, LCA research on battery recycling has been predicated on the idea that recycled materials are functionally equivalent to new materials. This makes sense because there is a dearth of or nonexistent empirical evidence to back up any other theory. The creation of a convincing business-as-usual case to serve as a benchmark for comparison presents another difficulty. There are still few stationary and electric vehicle LIBs that reach their end of life, and country-specific recycling and disposal methods exist. 95% of LIBs worldwide are thought to not be recycled. Further research will be required to track the date of these batteries and establish a clearer understanding of what processes are being used and what materials are ultimately recovered, particularly as the demand for energy storage in EV and stationary energy storage applications grows and batteries continue to reach their EOL (Nordelöf et al., 2014; Ayodele and Mustapa, 2020).

2.4.2. Recommendations for Battery LCA

It is astonishing that we cannot name a single study that establishes the benchmark for best practices in battery LCAs, even with the publication of LIB LCAs going back over 10 years. This is not to argue that earlier research has not been valuable and insightful for the scientific community; the most well-cited studies frequently exhibit strong performance in one or more dimensions, but they all have limitations. The majority of LCAs that have been published have included comprehensive information about how raw material extraction and processing affect the environment. There are two areas where our knowledge of the extraction and processing of raw materials is deficient. First, differences in the exposure risk for adjacent populations and geographic variation in mining practices were not sufficiently taken into account in the studies we reviewed. Due to regional circumstances and customs, certain mining operations that account for a small portion of total production are probably responsible for an excessive portion of the overall environmental effects. These operations, which might be unregulated or loosely regulated, are comparable to the idea of "superemitters" in the natural gas sector. Second, studies rely on regional or global average factors that are likely to be one or more orders of magnitude different from the actual values because no LCA can conduct detailed fate and transport modeling for every emission to air, soil, and water. These inherent limitations in the underlying midpoint and impact methodologies make this impossible. The lack of documentation regarding the precise locations of waste streams released from mining and material processing activities exacerbates these uncertainties.

Previous studies mostly fail to reach a consensus and provide clarity regarding the energy use and impacts of battery manufacturing. Since there is little evidence of significant non-combustion emissions to the air, water, or soil during the manufacturing process, almost all direct environmental effects from this stage are anticipated to be directly related to emissions from the production of electricity and on-site fuel combustion. Dunn and associates gave the first convincing and understandable explanation of dry room conditioning in an LCA context, as well as the causes of significant discrepancies in reported energy use. However, it did not offer a similarly thorough examination of the energy required for NMP recovery. Despite having a narrower focus that was exclusive to NMC-graphite LIBs, Dai et al. did offer a more comprehensive investigation of NMP recovery as well as dry room conditioning. In contrast to battery manufacturing, battery recycling has been extensively researched and has garnered greater consensus in the literature. However, due to the small quantity of LIBs that are recycled, battery recycling LCAs heavily depend on estimated or simulated mass and energy balances. When primary data is made more accessible, it might show discrepancies between simulations and primary data that are comparable to those observed in the battery manufacturing process. We hope to offer suggestions for addressing these uncertainties in a way that makes each study easier to interpret, replicate, and update as battery technologies and the infrastructure supporting their production continue to advance. After all, disagreements between primary and secondary data sources will persist and battery technologies will continue to evolve.

2.4.2.1. Defining Appropriate System Boundaries

Researchers must balance the benefits of comprehensiveness against the drawbacks of adding more assumptions that are not essential to battery technology to define the boundaries of the system. The most thorough evaluation of the net environmental impacts will come from expansive system boundaries that encompass the use- and end-of-life phases. When these stages are taken into account, comparisons of battery technologies (or recycling technologies) can yield different results. For example, a cradle-to-grave analysis can identify significant differences in battery cycle life that arise from using less resource-intensive battery technologies or directly recycled cathode materials. In this instance, a cradle-to-gate analysis that solely used kWh of battery capacity as the functional unit would be deceptive. In a cradle-to-gate analysis, a battery that depends more on Co might also seem to be at a disadvantage, but Co is more likely to be recovered, and this recovered material can counteract the effects of extracting and processing raw Co.

A similarly strong case can be made for the fact that cradle-to-grave LCAs have significant drawbacks in certain situations. When results are tied to a particular use case, it can be very difficult to compare them between studies. This is particularly true for applications involving stationary energy storage, where different configurations and services are offered. An otherwise rigorous and precise study of the effects on battery production may be compromised by the layers of assumptions and uncertainty brought about by including the use and EOL phases. For instance, use phase performance may be unknown and the most feasible use case may not be known in analyses based on novel battery technologies. Regardless of the study's overall system boundaries, reporting cradle-to-gate results separately is beneficial for cross-study comparisons and enhancing the reproducibility of results.

2.4.2.2. Selecting Relevant Environmental Metrics

Above all, we advise that future research take into account the objectives and study scope, as well as the availability of sufficient high-quality data when choosing environmental metrics (inventory, midpoints, and endpoints). A lengthy list of effects presented without any context or uncertainty analysis is likely to cause confusion and provide little benefit to the larger research community. Although leaving out an impact category could give readers the impression that it is unimportant, doing so could also result in results that are erroneous or give the impression of false precision. It is also essential to put in more effort to identify the primary causes of each impact. It's critical to distinguish between the environmental effects of energy use (and the combustion required to produce that thermal or electrical energy) and the effects of other activities, such as the releases of non-combustion pollutants into the air, water, or soil or the mining industry's depletion of finite resources.

The primary source of GHG emissions and the 100-year GWP for LIB production is CO_2 related to combustion. The same is true for the effects of other air pollutants on human health, such as particulate matter (PM), SOx, and NOx, as well as the potential for terrestrial acidification, which is fueled by SOx emissions. It is highly recommended that future life cycle assessments attempt to distinguish between impacts related to energy use and impacts related to other activities. It will be simpler to modify the results if facilities switch from coal to lower-emission fuels (such as natural gas to renewable fuels) and note this change in published LCAs. Furthermore, future researchers may want to update underlying grid mixes or select mixes that are more representative of where production occurs. By indicating which impacts are dependent on assumed grid mixes, LCA data can be used for much simpler future studies. Additionally, changes in the location chosen for analysis may result in a tightening or loosening of emissions regulations. Considering how quickly many nations are decarbonizing their electricity, researchers must have the chance to update our knowledge of the effects of battery production in relation to both present and future grid configurations. The majority of the LIB life cycle's non-energy-related environmental effects are linked to the mining and material processing industries. A range of midpoint and endpoint metrics are available to help describe the depletion of non-renewable resources. We contend that these multipliers miss the subtleties of some of the important inputs used in the production of batteries, where the availability of the material itself may be of secondary or tertiary importance and a long-term shift toward more expensive and energy-intensive extraction techniques is more likely. We advise future LCAs to create a range of average, marginal, and incremental scenarios for the recovery and processing of a few critical material inputs (like lithium, for instance) instead of trying to quantify resource depletion using a single metric. These scenarios can then be used to show the long-term effects of extracting these materials further without recovering and recycling them when batteries reach the end of their useful lives. The potential for eutrophication is the other effect we have covered here that is most pertinent to the extraction of raw materials. We are hesitant to suggest that this metric be used in subsequent research, in part because it may not be possible to derive significant conclusions from the findings from the necessary data on pertinent waste stream discharges. Human toxicity does not play a significant role in non-energy-related impacts in the studies we surveyed, but as previously mentioned, published values frequently rely on data provided by large mining companies. Those datasets likely reflect best industry practices and fail to account for the impact of outliers, particularly in artisanal and small mining operations. Ozone depletion potential is not likely to be a useful metric to quantify given the ongoing phase-out of ozone-depleting substances like CFC-11. In emissions and environmental impact accounting, omitting such outliers is a recognized issue that needs to be addressed in future LCAs, especially when it comes to the extraction and processing of materials.

Although water withdrawals and consumption related to battery production can be significant, LCAs frequently ignore them. Over the course of their lifetime, battery-electric vehicles are linked to over 50% higher water consumption than vehicles with internal combustion engines. This is primarily related to the electricity used for charging cars, but the LIB itself is responsible for a significant portion of the water consumption, which can range from 5%–10% of the total depending on the chemistry of the battery. According to the estimate, 752 liters of water are used for every kWh of Li-ion battery pack produced. Approximately half of this amount is related to the housing made of aluminum and the remaining 30% is related to the cathode active material used in NMC-111 cathodes. If aluminum is produced using thermal power, then 65% of the water used in the process is due to electrolysis; for cathode active material, the majority of the water used in the process is due to mineral extraction. Specifically, coproduction accounts for half of the embedded water consumption in NMC-111 cathode active materials, making it the most water-intensive process. Although water use is not the primary focus of most battery

LCAs, it is still important to investigate further for studies that aim to go beyond energy use and greenhouse gas emissions. This is especially true if the values can be weighted according to a water stress index or other comparable metric that captures the impacts of local water scarcity.

2.4.2.3. Defining Functional Units for Analysis

It can be difficult to define functional units for battery LCAs. The more closely a functional unit resembles the real service a battery performs (such as enabling a vehicle to travel one kilometer), the more underlying assumptions—all of which are non-standardized—that must be made. On the other hand, results are frequently reported in studies per kilogram of battery mass, which is not directly related to the service a battery offers but does make comparisons between studies easier. Battery or material mass (kg), individual battery pack, energy capacity (kWh of battery capacity), energy throughput (kWh passed through the system over the battery lifetime), and distance driven (km) for battery electric vehicles are some of the functional units used in previous studies. Ellingsen and associates established a helpful benchmark by disclosing their findings in terms of several functional units (per pack, per kWh, and per kg of battery mass (Ellingsen et al., 2014).

Scientists propose that kg of battery mass should no longer be used as a functional unit in LCAs. Normalizing results in terms of kWh of battery capacity not only provides a more logical functional unit but also offers comparable cross-comparison potential because storage capacity is directly related to the primary function that rechargeable batteries provide. Another attractive option is to normalize LCA results in terms of lifetime energy throughput, which takes into account variations in cycle life and round-trip efficiency that are not taken into account by a per unit-capacity analysis. If phase-out of NMP has a negative impact on cycle life, one can imagine interesting analyses of the tradeoffs between, say, switching to aqueous solvents and the resulting impact on lifetime kWh throughput. Researchers propose that studies would be well served to report results in these two formats (per kWh of capacity and per kWh of lifetime throughput), provided that underlying assumptions about cycle life are clearly documented. Having said that, it is reasonable for studies that seek to quantify the impact of battery-related issues within the framework of a larger system to report their findings in different ways, such as at the level of individual packs or kilometers traveled (Nordelöf et al., 2014; Ayodele and Mustapa, 2020).

REFERENCES

1. Ayodele, B. V., & Mustapa, S. I. (2020). Life cycle cost assessment of electric vehicles: A review and bibliometric analysis. *Sustainability, 12*(6), 1–17. https://doi.org/10.3390/su12062387.
2. Ayoubi, M., Eilemann, A., Mankau, H., Pantow, E., Repmann, C., Seiffert, U., Wawzyniak, M., & Wiebelt, A. (2013). Fahrzeuphysik. In H. H. Braess & U. Seiffert (Eds.), *Vieweg Handbuch Kraftfahrzeugtechnik* (pp. 47–118). Springer.
3. Bauer, C., Hofer, J., Althaus, H. J., Del Duce, A., & Simons, A. (2015). The environmental performance of current and future passenger vehicles: Life cycle assessment based on a novel scenario analysis framework. *Applied Energy, 157*, 871–883. https://doi.org/10.1016/j.apenergy.2015.01.019.
4. Bicer, Y., & Dincer, I. (2018). Life cycle environmental impact assessments and comparisons of alternative fuels for clean vehicles. *Resources, Conservation and Recycling, 132*, 141–157. https://doi.org/10.1016/j.resconrec.2018.01.036.
5. Burchart-Korol, D., Jursova, S., Folega, P., Korol, J., Pustejovska, P., & Blaut, A. (2018). Environmental life cycle assessment of electric vehicles in Poland and the Czech Republic. *Journal of Cleaner Production, 202*, 476–487. https://doi.org/10.1016/j.jclepro.2018.08.145.
6. Del Pero, F., Delogu, M., & Pierini, M. (2018). Life cycle assessment in the automotive sector: A comparative case study of internal combustion engine (ICE) and electric car. *Procedia Structural Integrity, 12*, 521–537.
7. Diao, Q., Sun, W., Yuan, X., Li, L., & Zheng, Z. (2016). Life-cycle private-cost-based competitiveness analysis of electric vehicles in China considering the intangible cost of traffic policies. *Applied Energy, 178*, 567–578.
8. Ellingsen, L. A. W., Majeau-Bettez, G., Singh, B., Srivastava, A. K., Valøen, L. O., & Strømman, A. H. (2014). Life cycle assessment of a lithium-ion battery vehicle pack. *Journal of Industrial Ecology, 18*(1), 113–124.
9. Faria, R., Marques, P., Moura, P., Freire, F., Delgado, J., & de Almeida, A. T. (2013). Impact of the electricity mix and use profile in the life-cycle assessment of electric vehicles. *Renewable and Sustainable Energy Reviews, 24*, 271–287.
10. Geyer, R., Stoms, D., & Kallaos, J. (2013). Spatially-explicit life cycle assessment of sun-to-wheels transportation pathways in the U.S. *Environmental Science & Technology, 47*(2), 1170–1176.

11. Hawkins, T. R., Singh, B., Majeau-Bettez, G., & Strømman, A. H. (2013). Comparative environmental life cycle assessment of conventional and electric vehicles. *Journal of Industrial Ecology, 17*(1), 53–64.
12. Nordelöf, A., Messagie, M., Tillman, A. M., Ljunggren Söderman, M., & van Mierlo, J. (2014). Environmental impacts of hybrid, plug-in hybrid, and battery electric vehicles—what can we learn from life cycle assessment? *International Journal of Life Cycle Assessment, 19*, 1866–1890.
13. Notter, D. A., Gauch, M., Widmer, R., Wäger, P., Stamp, A., Zah, R., & Althaus, H. J. (2010). Contribution of Li-ion batteries to the environmental impact of electric vehicles. *Environmental Science & Technology, 44*(17), 6550–6556.
14. Oliveira, L., Messagie, M., Rangaraju, S., Sanfelix, J., Hernandez Rivas, M., & Van Mierlo, J. (2015). *Journal of Cleaner Production, 108*, 354.
15. Scown, C. D., Taptich, M., Horvath, A., McKone, T. E., & Nazaroff, W. W. (2013). *Environmental Science & Technology, 47*, 9044.
16. Yang, F., Wang, D., Zhao, Y., Tsui, K. L., & Bae, S. J. (2018). *Energy, 145*, 486.
17. Ziegler, M. S., Mueller, J. M., Pereira, G. D., Song, J., Ferrara, M., Chiang, Y. M., & Trancik, J. E. (2019). *Joule, 3*, 2134.

Chapter 3

Manufacturing and Supply Chain Impacts

Contents

INTRODUCTION

Without each other, the transportation sector and global supply chains could not function. One must grow alongside the other as it grows. As most automakers increase their production of electric vehicles, supply chain leaders need to adjust. The processes involved in producing batteries and other electric vehicle components, as well as the mining and refining of raw materials, make up the supply chain for electric vehicles.

3.1. THE EV SUPPLY CHAIN ECOSYSTEM

An intricate and diverse network, the EV Supply Chain Ecosystem is vital to the changing dynamics of the automotive sector. It includes a range of activities and parties, from the procurement of raw materials and battery production to the assembly of vehicles and end-of-life care. It plays a key role in promoting sustainability and innovation in the transportation sector, mirroring the movement worldwide toward more technologically sophisticated and environmentally friendly mobility options.

3.1.1. Scope of the EV Supply Chain Ecosystem

The term "ecosystem" refers to the intricate web of interdependent parties that manufacture, distribute, and maintain electric vehicles (EVs). It includes every stage of an EV's lifecycle, from the extraction of raw materials to the management of the vehicle's end-of-life. The ecosystem is broad and consists of the intricately linked network of different parties involved in every stage of an electric vehicle's lifecycle, from sourcing raw materials to the last phases of use and maintenance.

In order to source raw materials, mining operations must be carried out to extract vital elements like lithium and rare earth elements, which are needed to manufacture EV components. The first part of the supply chain is made up of processing facilities and the movement of these materials. The industrial factories that manufacture batteries comprise battery packs, modules, and cells that are assembled with an emphasis on electrical energy management and storage, which is essential to the operation of EVs. Engineering design rooms and vehicle assembly lines are part of the vehicle design and manufacturing process. It entails incorporating electric motors, batteries, and other essential parts into the conception and construction of EVs. The variety of suppliers supporting the EV manufacturing process is demonstrated by the parts and components suppliers, which are a collection of smaller establishments offering necessary components like semiconductors, electrical systems, power electronics, and thermal management systems. Infrastructure for EV recharging that is represented by a sequence of charging station icons is the result of infrastructure development, installation, and upkeep (Perera, Hewage, Sadiq, 2020) The EV ecosystem has components that recommend creating software solutions for vehicle operation in terms of connectivity and software. These comprise battery management systems, driver assistance systems, and platforms that help connect smart grids and the Internet of Things (IoT). Lastly, the sales and distribution network is represented by dealership buildings and other sales platforms; these spaces show how EVs are sold to consumers and businesses. All things considered, the EV supply chain ecosystem includes the complexity and range of everything that contributes to the EV lifecycle.

3.1.1.1. Scale of the EV Supply Chain Ecosystem

The total amount of money made from the sale of EVs worldwide is known as the global EV market size. It includes many categories, such as plug-in hybrid EVs (PHEVs), hybrid EVs (HEVs), and battery EVs (BEVs) (Rapson and Muehlegger, 2023). The number of EVs sold, their average selling price, the total cost of EV-related services, and the amount invested in EV infrastructure all have a role in determining the size of the market. It also takes into account the size of the consumer base, the scope of EV manufacturers' operations, government subsidies and incentives, EV technological advancements, and the worldwide movement towards sustainability and carbon emission reduction. The size of this market reflects how quickly EVs are being adopted. It is an important measure of the growth trajectory of the EV industry and its influence on the overall automotive market.

The global EV market is projected to expand at a compound annual growth rate (CAGR) of 21.26% from USD 206 billion in 2022 to USD 1717 billion in 2032 (*See* Figure 3.1). This growth is indicative of a trend toward more environmentally friendly transportation, which is being aided by improvements in EV technology, an expanding network of charging stations, and supportive policies from the government to cut carbon emissions. The EV industry's growing role in the global automotive sector is strongly indicated by this market size, which reflects rising consumer demand and the coordinated push for cleaner energy solutions.

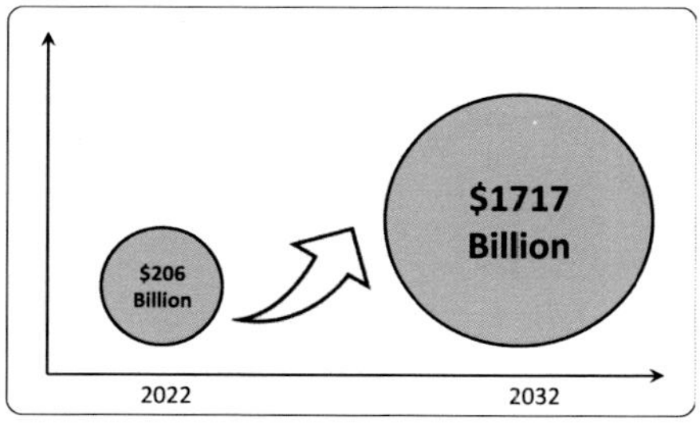

Figure 3.1. *Global EV market growth projection.*

Source: Jagani, S.; Marsillac, E.; Hong, P. The Electric Vehicle Supply Chain Ecosystem: Changing Roles of Automotive Suppliers. Sustainability 2024, 16, 1570. https://doi.org/10.3390/su16041570.

The breadth and depth of an EV supply chain ecosystem's operations and infrastructure needed to support EVs over their whole lifecycle can also be used to measure the ecosystem's scale (Morgan, 2020). The scope and effectiveness of battery manufacturing facilities; the reach and accessibility of the charging infrastructure; the sophistication of software and connectivity solutions for vehicle operation and integration with energy grids; the breadth of the parts and component supplier network; the capability of raw material extraction and processing facilities that supply critical materials like lithium and cobalt; and the coverage of sales, service, and recycling networks are all included. The ecosystem's capacity to expand and adjust in response to changing consumer needs and technical developments, assuring a steady supply chain to support an expanding global fleet of electric vehicles, is another indication of its scale.

The infrastructure for charging and maintenance, the sourcing of raw materials, and production capacity are three important metrics that can be used to assess the size of an EV supply chain ecosystem.

- **Production Capacity**: This metric would show how many EVs can be produced overall by manufacturers in a specific amount of time. It includes the ability to produce batteries, the availability of vital parts like electric motors and power electronics, and assembly lines for the manufacture of automobiles. This metric shows how well the ecosystem can satisfy consumer demand. It is affected by variables like the quantity and size of manufacturing facilities, the availability of labor, and the effectiveness of manufacturing procedures.
- **Raw Material Sourcing**: The volume and geographic distribution of raw materials, such as nickel, cobalt, and lithium, that are obtained for the manufacturing of batteries serve as indicators of the sourcing

process's scale. The quantity and capacity of mining operations, the longevity and variety of supply sources, and the alliances formed with suppliers of raw materials could all be used to evaluate this. With raw materials serving as the primary inputs for the production of electric vehicles, a robust and wide-ranging sourcing network is essential for maintaining stability in the supply chain.

- **Charging and Maintenance Infrastructure**: The size and density of the infrastructure for EV charging, such as public charging stations and service centers, are important markers of the ecosystem's size in the EV supply chain. The extent of the ecosystem's preparedness to accommodate the expanding EV fleet is reflected in the quantity, distribution, and accessibility of charging stations. Furthermore, maintaining the long-term operation of EVs on the road requires the ability to provide maintenance and after-sales services, including the availability of spare parts and the network of service centers.

With the use of these metrics, the EV supply chain ecosystem's capacity to sustain EV production, distribution, and maintenance as well as promote widespread adoption could be quantitatively evaluated.

3.1.2. Transitioning to the EV Supply Chain Ecosystem

Key components of its ecosystem must be carefully examined when analyzing the shift from the conventional internal combustion engine (ICE) supply chain ecosystem to the EV supply chain ecosystem. The scope of the EV supply chain network is depicted in Figure 3.2, taking into account multiple factors. The industrial policy initiative variables are included in the external macroenvironmental factor (e.g., strong government backing, rules, mandates for climate control) and the dynamics of the global market (e.g., market demand trends, trade agreements, geopolitical alliances, and investments made by multinational corporations). The elements that facilitate the electric vehicle supply chain are technological innovation (e.g., innovation in battery technology, digital transformation, artificial intelligence (AI) connectivity, smart grids, renewable energy, and ecosystem partnerships (e.g., networks of cooperation between suppliers and OEMs, public-private partnerships, coordination across industries, and circular economy systems).

Additionally, Figure 3.2 displays the elements of the power source and energy infrastructure, battery technology, materials, and the three main variables of the EV supply chain network factor: supply chain structure, supply chain process, and supply chain outcomes. Along with manufacturing and production processes, it also displays logistics, workforce level changes, supply chain network, skill diversification, customer demand levels, and satisfaction scores.

Power sources and energy infrastructure are included in the supply chain structure's variable elements (Das et al., 2020). The change in power sources is one of the most important aspects of the shift to the EV supply chain ecosystem. Gasoline, diesel, and other fossil fuels must be extracted, refined, and distributed for traditional fossil fuel supply chains to function. On the other hand, the main energy source in the EV supply chain is electricity. This necessitates the development of robust charging infrastructure, including charging stations at public places, workplaces, and homes, to support the widespread adoption of EVs. Technology and materials for batteries are also included in the supply chain structure. EVs are powered by advanced battery technologies, such as lithium-ion batteries, which differ significantly from internal combustion engines in traditional vehicles. The shift entails the creation of batteries with greater capacity and efficiency, improvements in battery recycling and reuse, and studies into substitute materials to lessen dependency on finite resources and improve sustainability.

Second, the supply chain network, logistics, and manufacturing and production processes are all included in the supply chain processes. Making the transition to the EV supply chain ecosystem necessitates changes to the manufacturing and production processes. To assemble EVs and their parts, like electric motors and batteries, traditional automakers and automotive suppliers must reorganize their production lines. Purchasing new equipment and technology, especially for EV manufacturing falls under this category. Every aspect of logistics and the supply chain network is impacted by the switch from ICE to EV. The supply chain must modify from raw materials to final goods to meet the unique needs of electric vehicle components. Suppliers must establish efficient transportation routes for batteries and other EV parts, which might differ from traditional fossil fuel counterparts. Batteries recycling and disposal also present special logistical difficulties.

Thirdly, workforce, skill diversification, customer demand, and satisfaction scores are all included in supply chain outcomes. With the automotive industry adopting electric mobility, there is an increasing need for qualified personnel with knowledge of technologies related to electric vehicles. It might be necessary for current auto suppliers to upskill employees or bring in experts in software integration, electric drivetrains, battery technology, and other EV-specific fields. New employment opportunities in the EV sector attract professionals with a variety of skill sets to meet the industry's evolving demands, which causes a simultaneous shift in the workforce. Changes in consumer trends have been observed on the demand side, which supplements changes in the workforce on the supply side. Consumer demand level and satisfaction ratings reflect the sales volume and market share of EVs in the automotive industry and report the growth and acceptance level among consumers. Customer satisfaction ratings take into account several variables, including battery life, driving range, accessibility to charging infrastructure, overall ownership experience, and vehicle performance. An extensive, widely dispersed network of charging infrastructure with high utilization rates suggests that the ecosystem of the EV supply chain supports consumer needs, encourages EV adoption, and raises customer satisfaction (Das et al., 2020).

These diverse factors of the EV supply chain ecosystem network—both external and internal—represent some of the key changes that occur during the transition from the traditional fossil fuel supply chain ecosystem to the EV supply chain ecosystem. Each of these changes presents challenges and opportunities for stakeholders in the automotive industry as they adapt to the evolving electric mobility landscape. Table 3.1 provides definitions and examples of the key factors, variables, and elements shown in Figure 3.2.

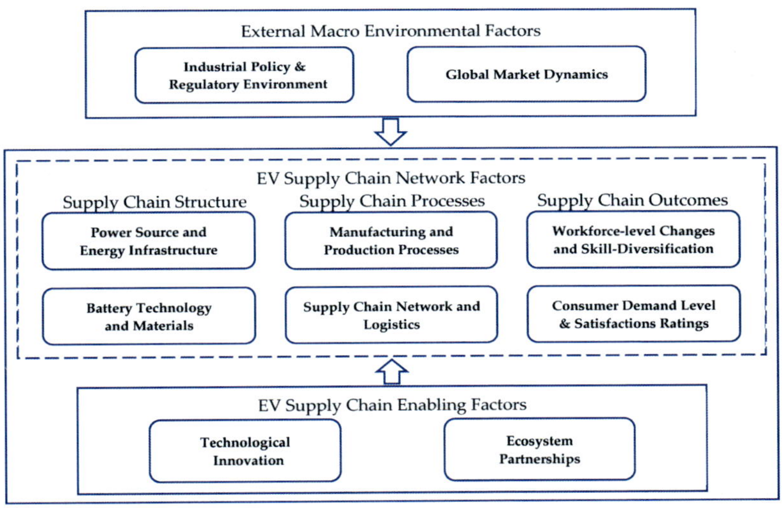

Figure 3.2. *Scope of EV supply chain ecosystem network.*

Source: Jagani, S.; Marsillac, E.; Hong, P. The Electric Vehicle Supply Chain Ecosystem: Changing Roles of Automotive Suppliers. Sustainability 2024, 16, 1570. https://doi.org/10.3390/su16041570.

Table 3.1. *Factors Affecting EV Supply Chain Ecosystem*

(1) **Macro-Environmental Factors** in the broader external environment can significantly impact the EV supply chain. They include macroeconomic, political, regulatory, and environmental influences. For example, government policies on emissions, fuel prices, and trade agreements can all affect the demand for EVs and the availability of key resources like batteries and rare earth metals. Environmental concerns and sustainability initiatives can also influence consumer preferences and drive changes in EV production and supply chain practices.		
Key Variable	**Definition**	**Examples/Indicators**
Industrial Policy and Regulatory Environment	The set of government policies, rules, regulations, and initiatives that impact industries, including the EV industry.	Tax incentives, subsidies, tax credits, and regulations related to emissions, safety, and manufacturing standards. They also encompass trade policies, tariffs, and international agreements that influence the import and export of EV components and vehicles.
Global Market Dynamics	Global market dynamics refer to the factors and forces shaping the global EV marketplace.	Consumer preferences, economic conditions, and competitive forces on a global scale. Market dynamics also encompass competition among automakers, technological advancements, and shifts in supply and demand that impact the global EV market's growth and profitability.
(2) **EV supply chain Network Factors** pertain to the structure and efficiency of the EV supply chain network itself. They encompass considerations such as the geographical distribution of suppliers, manufacturing facilities, distribution centers, and retail outlets. The design and optimization of the supply chain network impact cost efficiency, lead times, and the ability to meet customer demand. Factors like the location of charging infrastructure and the availability of recycling facilities for batteries are also part of this dimension.		
Key Variable/Element	**Definition**	**Examples/Indicators**
Power Source and Energy Infrastructure:	The extent of the energy sources used to power EVs and the infrastructure needed to support them.	The generation, distribution, and availability of electric power, as well as the development of charging stations, battery swapping stations, and renewable energy sources (such as solar and wind) integrated into the grid to support sustainable energy for EVs.

Battery Technology and Materials:	The specialized chemistries, structural components, and cutting-edge materials integral to the development of high-efficiency, high-density, and durable energy storage solutions for EVs.	It includes sourcing materials like lithium, cobalt, and nickel, as well as advancements in battery chemistry and manufacturing processes. Battery technology influences the performance, range, and cost-effectiveness of EVs.
Manufacturing and Production Processes:	The sequence of operations and techniques employed in creating EVs and their components through a systematic assembly and integration of materials into finished products.	It encompasses the assembly of EVs, battery production, and the sourcing of various parts. Efficiency and innovation in manufacturing processes can impact production costs and the speed at which EVs can be brought to market.
Supply Chain Network and Logistics:	The planning, coordinating, and managing of the physical flow of goods, including components, finished vehicles, and batteries, throughout the supply chain.	This includes decisions related to warehousing, transportation, inventory management, and distribution networks. Effective logistics and supply chain network design ensure timely delivery to customers and minimize costs.
Supply Chain Outcomes:	Consumer Demand Level and Satisfaction Rating are related to consumer behavior and preferences.	Consumer demand for EVs can significantly impact the supply chain, influencing production volumes and distribution strategies. Satisfaction ratings reflect how well the supply chain and EV products meet consumer expectations, affecting brand reputation and future market growth.
Workforce Level-Changes and Skill Diversification:	Workforce-level changes involve shifts in employment within the EV supply chain.	As the industry evolves, there may be increased demand for skilled workers in areas such as battery technology, EV manufacturing, and supply chain management. Skill diversification refers to the development of specialized skills needed to support the growing EV ecosystem.

(3) EV supply chain Enabling Factors encompass the technologies, processes, and strategies that enable a smooth and efficient EV supply chain. They include advancements in battery technology, supply chain automation, digitalization, and data analytics. For example, innovations in battery chemistry and manufacturing techniques can lead to longer-range EVs with improved performance. Supply chain enabling factors also involve practices like demand forecasting, inventory management, and sustainability initiatives that enhance the overall effectiveness and sustainability of the supply chain.

Key Variable	Definition	Examples/Indicators
Technological Innovation:	The continuous development, improvement, and application of new technologies and solutions that enhance the efficiency, performance, and sustainability of EVs and their related supply chains.	This includes innovations in battery technology, electric drivetrains, charging infrastructure, materials, software systems, and automation. Technological innovation drives advancements in EVs, making them more accessible, affordable, and environmentally friendly.
Ecosystem Partnerships:	Collaboration and strategic alliances between various stakeholders, including automakers, suppliers, charging infrastructure providers, energy companies, government entities, and other relevant organizations.	These partnerships aim to create a supportive ecosystem that facilitates the development, production, and adoption of EVs. They may involve joint ventures, research and development agreements, shared infrastructure investments, and initiatives to promote policy and regulatory changes that benefit the EV industry. Ecosystem partnerships help build a cohesive network of organizations working together to overcome challenges and accelerate the growth of the EV market.

Source: Jagani, S.; Marsillac, E.; Hong, P. The Electric Vehicle Supply Chain Ecosystem: Changing Roles of Automotive Suppliers. Sustainability ***2024****, 16, 1570. https://doi.org/10.3390/su16041570.*

3.1.3. Comparing ICE and EV Supply Chains

Because of where EVs are at in the product life cycle, their supply chain has a few unique characteristics. After their recent introduction, EV sales are rising quickly. About 25% of the EV supply chain will need to be rebuilt, but the majority of the ICE vehicle supply chain components can be modified to support the EV supply chain. EVs lack several crucial ICE vehicle parts, such as the cooling, exhaust, fuel, and transmission systems. To survive, some suppliers will therefore need to look into diversifying into EV components sooner rather than later.

3.1.3.1. Factors from the External Macro-Environment and Supply Chain Advancement

For the EV supply chain to potentially advance, three key factors are identified: the scope of innovation impact, resilience to disruption risk, and responsiveness to market changes. These variables help to prevent the dilution of focus that can happen when evaluating the robustness of the ecosystem. Each could either support or hinder the expansion of the ecosystem supporting the EV supply chain.

- **Innovation Impact Scope:** ICE vehicles are built with well-established engine designs. However, there are still certain things that can be done better, like fuel efficiency, emissions reduction, and autonomous driving. However, the efficiency and cleanliness of an ICE vehicle is limited by the inherent limitations of the technology and fuel utilized in them. Although the ICE vehicle supply chain offers little room for innovation, automakers such as Toyota are looking into the possibility of converting their current models to run on cleaner, alternative fuels. Even though EVs have been around for a while, mass production of them has only recently begun, so there is still plenty of room for advancements and innovations. Energy-related innovations in EVs have a lot of room to grow, including sustainability, recycling, and battery and charging technology.
- **Risk of Disruption and Resilience:** While there is a possibility of disruptions to the supply chains for both ICE and EV vehicles, the ICE supply chain is more resilient than the EV supply chain. The ICE supply chain is well-established, and its primary raw materials are widely available. For ICE vehicles, disruptions can still occur due to global crises like pandemics, geopolitical instability in regions that supply these commodities, and price fluctuations for commodities like steel and petroleum. However, the relatively recent and limited availability of EVs, geopolitical concerns surrounding raw materials like lithium, cobalt, and nickel used in batteries, production bottlenecks, technological advancements that could surpass current infrastructure, strict laws governing battery production and disposal, and any delays in the construction or operation of the charging infrastructure all pose risks to the EV supply chain. Although both supply chains have resilience strategies in place, their levels of maturity differ. While EV supply chains are still developing, they

are becoming more resilient due to rapid technological innovation and growing collaborations. In contrast, ICE supply chains benefit from established networks and a multitude of suppliers.

- **Market Change Responsiveness:** Market change responsiveness is crucial for EV suppliers to adapt to a fast-evolving automotive landscape. This ability allows them to swiftly respond to shifts in consumer preferences towards greener transport options, technological advancements like better battery life and faster charging, and regulation changes, such as emissions standards. Responsiveness also contributes to supply chain resilience, helping suppliers adapt to unforeseen disruptions like trade tensions or global pandemics. Being agile gives suppliers a competitive advantage by allowing them to capture a more significant market share and forge stronger relationships with automakers. Moreover, quick adaptability helps suppliers manage costs effectively, letting them adjust to fluctuating raw material prices or adopt new, cost-saving technologies. In essence, a high level of market change responsiveness is key for EV suppliers to remain competitive, align with regulatory shifts, and meet growing consumer demand.

3.1.3.2. Comparative Aspects of Automotive Supply Chains

In addition to these three critical elements, supply chains can be distinguished by eight key characteristics. The scope of innovation, market dynamics, raw materials, suppliers' roles, infrastructure, technology use, resilience and disruption risk, and regulations are some of these aspects. Every component is essential to forming and setting the EV supply chain apart from the ICE vehicle supply chain.

Table 3.2 presents a side-by-side comparison between ICE vehicle and EV supply chains across various features, highlighting the differences and nuances. Starting with innovation scope, ICE supply chains are labeled as "Limited," suggesting that possibilities for innovation within this mature technology are somewhat constrained. In contrast, the EV supply chain is marked as having a "Broad" scope for innovation, reflecting the nascent and rapidly developing nature of EV technology, which is ripe with opportunities for advancement, particularly in areas such as battery technology and charging solutions.

Table 3.2. *Comparison of ICE and EV Supply Chains*

Feature	ICE Supply Chain	EV Supply Chain
Innovation scope	Limited	Broad
Market dynamics	Mature	Growing
Raw materials	Steel, Aluminum, Plastic	Lithium, cobalt, nickel
Role of suppliers	Straightforward	Complex
Infrastructure	Well-established	Not as well-established
Technology use	Mature	Still in development
Risk of disruption and resilience	Resilient	Vulnerable
Regulations	Established	Evolving

Source: Jagani, S.; Marsillac, E.; Hong, P. The Electric Vehicle Supply Chain Ecosystem: Changing Roles of Automotive Suppliers. Sustainability 2024, 16, 1570. https://doi.org/10.3390/su16041570.

The ICE supply chain is regarded as "Mature" in terms of market dynamics, which suggests a well-established market with steady growth and predictable trends. The EV supply chain, on the other hand, is described as "Growing," indicating a dynamic and growing market as consumer demand rises and technology advances. The standard ICE vehicle raw materials are plastic, aluminum, and steel, all of which are easily obtainable and have well-established supply chains. On the other hand, because of their relative scarcity and the geopolitical complexities surrounding their mining, EVs depend on lithium, cobalt, and nickel, which may have more complicated sourcing requirements. Suppliers play a "Straightforward" role in the ICE supply chain, most likely as a result of established relationships and precise component requirements. But the role for EVs is "Complex," which might be because new supplier relationships and the integration of cutting-edge technology components are required. The "Well-established" ICE infrastructure refers to a vast network of gas stations and repair facilities. However, EVs are "Not as well-established," referring to the continuous construction of EV maintenance facilities and charging stations.

Technology usage in ICE vehicles is classified as "Mature," meaning that it has reached a plateau in development, whereas technology used in EVs is classified as "Still in development," indicating that ongoing advancements in electric drivetrains and battery management systems are possible. A crucial factor to consider is the risk of disruption and resilience. While EV supply chains are viewed as "vulnerable" because of their reliance on emerging technologies and less established supply chains, ICE supply chains are seen as "resilient" because they have withstood numerous market and geopolitical challenges over time. Last but not least, there are "Established" regulations for ICE that include precise standards and guidelines. In parallel, EVs are "Evolving," reflecting the quick adjustments being made to regulations and industry standards as they adjust to the new reality of electric mobility.

3.1.4. EV Supply Chain Ecosystem Transformation

The dynamic nature of the EV supply chain and the changing role of its suppliers define the ecosystem.

The environment that fosters innovation and the EV market's quick growth set it apart. It is impacted by things like changes in consumer preferences for environmentally friendly transportation, government regulations encouraging the use of EVs, technological developments (particularly in the areas of battery and charging infrastructure), and the worldwide effort to cut carbon emissions. This market is important because it sets the demand for EVs, which in turn affects investment in R&D, the need for new supply chain structures, and the rate at which the industry must change to keep up with these demands.

Because EV production has distinct technological requirements compared to traditional ICE vehicle production, suppliers are an essential part of the EV supply chain ecosystem. Suppliers have to deal with new manufacturing processes (for electric motors and battery packs), new materials (like lithium and cobalt for batteries), and shifting OEM (Original Equipment Manufacturer) expectations. Understanding suppliers is vital since they play a major part in the supply chain's overall sustainability and efficiency. They are essential to industry innovation and have a direct impact on the price, availability, and quality of EVs.

To forecast demand trends and the size of the infrastructure needed for the EV ecosystem, it is essential to have a thorough understanding of the global automotive market. It facilitates stakeholders' ability to foresee changes and adjust their plans in response to market expansion. Understanding suppliers is crucial because they form the foundation of the supply chain and because the industry's transition to electric vehicles depends on their capacity and readiness to move from traditional automotive supply chains to those needed for EVs. A resilient supply chain that can support the expansion of the EV market is ensured by a strong network of suppliers who can adjust to technological advancements.

3.1.4.1. A Typology of the EV Market

The automotive industry segment that focuses on creating, marketing, and developing electric vehicles is known as the EV market. Growing consumer preferences, technological advancements, and growing environmental concerns have made this market a crucial area for growth. Different EV markets may have distinct characteristics based on market factors such as:

- **Consumer Demand:** Varies based on environmental awareness, economic factors, and government incentives. High demand often correlates with higher sales and more rapid market growth.
- **Government Policies:** Strong incentives, such as subsidies, tax breaks, and emissions regulations, can accelerate the adoption of EVs. Conversely, a lack of support can stifle market growth.
- **Charging Infrastructure:** The availability of charging stations is crucial. Well-established networks encourage EV adoption, whereas poor infrastructure can be a barrier.
- **Technological Advancements:** Breakthroughs in battery technology, such as increased energy density or faster charging capabilities, can make EVs more attractive to consumers.
- **Economic Factors:** The cost of ownership, which includes the purchase price, maintenance, and operating costs, influences market characteristics. As battery costs decrease, EVs become more competitive with ICE vehicles.
- **Energy Prices:** Fluctuations in the price of electricity versus fossil fuels can make EVs more or less appealing to consumers.
- **Cultural and Social Factors:** Social norms and values, such as environmental consciousness, can drive market growth.

Technological availability and the impact factors that may determine different EV markets include:

- **Battery Technology:** The state of battery technology, including energy capacity, longevity, and charge times, is a primary driver of EV functionality and, therefore, market viability.
- **Power Electronics:** Advances in power electronics, which control the flow of electricity within the vehicle, can improve efficiency and performance.
- **Electric Motor Technology:** The efficiency and power of electric motors influence vehicle performance, impacting consumer adoption rates.
- **Software and Connectivity:** Integration of EVs with smart grids

and the IoT for better energy management and user experience.

- **Materials Science:** Developing lightweight materials and rare-earth elements for motors can impact vehicle efficiency and production costs.
- **Recycling and Sustainability:** Technologies for recycling batteries and sustainable sourcing of materials can affect EVs' long-term viability and environmental impact.

The EV market's features are shaped by the interplay of these market factors and technological accessibility. For example, markets with high levels of technological innovation but poor infrastructure for charging could prioritize hybrid vehicles or high-tech battery technology with extended ranges. On the other hand, markets with less developed technology but stronger infrastructure might place more value on integration with current power grids and accessibility. Because of the ways in which these and other factors interact, every market will have a distinct profile.

A matrix that classifies various EV markets according to two dimensions—the Innovation Impact Scale and the Market Transformation Scope—is shown in Figure 3.3. The market transformation scope, represented by the horizontal axis, indicates how much the EV market affects or modifies the larger automotive market. The range of values is from low (little change from traditional markets) to high (significant market evolution due to the adoption of EVs). The Innovation Impact Scale, which measures how innovations in the EV market affect local or regional economies and industries, is represented by the vertical axis. This scale goes from low, where innovations barely affect the market, to high, where they drastically change the market. The four EV market types are defined by the quadrants.

Low Market Transformation Scale and low Innovation Impact Scope define Legacy Holder markets. These providers, who concentrate on established, conventional technologies, frequently operate in markets that are stagnant or in decline. Their challenge is to find niche markets where their knowledge is still valued or to adjust to a landscape that is changing quickly. Their entry into the burgeoning EV market may be severely hampered by the fierce competition already in place and the rapid advancement of technology. These organizations have to change their business strategies to incorporate new technologies or stick to catering to a shrinking market niche where their present products and services are still useful.

Innovation Impact Scale		
High	Pioneers Example: US market. Charging infrastructure is regional but growing. General US EV new car sales are at ~8%, but sales are not equitable across the US.	Leading Edge Example: China, Japan, South Korea, Germany, Norway markets. Leaders in technological advancements and EV adoption. Significant subsidies underpin a strong EV market presence. EVs near 50% of sales in some cases.
Low	Legacy Holders Example: African market, primarily Kenya and South Africa. Ranked lowest globally. Virtually no current charging infrastructure and <1% of the market share, but market development opportunities are vast.	Optimizers Example: Indian market. Charging infrastructure is currently geographically limited. One of the fastest growing EV markets.
	Market Transformation Scope **Low**	**High**

Figure 3.3. *EV market typology.*

Source: Jagani, S.; Marsillac, E.; Hong, P. The Electric Vehicle Supply Chain Ecosystem: Changing Roles of Automotive Suppliers. Sustainability 2024, 16, 1570. https://doi.org/10.3390/su16041570.

Pioneer markets have a high scope for innovation impact but a low scale for market transformation. They are trailblazers in sluggish markets, setting the stage for the eventual shift in market dynamics in favor of electric vehicles. Pioneers like the US EV market in the Midwest indicate a high impact on innovation but a limited scope for transformation. Although these markets are innovators or early adopters of EV technology, the general market shift toward EVs is still in its infancy. Despite not being at the forefront of innovation, markets designated as "Pioneers" are anticipated to be crucial to the adoption of EVs in the region. Additional investigation is required to comprehend the tactics utilized by these markets to encourage early adoption and to evaluate the potential impact of these tactics on more significant market transformations.

Market transformation and innovation impact scope are both high in leading-edge markets. They are leaders in the EV industry, offering technological advancements, social and political backing, and a track record of approving and enacting significant market reforms that spur ongoing expansion.

Optimizer markets exhibit a high scope for market transformation, but a low scope for innovation impact. Innovation in technology isn't advancing progress in this area. However, macro initiatives like financial subsidies and social and political support are being used to force market expansion vigorously, almost as if to bypass any technological constraints.

3.1.4.2. A Typology of EV Automotive Suppliers

From tiny, specialized businesses to enormous conglomerates that manufacture a wide range of parts and materials for automakers, automotive suppliers make up a sizable and varied portion of the global manufacturing industry. These suppliers, who offer everything from simple fasteners to intricate electronic systems, fundamental raw materials, and cutting-edge drivetrains, are essential to the automotive sector. Their activities cover several continents and are frequently entwined with complex sub-supplier networks, all of which contribute to the final vehicle assembly. Not only do they differ in terms of size and product offering, but also in terms of their technological prowess, operational approaches, and market presence.

The key characteristics of the Market Change Scope typically include:

- **Adaptability:** The ability to respond to new market demands, technology shifts, and evolving consumer preferences.
- **Innovation:** Developing new products, services, or processes that meet emerging market trends.
- **Agility:** Quickly adjusting business strategies and operations to capitalize on market opportunities.
- **Diversification:** Offering a range of products or services that cater to various market segments to mitigate the risk associated with market volatility.

Disruption Risk Scale factors include:

- **Dependency on Specific Technologies or Materials:** Relying heavily on certain inputs that could be subject to shortages or price volatility.
- **Regulatory Compliance:** The capacity to meet changing regulatory requirements without significant difficulty or delay.
- **Operational Robustness:** The overall stability of operational processes to handle disruptions without major losses.
- **Financial Stability:** Sufficient capital reserves or access to credit to survive periods of disruption without catastrophic consequences.

For suppliers to comprehend their positioning and create strategies to improve their competitiveness and sustainability in the automotive industry, they must consider the extent of market change and the risk of disruption. To sustain or enhance their market position, suppliers need to strike a balance between their inventiveness, flexibility and effective risk management. Based on their Disruption Risk Scale and Market Change Scope, the automotive supplier typology shown in Figure 3.4 distinguishes four different categories that provide insight into their competitive positioning and industry adaptability.

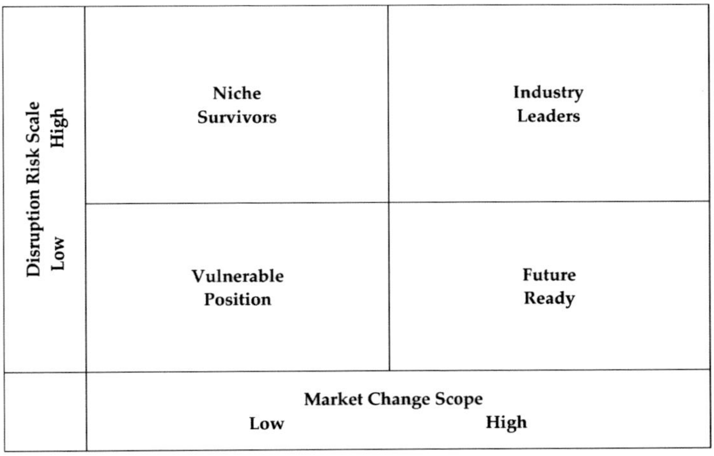

Figure 3.4. *Automotive supplier typology.*

Source: Jagani, S.; Marsillac, E.; Hong, P. The Electric Vehicle Supply Chain Ecosystem: Changing Roles of Automotive Suppliers. Sustainability ***2024****, 16, 1570. https://doi.org/10.3390/su16041570.*

Suppliers in a vulnerable position are those who are not very resilient to disruptions and are not very responsive to changes in the market. Their main advantage is rooted in their use of established procedures and conventional technologies, which are becoming less and less relevant as the industry moves toward greater innovation. These suppliers face an immediate risk of going out of business in a market that is changing quickly. If they want to become more competitive, they have to make innovation investments and adjust to the digital revolution. They also need to completely rethink their business plans in order to accommodate the latest developments in technology and consumer behavior.

Future Ready suppliers are distinguished by their flexibility and ability to quickly adjust to changing market conditions, a quality that puts them in a favorable position to seize new opportunities. They are not as resilient to unforeseen setbacks, though. These suppliers need to make investments in sustainable operations and solid risk management procedures to keep their competitive edge. Their long-term viability and success in the sector will depend greatly on their capacity to maintain flexibility while erecting a strong foundation. Niche Survivors are distinguished by their strong resilience to disruption risk, which is rooted in their specialized product offerings and devoted clientele within specific market niches. These suppliers face a challenge in that their growth potential may be constrained by their reduced responsiveness to broader market shifts. They should think about strategic diversification and pursue incremental innovation that can open doors to new market segments without compromising their niche authority if they want to secure their future and increase their market reach. Prominent industry leaders exhibit remarkable resilience to disruption risks in addition to taking a proactive stance towards market developments. They are the trendsetters, frequently directing the market's course with creative fixes. These leaders need to keep making significant investments in research and development if they want to maintain their position as industry benchmarks. To stay ahead of the curve, they should foster a culture that not only reacts to but also anticipates changes in the market and customer demands. All things considered, this typology emphasizes how crucial it is to be flexible and proactive in a field that is prone to sudden changes and technological upheavals. Every supplier category has a different set of tactics to use to stay competitive and relevant in the ever-changing automotive market (Jagani et al., 2024).

3.2. MINING AND PROCESSING OF RAW MATERIALS (E.G., LITHIUM, COBALT)

One of the main initiatives to lower greenhouse gas emissions from the transportation sector is the substitution of electric vehicles for gasoline and diesel-powered ones. Several essential minerals are needed for the batteries used in electric vehicle designs, including cobalt, nickel, manganese, lithium, and natural graphite.

Modern life would not be possible without lithium-ion batteries, which power everything from cars and mining trucks to consumer electronics like phones and laptops. They are essential to decarbonizing society and weaning the world economy off of fossil fuels at a time when environmental protection is the top priority. Sales of EVs increased from 1.2 million in 2017 to over 2 million in 2018 and are projected to reach 4 million in 2020. By 2025, sales should reach 12 million, and by 2030, they should nearly double to 23 million. According to Deloitte, EVs will no longer be prohibitively expensive to buy when they reach cost parity with ICE vehicles in 2022. It is anticipated that after the COVID-19 pandemic, EV sales will increase. But the industry is having trouble getting the raw materials needed to make EV batteries, which is causing supply problems. The supply chain is beset by challenges such as exorbitant expenses and ecological issues arising from the extraction and refinement of essential metals (Cohen, 2020).

3.2.1. Raw Materials: Lithium

Lithium carbonate and lithium hydroxide are used as the cathode materials in lithium-ion batteries. As more people switch to electric vehicles, there is an increasing demand for these raw materials and a corresponding increase in their cost. According to a UNCTAD report, the global lithium market, the majority of which is mined in China, was valued at $7 billion in 2018 and is expected to grow to $58.8 billion by 2024. Although there is no shortage of lithium, the world's demand for EVs has declined because of a weak market and fewer sales in China, the country that consumes the most EVs. As a result, mining and plant construction have slowed down. While a temporary pause in lithium mining due to the pandemic has allowed many manufacturers to resume operations, there is currently a shortage of the metal, especially in the West. The process could be accelerated with the use of new lithium mining and extraction techniques. Nanotech membranes, which allow for quick and effective separation of lithium during the refining process, are being tested by Puerto Rico-based EnergyX. The global supply chain will be greatly enhanced if this process can be scaled up.

3.2.2. Raw Material: Cobalt

About one-fifth of the bulk of cobalt is used in the cathode of lithium-ion batteries. Within the next 10 years, demand is predicted to reach up to 430,000 tons, and current trends point to a future shortage of raw cobalt because current capacity cannot keep up with the growing demand. Approximately half of the global cobalt reserve, and consequently, mining operations, are centered in the Democratic Republic of Congo (DRC). Here, in one of the least developed nations on Earth, cobalt is mined to supply about 65% of the world's total supply. The cobalt value chain lacks transparency, and mining is a risky industry with hazardous working conditions, exploitation, and low pay. Moreover, a supply chain that is so highly concentrated is vulnerable to supply chain instability due to political unrest and unfavorable environmental effects. This could have an impact on the global transition to electric vehicles by creating more competition in the market, driving up prices, and raising the cost of batteries.

3.2.3. Transition Minerals and Battery Supply Chains

The extraction and processing of five minerals—lithium, nickel, cobalt, manganese, and graphite—that are used in many of the top EV batteries on the market today present numerous supply chain challenges. These chemical components give lithium-ion battery cells their fundamental strength and enable them to store and release energy needed to power electric cars. Because batteries and EVs are becoming more and more important to the U.S. economy, the Department of Interior has listed these substances as "critical minerals" on its list in recent years, expressing worries about ensuring supply in the future. The "transition minerals" that we need for a zero-emission vehicle (ZEV) economy include these five battery minerals, a few other critical minerals, a few rare earth elements, and other non-critical minerals like copper. Over the next several decades, the need for transition minerals in electric vehicle batteries will be a major factor in driving the overall market demand for these minerals (Figure 3.5).

Figure 3.5. *An electric vehicle battery.*

Source: By United States Department of Energy – https://www1.eere.energy.gov/vehiclesandfuels/pdfs/success/nimh_batteries_mar_2001.pdf, Public Domain, https://commons.wikimedia.org/w/index.php?curid=91271723.

The battery supply chain is made up of numerous parties who labor together to convert unprocessed mineral building blocks into the complex gadgets we use on a daily basis to power our laptops, cellphones, and electric cars. The upstream part of battery supply chains is handled by the mining sector, which is also in charge of finding and exploring mineral reserves as well as removing ores—sediments combined with valuable minerals—from these resources. Following their transportation, these ores are processed to remove unnecessary materials and refined to a quality appropriate for batteries at a facility. After these materials are refined, one manufacturer uses them to create cathodes and anodes, or the "positive" and "negative" sides of batteries, respectively. They then send these materials to facilities further down the supply chain that create battery cells. The battery cells are then assembled into sizable packs that can be utilized in electric cars at yet another manufacturing facility. Batteries are ideally recycled or reused at the downstream end of this supply chain so that their materials can be recovered and used to make new batteries.

3.2.4. Supply Challenges

This supply chain is very intricate, and difficulties arise at every stage. Let's begin with the extraction and refining processes that occur upstream.

A common concern is the low domestic supply of minerals and recycled materials needed to manufacture EV batteries. Transition mineral reserves are concentrated outside of the United States; the DRC and Chile account for half of the world's cobalt and lithium reserves, respectively. Geographic concentration of mineral reserves is determined by nature rather than necessity, but problems with concentration also affect other stages of the battery supply chain. Midstream supply chain operations, such as the production of battery cells and the refining of minerals, are also concentrated in a small number of nations, mostly outside the U.S. Thus, even in the event that the U.S. Currently, after its mineral resources are mined, they must be sent to other nations for processing. Furthermore, it may take more than 10 years before minerals are extracted from a reserve in the United States due to protracted discovery and exploration phases, poor industry data quality, and a shortage of federal agency resources. Because of their concentrated geographic location, mineral and battery supplies have the potential to become a significant source of geopolitical risk, if not outright conflict.

Extraction is being moved to the U.S. might lessen these geopolitical risks and provide workers with better health and safety protections than many other nations. But the combination of antiquated mining laws and inadequate National Environmental Policy Act, or NEPA, procedures in the U.S. prioritize extraction over other land uses, keep a close eye on water use and contamination without independent oversight, and don't enforce strict enough regulations for the handling of mining waste and tailings or give communities enough information about possible effects. Thus, the largest single source of toxic waste in the US is the metals mining sector. Given that the United States has reserves of 97% nickel, 89% copper, 79% lithium, and 68% cobalt, native communities are probably disproportionately affected by these regulatory gaps are located 35 miles from reservations held by Native Americans. The environment and Indigenous communities' right to free prior and informed consent must not be sacrificed in the name of metal security.

3.2.5. Congressional and Executive Supply Chain Engagement

Through congressional hearings, the federal government is trying to better understand these domestic supply issues. To help with this, President Biden has issued a Presidential Determination that uses the Defense Production Act to secure domestic supply chains for critical minerals. Additionally, the Bipartisan

Infrastructure Law has provided $7 billion in grant funding to support domestic battery supply chains. Further, the Department of Interior launched a joint agency collaboration to improve mining and land use regulations, and the Clean Energy Minerals Reform Act has been introduced in Congress in both the House and the Senate to reform the Mining Law of 1872 which the U.S. is still operating under 150 years later. Strong cultural, environmental, and due diligence standards must be implemented by these federal actions as well as any subsidized activities. It is also crucial to promote the use of less damaging and inefficient extraction techniques like direct lithium extraction and mineral recovery from waste treatment.

3.2.6. Material Substitution and Technological Improvements

Supply chains are finding it difficult to keep up with the rapidly increasing demand for transition minerals. It is particularly difficult to meet supply needs sustainably when geopolitical, resource location, environmental protection, and pressure to meet rising demand are combined. In order to solve these issues and alleviate supply concerns, less reliance on new extraction must be used. The two main ways to lower the demand for minerals are material substitution and technological advancements. Better battery chemistries can store the same amount of energy with far less mineral inputs or with different minerals that are more prevalent and less significant. By increasing material efficiency during the battery production process, advanced manufacturing techniques can lower the amount of inputs required.

3.2.7. Recycling Lithium-Ion Batteries

In order to lessen global reliance on the aforementioned raw material producing nations, it will be more crucial than ever to set up an extensive recycling system. Partially developed procedures are already in place for extracting raw materials from tiny lithium-ion batteries, like those found in cell phones. But because car batteries are bigger, heavier, and more potent than ever, industrializing the recycling process is more difficult. The Libero research project at RWTH Aachen University is supported by the Central Innovation Program for SMEs (ZIM) of the German Federal Ministry for Economic Affairs and Energy (BMWi) and Vinnova, the innovation agency of Sweden. The German-Swedish consortium is working on creating a reliable, adaptable, and largely waste-free battery recycling process. It is made up of two partners from industry and two from research in each nation. The project started in 2019 and aims to design a plant that can recycle 25,000 t of battery mass annually. A method for recycling lithium-ion batteries from electric vehicles has already been developed by the partially state-owned Finnish company Fortum.

Umicore is a trailblazer in the realm of commercial battery recycling. The company's process is divided into two phases: hydrometallurgical and pyrometallurgical. The alloy that is produced during the first thermal processing stage includes a slag fraction along with cobalt, nickel, and copper. In the process's subsequent hydro-metallurgical step, the metals are recovered. The first recycling facility owned by Umicore can process 7000 t of battery mass annually, or about 35,000 electric vehicle batteries.

Volkswagen started operating a high-voltage vehicle battery recycling pilot plant at its Salzgitter, Germany, location in the beginning of 2021. In addition to recovering 90% of the aluminum, copper, and plastic, the plant will recover 100% of the lithium, nickel, manganese, and cobalt. As of right now, the plant is built to recycle up to 3600 battery systems annually, or about 1500 t of battery mass. However, when more used batteries become available, the system can be expanded to handle higher volumes.

Table 3.3. *Examples of Lithium-Ion Battery Recycling Details*

Company	TRL level	Country	Technology	Capacity (tonnes)
Akkuser Oy	Commercial	Finland	Mechanical	1 000
Fortum	Commercial	Finland	Mechanical, thermal treatment and hydro	
Umicore	Commercial	Belgium	Pyro and Hydro	7 000
Lithion	Commercial	Canada	Mechanical	2 500
Li-cycle	Pilot	Canada	Mechanical	7 500
Retriev Technologies	Pilot	United States	Mechanical	4 500
Accurec	Commercial	Germany	Mechanical and thermal Treatment	2 500
Valdi	Commercial	France	Mechanical and hydro	20 000[18]
GEM High-Tech	Commercial	China	Mechanical and hydro	10 000
Brunp	Commercial	China	Mechanical and hydro	25 000 - 30 000
JX Nippon Mining and Metals	Commercial	Japan	Pyro and hydro	5 000
NorthVolt	Pilot	Sweden	Mechanical and hydro	

Source: By Awan, A – Global EV Outlook 2020: Entering the decade of electric drive?, *https://commons.wikimedia.org/w/index.php?curid=105155381.*

Volkswagen claims that there is no blast furnace smelting involved in the recycling process, which would require a significant energy consumption. The plant receives used battery systems that have been dismantled and deep discharged. Granulate is created by shredding the individual components, which is subsequently dried. Aluminum, copper, plastics, and—most importantly—a black, powdery mixture containing the necessary raw materials for batteries—lithium, nickel, manganese, cobalt, and graphite—are produced during the process. Volkswagen's specialized partners are then in charge of using chemicals and water in hydro-metallurgical processes to separate and process each component (Table 3.3).

3.2.8. Second-Life Approach

Old car batteries may have more useful life before they need to be recycled if they are reused in stationary applications. There isn't any real-world experience yet regarding how many batteries would suffice in terms of service life and remaining storage capacity for a second use. The second-life idea is generally limited to uses that allow for the use of outdated batteries with low energy densities. Furthermore, problems like warranties and standardization must be fixed.

Decentralized battery storage systems for residential buildings, for example, cannot guarantee the high levels of reliability required; higher failure and replacement rates are to be expected compared to new batteries. The number of cells required and, consequently, the cost of the batteries, would be higher due to the required redundancy levels. In reality, only a small portion of the outdated traction batteries could be revived.

3.2.9. Reducing Demand and Limiting Impacts

To prevent the worst effects of the climate crisis and to safeguard the communities that are most affected, we must keep advocating for a net-zero economy. One significant component of this puzzle is electric vehicles. By giving priority to demand reduction, recycling, and material reuse, the shift to zero-emission transportation can prevent the mistakes made during the fossil fuel era from being repeated. When mining is necessary, the effects on the community, Indigenous rights, and the environment should all be carefully considered. We need to back laws and initiatives that guarantee the safety, sustainability, and limitation of harm to the environment and the populations we work to safeguard in the supply chains for electric cars and their batteries.

3.3. INNOVATIONS IN SUSTAINABLE MANUFACTURING OF EVS

The need to address environmental issues, lessen reliance on fossil fuels, and lessen the effects of climate change is driving a massive global shift toward sustainable transportation. With their potential to completely change the way we travel and the way our transportation systems are designed, electric vehicles, or EVs, have come to light as a promising answer to these problems. The design of electric vehicles, their charging infrastructure, and battery performance have all improved significantly in recent years. Smart features and autonomous capabilities have also been integrated. These developments are expected to have a significant impact on society, the economy, and the environment in addition to quickly altering the transportation landscape.

Since they were first introduced in the late 19th century, electric vehicles have advanced significantly. A number of EV technology areas have seen tremendous breakthroughs over time, improving overall user experience, performance, range, and charging speed. Battery technology is one of the main areas of innovation in electric vehicle technology, as it is essential to the functionality and viability of EVs. Recent years have seen a rapid advancement in battery technology due to growing demands for increased safety, faster charging times, and longer ranges. The most popular kind of batteries used in EVs, lithium-ion (Li-ion) batteries, have seen notable advancements in energy density that enable longer ranges between charges. These developments have been made possible by advancements in manufacturing techniques, electrode materials, electrolytes, and cell designs. Battery technology innovation has also focused on charging speed in addition to energy density. EVs can now charge much faster and require shorter charging times thanks to the development of fast-charging technologies like DC fast charging and ultra-fast charging. For example, some of the newest EV models can now reach an 80% charge in under 30 minutes, which increases the practicality and convenience of EVs for long-distance driving.

Another important area of EV technology research and development has been battery safety. An attempt has been made to increase the safety of EV batteries due to worries about the possibility of thermal runaway and battery fires. The safety of EVs has been improved through innovations like thermal management systems, solid-state electrolytes, and fire-resistant battery designs, which make them more dependable and secure for daily use. Electric vehicle supply equipment (EVSE), commonly referred to as fast-charging stations, has made great strides in terms of user-friendliness, dependability, and charging speed. EVs can be quickly charged at these stations thanks to the high-power chargers they use, which deliver electricity at a rate significantly faster than traditional chargers. Global expansion of fast-charging networks has made EV owners' access to more convenient and readily available charging options. Examples of these networks include Tesla's Supercharger network and other third-party networks. Furthermore, barriers to EV adoption have decreased due to improvements in charger designs, such as plug-and-play systems and integrated payment solutions, which have made charging more convenient and seamless.

Significant advancements in home and workplace charging solutions have been made possible by the creation of smart chargers that can be remotely controlled and integrated with the grid. With the help of these solutions, EV owners can maximize their charging schedules and take advantage of off-peak electricity rates by charging their cars at home or at work. In addition to making charging easier for EV owners, this has helped EVs become more integrated into the power grid and enable vehicle-to-grid (V2G) capabilities, which allow EVs to act as mobile energy storage units and return electricity to the grid during periods of high demand.

These days, an extensive variety of vehicle types, such as SUVs, crossovers, trucks, and buses, are considered EVs. They are no longer just tiny hatchbacks or sedans. Customers now have more options thanks to these new designs, which increases the appeal of EVs to a wider spectrum of buyers. Enhancing aerodynamics, reducing weight, and increasing overall energy efficiency have also been major focuses of vehicle design advancements. For instance, the weight of EVs has decreased thanks to the use of lightweight materials like carbon fiber, aluminum, and composites, which has improved their performance and extended range. To lower drag and increase energy efficiency, aerodynamic advancements like streamlined body designs and active aerodynamics have also been implemented. Technological developments in connected and autonomous vehicles have also impacted the electric vehicle market. Advanced driver-assistance systems (ADAS) are now standard on a lot of EVs. These systems use cameras, sensors, and other technologies to enable features like autonomous parking, lane-keeping assistance, and adaptive cruise control. These innovations not only improve driving pleasure but also hold promise for increased safety, less traffic, and optimized energy use, all of which will increase consumer interest in EVs.

3.3.1. Improved Battery Technology

With the potential to completely transform the performance and affordability of EVs, improved battery technology has emerged as a crucial component in the development of EVs. The battery is one of the most important parts of an EV, and new developments in battery technology are changing the face of electric mobility. The most popular battery type for EVs at the moment, lithium-ion batteries, are seeing major advancements in terms of cost, size, and energy efficiency. Thanks to these developments, EVs are becoming more competitive and useful with internal combustion engine cars. Additionally, new battery chemistries like lithium-air, lithium-sulfur, and solid-state batteries are being investigated by researchers because they have the potential to provide even higher energy density, longer driving ranges, and quicker charging times. It is impossible to overestimate the significance of improved battery technology since it could allay worries about range anxiety, shorten charging times, and cut costs overall, opening up EVs to a larger spectrum of consumers (K. T. Chau, Y. S. Wong, and C. C. Chan, 1999).

Enhancements in battery technology can also support the sustainability of electric vehicles and the larger energy system. Greater energy density batteries can allow EVs to go farther between charges, which lessens the need for frequent recharging and lessens the vehicle's dependency on fossil fuels. Faster charging times can also improve an EV's usability and convenience, which will entice buyers who might otherwise be hesitant because of worries about the infrastructure needed for charging. By bringing down the total cost of batteries, EVs may become more accessible to consumers, which could hasten their adoption and cut down on greenhouse gas emissions from conventional combustion engines. Better battery technology has an impact that goes beyond how well and how cheaply EVs operate. It can also spur innovation in fields like energy management, grid stability, and the integration of renewable energy sources. Through vehicle-to-grid (V2G) communication, EVs with cutting-edge batteries can act as mobile energy storage units and enhance the stability of the electrical grid. Because of this two-way energy flow, EVs can feed power back into the grid during times of peak demand, which helps to better integrate renewable energy sources and balance the load. Technological developments in batteries can lead to more intelligent and efficient energy management systems, which will optimize EV battery charging and discharging and raise the transportation and energy sectors' overall sustainability.

Advancements in battery technology hold promise for stimulating economic expansion and employment generation. The demand for batteries and related technologies is predicted to rise sharply as EV adoption continues to rise. This offers chances for funding, R&D, and production in the battery sector, resulting in the expansion of the economy and the creation of jobs in sectors like battery production, component manufacturing, and electric vehicle assembly. By lowering dependency on finite resources and minimizing environmental effects, the use of recycled materials and circular economy techniques in battery production can also further support sustainable economic development.

3.3.2. Wireless Charging

Recent years have seen a significant increase in interest in wireless charging, a promising technology that has the potential to completely change how EVs are charged. Wireless charging systems have the power to completely change the way that electric vehicles are used since they can make charging more convenient and effective by doing away with the need for physical connectors. Instead of requiring bulky cables or plugs, these systems work by transferring energy between the EV and the charging infrastructure via electromagnetic fields. This discovery has important ramifications since it could streamline the charging procedure, lessen urban clutter, and eventually boost EV adoption by making charging more convenient and accessible for all users (Liu et al,, 2013).

Precise alignment of connectors is often necessary for traditional charging methods, which can be difficult and time-consuming, particularly in bad weather. Conversely, EV owners can enjoy a more effortless and convenient experience with wireless charging, as they only need to park their cars over a charging pad or station, saving them the trouble of having to connect cables. This simplicity of use has the potential to significantly improve charging efficiency and user-friendliness, making EV adoption more rapid and appealing to a wider variety of drivers. Urban clutter may decrease if wireless charging becomes more commonplace. Conventional charging stations can clog parking lots, walkways, and other public areas because they often need a large amount of room for numerous charging cables and connectors. On the other hand, wireless charging systems don't require extra wires or connectors because they can be incorporated into the infrastructure that already exists, like parking lots or roadways. This could lessen the visual pollution that charging stations cause, improving the aesthetics and accessibility of urban areas for pedestrians and other users of the roads.

Conventional charging stations are usually found in specific locations, which may not always be handy or reachable for electric vehicle owners, especially those who live in cities or don't have access to private infrastructure for charging. This restriction may be addressed by wireless charging, which allows for greater flexibility in charging locations. To give EV owners more options for charging their cars, wireless charging pads or stations could be added to parking lots, shopping centers, or other public areas. With greater accessibility, one of the obstacles to EV adoption may be lessened, enticing more drivers to convert to electric cars. Conventional charging techniques frequently lose energy because of resistance in the connectors and cables, which makes charging less effective. On the other hand, because wireless charging systems minimize energy loss during the charging process by transferring energy through electromagnetic fields, they can achieve higher efficiencies. Because of their potential to save energy waste and speed up charging times, electric vehicles may become even more desirable and useful to consumers. Moreover, wireless charging might provide EV owners with even more convenience. It can be inconvenient to physically plug and unplug the charging cable when using wireless charging, especially in inclement weather. Rather, EV owners just need to park their cars over the station or pad, and the charging process will start on its own. For EV owners, this hands-free method could save time and effort, improving the convenience and usability of charging (Ahmad et al., 2018).

3.3.3. Vehicle-to-Grid (V2G) Technology

The technology known as Vehicle-to-Grid (V2G) holds great promise for the future of electric vehicles and has the potential to completely transform the energy sector. EVs can now return excess energy to the grid through V2G technology, enabling them to do more than just consume electricity from it. EVs could become mobile energy storage units as a result of this two-way energy flow, which could help create a more flexible and sustainable energy system. V2G technology has the potential to mitigate the strain on the grid during peak periods and improve the integration of renewable energy sources by mitigating fluctuations in electricity demand and supply by utilizing the battery capacity of electric vehicles. As a result, EVs can actively contribute to the balancing of the electrical grid and aid in the shift to a more sustainable and clean energy source (Figure 3.6).

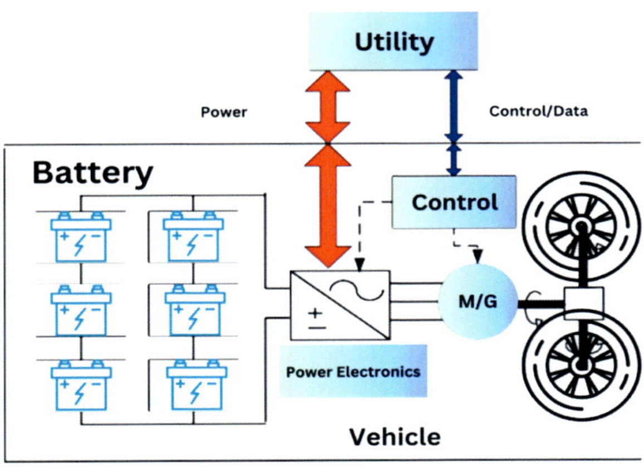

Figure 3.6. *An EV with vehicle -to grid technology.*

Source: Patil, P. (2019) "Innovations in Electric Vehicle Technology: A Review of Emerging Trends and Their Potential Impacts on Transportation and Society," Reviews of Contemporary Business Analytics, 2(1), pp. 20–32.

EV owners may see financial advantages from V2G technology. EV owners may be able to make money by selling electricity back to the grid during times of high demand or by participating in energy markets if they have the capacity to return excess energy to the grid. By offsetting the costs of EV ownership, this additional revenue stream may increase consumer interest in buying electric cars and offer a financial incentive for adoption. This could hasten the shift to a greener transportation system and lower greenhouse gas emissions from conventional combustion engines by encouraging more customers to think about EVs as a practical option for their transportation needs. V2G technology could make new services and business models possible. With the flexibility of V2Genabled EVs, energy aggregators and grid operators can optimize energy management by providing services like virtual power plants, demand response, and grid ancillary services. In the developing industries of e-mobility and energy services, this may open up new avenues for generating income, employment, and economic expansion. The growth of microgrids and local energy communities, where EVs can play a critical role in supplying local energy needs, boosting energy resilience, and encouraging community-based energy solutions, can also be supported by V2G technology.

Vehicle-to-Grid (V2G) technology has the power to completely change our perception of electric vehicles and the energy landscape. V2G-enabled EVs can help create a more flexible, sustainable, and reasonably priced energy system by having the capacity to both consume and return excess energy back to the grid. Energy can move both ways, which can help stabilize the electrical grid, integrate renewable energy sources, develop new services and business models, and give EV owners new sources of income. V2G technology has the ability to spearhead the shift to a cleaner, more resilient, and sustainable energy future as it develops and becomes more widely used.

3.3.4. Lightweight Materials

Creating lighter materials has become a viable way to enhance the efficiency and environmental friendliness of electric cars. A car's energy efficiency is directly impacted by its weight because lighter cars need less energy to accelerate, decelerate, and maintain speed. Technological developments in lightweight materials, like carbon fiber, aluminum, and advanced composites, could result in a substantial weight reduction for EVs, improving their energy efficiency and extending their range.

EVs may become safer and more durable with the use of lightweight materials. For example, materials with exceptional strength-to-weight ratios, such as carbon fiber, are perfect for structural elements in EVs that can enhance crash safety and shield occupants in the event of an accident. Additionally, since they can lower the vehicle's overall mass and increase its agility and responsiveness on the road, lightweight materials can help with handling and performance. One way that lightweight materials could lessen the environmental impact of vehicle manufacturing is through their use. Conventional car manufacturing usually uses a lot of energy and resources, which leads to high greenhouse gas emissions and pollution of the environment. On the other hand, when compared to traditional materials like steel, lightweight materials like carbon fiber and aluminum usually require less energy during production and have lower carbon footprints. This can contribute to a more sustainable transportation sector by lowering the overall environmental impact of the vehicle manufacturing process.

Although lightweight materials have been effectively employed in specialized markets, like aerospace and high-end sports cars, scaling up production for mass-market vehicles necessitates a large investment in R&D and manufacturing infrastructure. This entails creating dependable and affordable supply chains for lightweight materials and creating effective production techniques to satisfy the automotive industry's high demand. The adoption of lightweight materials in EVs have broader societal benefits. Lighter vehicles require less energy to operate, which can result in reduced dependence on fossil fuels and lower greenhouse gas emissions, contributing to mitigating climate change. Moreover, increased energy efficiency and extended range offered by lightweight materials can help address the issue of "range anxiety" – the fear of running out of battery power – which is a common concern among potential EV buyers. This can help boost consumer confidence in EVs and accelerate their adoption.

Innovation and technological advancements in the automotive industry can be stimulated by the use of lightweight materials in EVs. In an effort to make cars lighter, automakers are spending money on R&D to develop new, more sustainable and efficient materials and manufacturing techniques. This may encourage innovation in manufacturing processes, supply chain management, and materials science, resulting in new products and services that can help other industries in addition to the automobile one. Technological developments in lightweight materials have enormous potential to enhance the efficiency, durability, and security of electric automobiles. These materials may help EVs become lighter, which would increase their energy efficiency, increase their range, and lessen their environmental effect. But before lightweight materials are widely used in the automotive sector, issues with cost and scalability must be resolved. In order to accelerate the adoption of lightweight materials in EV production and ultimately contribute to a more efficient and sustainable transportation system, it is imperative that research, development, and investment in lightweight materials continue.

3.3.5. Autonomous Driving

Self-driving or driverless technology, commonly referred to as autonomous driving technology, has the potential to completely transform the transportation industry and could have a big impact on EVs. Those that are capable of autonomous driving can reduce traffic, optimize routes, and boost overall efficiency, resulting in more environmentally friendly and effective modes of transportation. In order to provide optimal route planning and navigation, self-driving EVs can employ sophisticated algorithms and sensors to evaluate traffic patterns, road conditions, and other data in real-time. This can maximize the range and efficiency of EVs by reducing travel time, decreasing idle time, and improving energy management (Nikitas et al., 2017).

Technology that allows for autonomous driving can facilitate on-demand and shared mobility services, like ride-hailing and car-sharing, which can decrease the number of vehicles owned and increase the use of electric vehicles. By strategically placing shared EVs to serve multiple passengers on distinct routes, the total number of vehicles on the road can be decreased, which can help to relieve traffic congestion. Because shared EVs can be used more efficiently than privately owned vehicles, which frequently sit idle for long periods of time, this may lead to lower emissions. The development of autonomous driving technology holds promise for increasing accessibility to public transportation for all demographics, including the elderly, those with disabilities, and those without access to private vehicles. To offer inclusive mobility options, self-driving EVs can be outfitted with accessibility features like lifts, ramps, and customized seating arrangements. For those who might have mobility issues, this can mean more independence and access to transportation, improving their quality of life and lowering transportation disparities.

The incorporation of autonomous driving technology into EVs has the potential to mitigate traffic accidents and fatalities by eliminating human error, including distracted or impaired driving. This has the potential to save lives and increase road safety, which will make transportation safer for all users of the road, including cyclists and pedestrians. In EVs, autonomous driving can also be profitable. Autonomous EVs have the potential to reduce transportation costs for both individuals and businesses through route optimization, increased efficiency, and decreased vehicle ownership. Users may find more affordable transportation options with shared and on-demand mobility services, which eliminate the need for private vehicle ownership and all of its costs, including upkeep, insurance, and parking. Furthermore, the development, production, and maintenance of autonomous EVs can generate new employment opportunities due to autonomous driving technology, thereby promoting economic growth and job creation.

The fact that emissions from electric vehicles are lower than those from conventional internal combustion engines is already well-known. EVs may minimize emissions even more when paired with autonomous driving technologies by streamlining driving habits, cutting down on energy waste, and utilizing renewable energy sources. By reducing climate change and enhancing air quality, this can help create communities that are healthier and more sustainable. Technology related to autonomous driving has the potential to have a big impact on the transportation industry, including EV use. It can lead to more sustainable and effective transportation by streamlining routes, easing traffic, and boosting overall efficiency. It can also facilitate on-demand and shared mobility services, which will decrease the number of vehicles owned and increase the use of electric vehicles. To fully utilize autonomous driving in EVs and create a more sustainable and accessible transportation future, it is imperative to address issues pertaining to safety, regulations, and technology.

3.3.6. Vehicle-to-Everything (V2X) Communication

A state-of-the-art technology called Vehicle-to-Everything (V2X) communication has the potential to completely change how EVs interact with their surroundings. By connecting EVs to the grid, other cars, and infrastructure through V2X communication, a smooth network of connectivity is created that can improve a number of EV performance factors (Habib et al., 2015).

V2X communication has the potential to enhance EV charging management. EVs and charging stations can coordinate charging so that the owner can maximize charging costs by scheduling charging for off-peak hours when electricity is less expensive. To help grid operators better manage the electrical grid, EVs can also communicate with the grid to provide information about their charging status, energy availability, and demand. This has the potential to enhance grid stability, load balancing, and the integration of renewable energy sources. Additionally, V2X communication can improve the overall effectiveness of the transportation system. V2X communication can enable intelligent transportation systems that optimize resource use, minimize energy waste, and lessen environmental impacts by connecting EVs, charging stations, and the grid. By enabling smart car-sharing programs, for instance, where EVs can be effectively shared among several users, V2X communication can lessen the need for private vehicle ownership and promote sustainable mobility options.

EVs and other cars on the road can converse and exchange data about direction, speed, and location. This can help avoid collisions in real time and lessen the likelihood of accidents. For instance, an EV may be able to avert a possible collision by acting evasively in response to a nearby car that is about to run a red light. This can greatly improve traffic safety and lower the likelihood of collisions, making driving an EV even safer. EV efficiency can be increased through V2X communication, which optimizes traffic flow. In order to get real-time traffic conditions information, EVs can communicate with other infrastructure such as road signs, traffic lights, and other devices. By doing so, EVs may be able to modify their speed and path, avoiding traffic and cutting down on travel time. For example, an EV can automatically reroute to a different route when it detects an impending traffic jam, preventing delays and increasing overall transportation efficiency.

For EV owners, V2X communication opens up new revenue streams and business models. EV owners could, for instance, take part in demand response programs, which allow them to resell any extra energy that their EVs store to the grid during periods of high demand and make money doing so. As a result, a more sustainable energy ecosystem may be created, encouraging the purchase of electric vehicles and encouraging the use of renewable energy sources. The accessibility of EVs for individuals with disabilities can be improved through V2X communication. For instance, EVs can exchange real-time data about availability, accessibility features, and charging options with infrastructure like parking lots and charging stations. This can promote inclusivity and accessibility in transportation by making it simple for individuals with disabilities to find and use parking spaces, charging stations, and other EV-related services. V2X communication can improve many facets of EV performance and help create a more environmentally friendly and intelligent transportation future. These benefits range from enhancing safety and efficiency to opening up new business opportunities and encouraging the integration of renewable energy. To guarantee the responsible and extensive adoption of V2X communication in the automotive industry, it is imperative to tackle issues pertaining to standardization, cybersecurity, privacy, and infrastructure deployment.

3.3.7. Circular Economy

A product's entire lifecycle, including its design, manufacture, use, and end-of-life phases, should be optimized for resource utilization and waste reduction, according to the circular economy concept. When it comes to EVs, the circular economy approach can be extremely helpful in reducing the environmental effects of both EV production and disposal. Designing EVs with recycling and reuse in mind is the context of EVs. This entails the use of easily recycled or repurposed materials and the design of components to make it easier to disassemble and separate them for recycling. For instance, since steel and aluminum are readily recyclable and can be used to make new cars or other products, using them in the construction of EVs can help with recycling. Furthermore, modular component design for EVs can make it simple to upgrade or replace particular parts, increasing the vehicle's lifespan and lowering the need for total vehicle replacement. EV encourages manufacturers to use recycled materials in their products. Recycling elements from electric vehicle batteries, such as lithium, cobalt, and nickel, can lessen the need for raw materials, lessen the effects of mining on the environment, and preserve natural resources. Reducing the environmental impact of EV manufacturing can be achieved by implementing closed-loop recycling systems, in which materials from end-of-life EVs are recovered and repurposed in the creation of new vehicles.

In the context of the circular economy, formulating plans for the recycling and end-of-life repurposing of electric vehicle components is essential. For instance, after being used in cars, EV batteries can still hold a sizable portion of their useful capacity, which can then be used for stationary energy storage or other purposes. Other EV parts, like motors, inverters, and charging apparatus, can also be repurposed to increase their longevity and decrease waste. Furthermore, by recycling EV components properly, you can minimize the environmental impact of their disposal by recovering valuable materials and keeping them out of landfills or incinerators. Strategies to increase EV longevity and encourage circular consumption patterns can also be included in the circular economy approach for EVs. For example, putting in place programs for remanufacturing, upgrading, or refurbishing older EVs can help them last longer, which in turn lowers the need for new cars and reduces waste. Increasing the efficiency and performance of older EVs may involve upgrading antiquated technology or swapping out worn-out parts for refurbished or remanufactured ones. Encouraging EV batteries to be used for stationary energy storage or grid stabilization, for example, can increase their useful life and lessen the need for premature disposal.

Promoting the circular economy for EVs also requires raising awareness and educating people. Customers' demand for more environmentally friendly EVs can be increased by educating them about the advantages of circular consumption patterns, such as purchasing reconditioned or remanufactured EVs or taking part in take-back and recycling programs. This can also promote appropriate end-of-life disposal practices. Similarly, educating producers, decision-makers, and other interested parties about the financial and environmental advantages of circular economy strategies can encourage creativity and cooperation in the direction of more environmentally friendly EV production and lifecycle management techniques. A variety of tactics are included in the circular economy approach for EVs with the goals of cutting waste, increasing resource efficiency, and prolonging the life of EVs. The circular economy for EVs includes several key components, such as designing

EVs with recyclability and reusability in mind, encouraging the use of recycled materials in manufacturing, creating strategies for end-of-life recycling and repurposing, extending the lifespan of EVs through refurbishing and remanufacturing, putting in place efficient take-back and recycling programs, and raising awareness and education. The EV industry can transition to a more resource-conscious and sustainable transportation system that reduces waste, conserves energy, and lessens the environmental effects of EV production and disposal by embracing the principles of the circular economy.

In summary, battery technology has advanced significantly since its inception, but many issues still prevent it from being widely used. The low energy density of batteries, which reduces their range and adds weight, is one of the biggest problems. This is particularly true for electric cars since a short range worries a lot of prospective purchasers. Furthermore, batteries need to be changed after a specific number of charging cycles due to their limited lifespan. Particularly in large-scale applications like electric vehicles or grid-scale energy storage, this can be expensive and time-consuming. It is well known that lithium-ion batteries are vulnerable to thermal runaway, which can cause explosions or fires. This poses a serious risk to public safety, particularly for large-scale energy storage systems and electric cars. Even though battery safety has improved over time, much more needs to be done in this area. Solid-state batteries, one of the new battery chemistries, have the potential to completely change the market by offering safer and higher energy densities. By using a solid electrolyte rather than a liquid one, solid-state batteries can increase security and lower the possibility of thermal runaway. Furthermore, the environmental impact of battery production may be decreased by using recycled materials and environmentally friendly manufacturing techniques.

The environmental effects of battery production continue to raise concerns, particularly in light of the rising demand for raw materials like cobalt and lithium. These materials can raise ethical questions regarding their extraction because they are frequently mined in ways that harm the environment. To allay these worries, scientists are looking into sodium-ion batteries as an alternative to conventional battery chemistries, which may employ more readily available and environmentally friendly materials. Furthermore, the development of environmentally friendly manufacturing techniques and recycling technologies may contribute to a decrease in the negative environmental effects of battery production. Energy losses that occur during the transfer of power from the charging pad to the EV's battery can make wireless charging systems less effective. Consequently, drivers who need to swiftly charge their electric vehicles may find that wireless charging takes longer than plug-in charging. The need for specialized infrastructure, which can be more expensive to install than plug-in charging stations, is another drawback of wireless charging. This is since specific requirements for the wireless transmission of power must be met during the design and installation of wireless charging systems. Furthermore, wireless charging systems may require more frequent maintenance and repairs due to their increased complexity. Because of this, the initial expense of setting up a wireless charging system could be a major deterrent to its adoption, particularly for smaller companies and individuals who might not have the funds to purchase the technology.

Wireless charging systems raise additional safety concerns. Wireless charging necessitates high-power electromagnetic fields, which can be harmful to humans and animals if they are exposed to them for long periods of time. Nonetheless, research indicates that the electromagnetic radiation levels released by wireless charging systems are generally safe for both people and animals. However, safety concerns continue to be a major deterrent to the widespread use of wireless charging, and further study is required to address these issues and create safety guidelines for wireless charging infrastructure. Technological developments may result in quicker and more effective charging times, which would make wireless charging a more viable choice for EV users. Furthermore, EV manufacturers may find it simpler to integrate wireless charging into their vehicles if standardized wireless charging protocols are developed. A greater variety of EVs that are compatible with wireless charging systems may arise from this. To further improve accessibility and convenience, wireless charging might be incorporated into the infrastructure that already exists, such as parking lots and roadways. Although some nations and areas have begun putting V2G projects into action, they are still in the planning stages, and the necessary infrastructure is still not generally accessible. Furthermore, regulatory frameworks need to be established in order to guarantee the technology's safe and efficient integration into the grid. Grid stability and reliability become major concerns as EVs proliferate and their capacity to supply the grid with energy grows. To maintain the grid's stability and dependability while controlling the flow of electricity from EVs, a sophisticated energy management system is needed. Concerns exist over how much excess energy EVs can contribute to the grid, as this may be constrained by elements like battery capacity and charging habits.

The utilization of energy-intensive materials during the manufacturing process, like aluminum or carbon fiber, can raise the carbon footprint of the vehicle. Furthermore, the creation of lightweight materials might call for the use of hazardous chemicals, which, if not handled carefully, could have a negative impact on the environment. Modifications to safety regulations and vehicle design may also be necessary as a result of the use of lightweight materials.

Because lightweight materials have different qualities from traditional materials, it might be necessary to modify a vehicle's design in order to comply with safety regulations. Furthermore, the durability and dependability of vehicles may be impacted by the distinct qualities of lightweight materials. The requirement for standardization to guarantee compatibility and interoperability across various vehicles and infrastructure is one of the challenges facing V2X technology. The broad acceptance of this technology depends on the creation of uniform standards and protocols for V2X communication. Another issue that must be resolved is guaranteeing data security and privacy. Many technical obstacles still need to be addressed, such as creating algorithms that can consistently identify and react to intricate and unpredictable driving scenarios. In addition, there exist regulatory obstacles concerning insurance and liability, which might necessitate considerable modifications to laws and policies to promote the integration of autonomous driving technology. The environmental impact of EVs could be greatly decreased by incorporating the principles of the circular economy into the design and production process. Adoption of circular economy concepts also requires a shift in mindset and business models, which can be challenging for some companies. But a lot of companies have already incorporated circular economy strategies—like recycling batteries and reusing parts—into their manufacturing processes (Nikitas et al., 2017).

REFERENCES

1. Ahmad, A., Alam, M. S., & Chabaan, R. (2018). A comprehensive review of wireless charging technologies for electric vehicles. *IEEE Transactions on Transportation Electrification, 4*(1), 38–63.
2. Boyden, A., Kie Soo, V., & Doolan, M. (2016). The environmental impacts of recycling portable lithium-ion batteries. *Procedia CIRP, 48*, 188–193. https://doi.org/10.1016/j.procir.2016.03.100.
3. Calvão, F., McDonald, C., & Bolay, M. (2021). Cobalt mining and the corporate outsourcing of responsibility in the Democratic Republic of Congo. *The Extractive Industries and Society, 8*(4). https://doi.org/10.1016/j.exis.2021.02.004.
4. Chau, K. T., Wong, Y. S., & Chan, C. C. (1999). An overview of energy sources for electric vehicles. *Energy Conversion and Management, 40*(10), 1021–1039.
5. Cohen, A. (2020). Manufacturers are struggling to supply electric vehicles with batteries. *Forbes*. Retrieved from: https://www.forbes.com/sites/arielcohen/2020/03/25/manufacturers-are-struggling-to-supply-electric-vehicles-with-batteries/?sh=3b68d8a71ff3 (accessed on 9 November 2024).
6. Das, H. S., Rahman, M. M., Li, S., & Tan, C. W. (2020). Electric vehicles standards, charging infrastructure, and impact on grid integration: A technological review. *Renewable and Sustainable Energy Reviews, 120*, 109618.
7. Deberdt, R., & Le Billon, P. (2021). Conflict minerals and battery materials supply chains: A mapping review of responsible sourcing initiatives. *The Extractive Industries and Society, 8*(4), 100935. https://doi.org/10.1016/j.exis.2021.100935.
8. Dong, D., van Oers, L., Tukker, A., & van der Voet, E. (2020). Assessing the future environmental impacts of copper production in China: Implications of the energy transition. *Journal of Cleaner Production, 274*, 122825. https://doi.org/10.1016/j.jclepro.2020.122825.
9. Farjana, S., Huda, N., & Mahmud, P. (2019). Life cycle assessment of cobalt extraction process. *Journal of Sustainable Mining, 18*(3), 150–161. https://doi.org/10.1016/j.jsm.2019.03.002.
10. Habib, S., Kamran, M., & Rashid, U. (2015). Impact analysis of vehicle-to-grid technology and charging strategies of electric vehicles on distribution networks—A review. *Journal of Power Sources, 277*, 205–214.
11. Henderson, J. (2020). EVs are not the answer: A mobility justice critique of electric vehicle transitions. *Annals of the American Association of*

Geographers, 110(6), 1993–2010. https://doi.org/10.1080/24694452.2020.1744422.

12. Jagani, S., Marsillac, E., & Hong, P. (2024). The electric vehicle supply chain ecosystem: Changing roles of automotive suppliers. *Sustainability, 16*, 1570. https://doi.org/10.3390/su16041570.
13. Jerez, B., Garcés, I., & Torres, R. (2021). Lithium extractivism and water injustices in the Salar de Atacama, Chile: The colonial shadow of green electromobility. *Political Geography, 87*, 102382. https://doi.org/10.1016/j.polgeo.2021.102382.
14. Kara, S. (2023). *Cobalt Red: How the Blood of the Congo Powers Our Lives*. St. Martin's Publishing Group.
15. Li, G., & Zhang, X. P. (2012). Modeling of plug-in hybrid electric vehicle charging demand in probabilistic power flow calculations. *IEEE Transactions on Smart Grid, 3*(1), 492–499.
16. Liu, Z., Wen, F., & Ledwich, G. (2013). Optimal planning of electric-vehicle charging stations in distribution systems. *IEEE Transactions on Power Delivery, 28*(1), 102–110.
17. Morgan, J. (2020). Electric vehicles: The future we made and the problem of unmaking it. *Cambridge Journal of Economics, 44*, 953–977.
18. Nikitas, A., Kougias, I., Alyavina, E., & Njoya Tchouamou, E. (2017). How can autonomous and connected vehicles, electromobility, BRT, Hyperloop, shared-use mobility, and mobility-as-a-service shape transport futures for the context of smart cities? *Urban Science, 1*(4), 36.
19. Perera, P., Hewage, K., & Sadiq, R. (2020). Electric vehicle recharging infrastructure planning and management in urban communities. *Journal of Cleaner Production, 250*, 119559.
20. Rapson, D. S., & Muehlegger, E. (2023). The economics of electric vehicles. *Review of Environmental Economics and Policy, 17*, 274–294.
21. Schreiber, A., Marx, J., & Zapp, P. (2021). Life cycle assessment studies of rare earths production–Findings from a systematic review. *Science of the Total Environment, 791*, 148257. https://doi.org/10.1016/j.scitotenv.2021.148257.
22. Srivastava, P., Bolan, N., Casagrande, V., & Joshua, B. (2022). Appraisal of metal(loids) in the ecosystem. In *Elsevier* (pp. 81–104).
23. Zeng, A., Chen, W., Rasmussen, K. D., Zhu, X., Lundhaug, M., Müller, D. B., Tan, J., Keiding, J. K., Liu, L., Dai, T., Wang, A., & Liu, G. (2022). Battery technology and recycling alone will not save the electric mobility transition from future cobalt shortages. *Nature Communications, 13*(1), 1341. https://doi.org/10.1038/s41467-022-29022-z.

Chapter 4

Battery Production and Environmental Concerns

Contents

INTRODUCTION

The production of batteries presents serious environmental challenges despite being essential to contemporary technology and the shift to renewable energy sources. The process of extracting raw materials like nickel, cobalt, and lithium frequently uses methods that are harmful to the environment. For example, the extraction of lithium may result in contamination and water depletion, especially in arid areas like the Atacama Desert. The mining of cobalt, which is primarily done in the Democratic Republic of the Congo, has been connected to serious violations of human rights, including child labor, and environmental degradation. Because of the release of hazardous materials, these mining operations may cause habitat destruction, biodiversity loss, and pollution of the soil and water. Large amounts of energy are used in the manufacturing process to produce batteries, which frequently come from fossil fuels and increase greenhouse gas emissions. Hazardous waste from the process is also produced, including chemicals and heavy metals, which need to be managed carefully to prevent environmental contamination. Furthermore, improper regulation of the production stage may result in pollution of the air and water. During the manufacturing process, volatile organic compounds (VOCs) can be released into the atmosphere by the use of solvents and other chemicals. This can lead to air pollution and health hazards for workers and the surrounding communities.

Batteries have an ongoing environmental impact for the duration of their lifecycle after production. Because batteries contain metals and chemicals that can corrode soil and groundwater over time, improper battery disposal can result in the release of hazardous materials into the environment. Although recycling can lessen some of these effects, lithium-ion battery recycling rates are still low at the moment. If recycling is not done correctly, it can be an energy-intensive process that puts the environment at further risk. Furthermore, the need for sustainable battery production has increased due to the growing demand for renewable energy storage and electric vehicles (EVs). Although EVs emit fewer greenhouse gases than traditional cars, it is important to consider the environmental impact of their batteries. The transition to renewable energy storage emphasizes how crucial it is to create batteries with greater efficiency and longer lifespans to reduce their negative effects on the environment. More focus is being placed on creating environmentally friendly and sustainable battery technologies to allay these worries. There is research being done on innovations like solid-state batteries, which offer greater energy density, safety, and fewer environmental hazards. Furthermore, attempts are underway to enhance recycling technologies and augment the utilization of recycled materials in the battery manufacturing process. Industry standards and regulatory frameworks are changing as well to guarantee more environmentally friendly practices all the way through the battery lifecycle.

4.1. BATTERIES: ELECTRICITY THROUGH CHEMICAL REACTIONS

One or more electrochemical cells, used in batteries, store chemical energy that can subsequently be converted to electrical energy. Numerous everyday gadgets, including cars, laptops, cell phones, and clocks, depend on batteries for operation. An electrochemical cell, which is used to produce and store energy, makes up a minimum of one battery. Batteries typically contain at least one voltaic cell, despite the existence of a wide variety of electrochemical cells. Galvanic cells is another term that can be used to refer to voltaic cells. In contrast to the reactions found in fuel cells and electrolytic cells, voltaic cells undergo spontaneous chemical reactions and electrical energy production.

The term "battery" was originally used by Benjamin Franklin in 1749 to refer to linked capacitors during his electrical experiments. Though it was the first battery ever mentioned as such, his was not the original battery. Though their precise function is still up for debate, it is thought that the Baghdad Batteries, which were invented in 1936 and date back more than 2,000 years, were among the first batteries ever made. In 1780, Luigi Galvani—the man who gave the name to the galvanic cell—described animal electricity for the first time by passing an electrical current through a frog. This was a battery, even though he was unaware of it at the time. The voltaic cell and voltaic pile are named for Alessandro Volta, who was a contemporary of his and was adamant that the animal electricity was not coming from the frog but rather from another source. The voltaic pile, the first true battery, was created in 1800. John Frederic Daniell invented the Daniell cell in 1836 while attempting to find solutions for some of the issues surrounding Volta's voltaic pile. Following this discovery were the creation of the Grove cell by William Robert Grove in 1844, the gravity cell by Callaud in the 1860s, the Leclanche cell by Georges Leclanche in 1866, and the first rechargeable battery by Gaston Plante in 1859, which was made of a lead-acid cell.

Every battery used up until this point were wet cells. Then, using a zinc-carbon cell, Carl Gassner invented the first dry cell battery in 1887. Waldemar Jungner introduced the nickel-iron battery and the nickel-cadmium battery in 1899. Thomas Edison patented a slightly altered version of the nickel-iron battery in 1903 after Jungner was unable to secure a patent for his invention. The invention of the common alkaline battery by Lewis Urry, a worker at the company that is now known as Energizer, in 1955 marked a significant advancement. The nickel hydrogen battery was developed in the 1970s, and the nickel metal-hydride battery followed in the 1980s. Although the first lithium batteries were developed in 1912, the most popular kind—the lithium ion polymer battery, which is now found in the majority of portable electronics—was not commercially available until 1996 (Bitenc et al., 2020).

4.1.1. Voltaic Cells

The two half-cell reactions (oxidation-reduction) that make up voltaic cells are connected by a wire and a semipermeable membrane, usually a salt bath (Figure 4.1). There is a metal on each side of the cell that serves as an electrode. We refer to one electrode as the cathode and the other as the anode. By gaining electrons, the cathode-containing side of the cell becomes reduced and serves as an oxidizing agent for the anode. Oxidation, or the loss of electrons, happens on the side of the cell that houses the anode and serves as a reducing agent for the cathode. An electrolyte, which is an ion-containing compound, is immersed in each of the two electrodes.

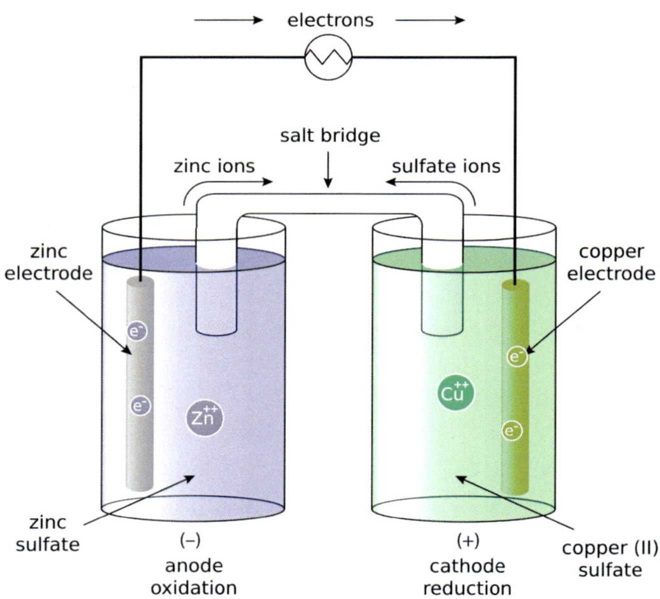

Figure 4.1. *A Zinc–Copper voltaic cell. The voltaic cell provides the electricity needed to power the light bulb.*

Source: By Rehua – Own work, CC BY 3.0, https://commons.wikimedia.org/w/index.php?curid=17902682ight=410.

To help with the electron transfer process through the wire, this electrolyte serves as a concentration gradient on both sides of the half reaction. The energy produced by these electrons moving back and forth powers the battery. Because the chemical reaction is spontaneous, the cell is divided into two compartments. Without this separation, heat would be released as a result of the reaction, rendering the battery ineffective.

4.1.1.1. Types of Batteries

- **Primary Batteries:** Primary batteries are disposable and not rechargeable. These batteries' electrochemical processes are irreversible. Since all of the materials in the electrodes have been used, electricity cannot be generated. Because primary batteries discharge much more slowly than secondary batteries, they are frequently used when extended periods of storage are needed. Flashlights, smoke detectors, and the majority of remote controls are examples of devices that use primary batteries.

Figure 4.2. *Primary versus secondary Batteries. Primary batteries are non-rechargeable and disposable. Secondary batteries are rechargeable, like a cellular phone battery.*

Source: By Lead holder – Own work, CC BY-SA 3.0, https://commons.wikimedia.org/w/index.php?curid=15389714.

- **Secondary Batteries:** Rechargeable secondary batteries are available. These batteries go through easily reversible electrochemical reactions. Secondary batteries have reversible chemical reactions because the reacting components are not entirely consumed. After their energy is depleted, rechargeable batteries require an external electrical source to replenish their energy. Car batteries and portable electronic devices are two examples of devices that use secondary batteries (Figure 4.2).

4.1.2. Battery Cell Types

- **Wet Cells:** Wet cell batteries have an electrolyte that is liquid. These may be classified as secondary or primary batteries insulator sheets are used to keep the anode and cathode apart in wet cells because of their liquid nature. Daniell cells, Leclanche cells (which were first employed in dry cells), Bunsen cells, Weston cells, Chromic acid cells, and Grove cells are examples of wet cell types. Wet cells are lead-acid cells found in car batteries (Figure 4.3).

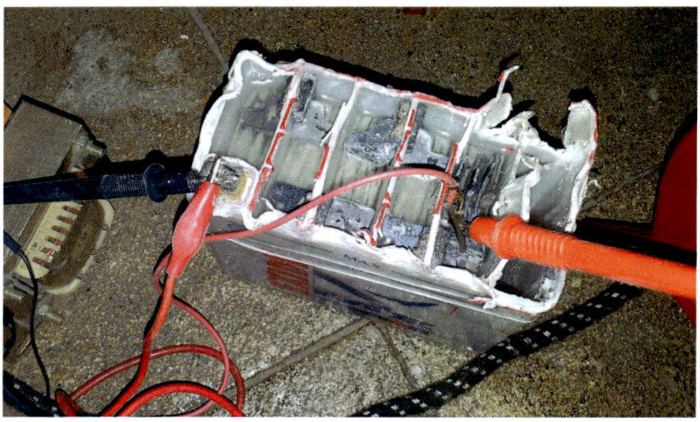

Figure 4.3. *A lead-acid battery in an automobile.*

Source: By Sakar.solanki – Own work, CC BY-SA 3.0, https://commons.wikimedia.org/w/index.php?curid=23692756.

- **Dry Cells:** There is no free liquid in dry cell batteries. Rather, the electrolyte is a paste that is just moist enough to permit the flow of current. Because of this, the dry cell battery can be used in any orientation without having to worry about its contents spilling out. For this reason, dry cell batteries are frequently found in items like portable electronic devices that are handled and turned around a lot. Primary or secondary batteries are both possible with dry cell batteries. The Leclanche cell is the most widely used type of dry cell battery.

4.1.3. Battery Performance

The amount of electrode and electrolyte material within a battery directly affects its capacity. When primary batteries are not used for a year, they can lose between 8% and 20% of their charge. Side chemical reactions that don't result in current are the cause of this. Lowering the temperature can slow the rate of side reactions. Because the side chemical reactions speed up in warmer temperatures, the battery's performance may also be reduced. With use, primary batteries polarize. At this point, hydrogen builds up at the cathode and lowers the battery's efficiency. Depolarizers can be used to get rid of this hydrogen accumulation. Even faster self-discharge occurs in secondary batteries. Each month, they typically lose roughly 10% of their charge. After each recharge cycle, rechargeable batteries gradually lose capacity as a result of deterioration. Electrolytes moving away from the electrodes or active materials falling off the electrodes are the causes of this. The approximate relationships between current, capacity, and discharge time can be found using Peukert's law. This is what the equation looks like.

$$t = \frac{Q_p}{K^k}$$

where I is the current, k is a constant of about 1.3, t is the time the battery can sustain the current, and Q_p is the capacity when discharged at a rate of 1 amp.

4.1.4. Current, Voltage, and Standard Reduction Potential

The voltage and current of a cell are significantly correlated. As the name suggests, current is the movement of electrical charge. Voltage is the maximum amount of current that a system is capable of handling. Figure 4.4 shows how voltage and current differ from one another.

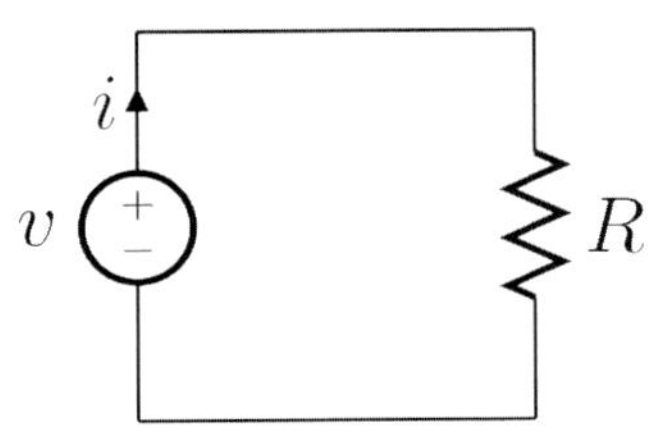

Figure 4.4. *The difference between voltage and current.*

Source: By GorillaWarfare – Own work, CC0, https://commons.wikimedia.org/w/index.php?curid=22824659.

A waterwheel is being turned by water pouring from a hose onto it. The amount of water that is passing through the hose is known as the current. The strength or pressure of the water flowing through the hose can be compared to the voltage. Because there is not enough water or pressure passing through the first hose, the waterwheel cannot be turned very successfully. The second hose has a lot of current flowing through it because there is a lot of water flowing through it. The third hose has less water running through it, but there is an obstruction obstructing a large portion of the hose. This gives the water hose a higher pressure as it exits, increasing its voltage and enabling it to strike the waterwheel with greater force than it could have with the first hose.

Standard reduction potential, E^o, is a measurement of voltage. Standard reduction potential can be calculated with the knowledge that it is the difference in energy potentials between the cathode and the anode: $E^o_{cell} = E^o_{cathode} - E^o_{anode}$. For standard conditions, the electrode potentials for the half cells can be determined by using a table of standard electrode potentials.

4.1.5. Different Sizes of Batteries and Some Additional Facts

Because of their internal composition and chemical characteristics, batteries differ in size as well as voltage. Nonetheless, the voltage of batteries with varying sizes might be the same. This phenomenon is caused by the fact that a battery's internal content, rather than its size, determines its standard cell potential. Batteries of various sizes can therefore have the same voltage (Figure 4.5). Furthermore, batteries can increase their voltage and current in a few different ways. Batteries can have their voltage increased by arranging them in rows, and their current can be increased by arranging them in columns.

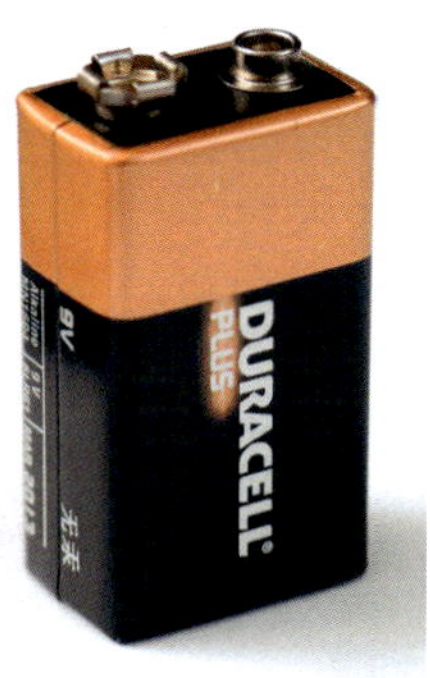

Figure 4.5. *A 9V battery.*

Source: By Ashley Pomeroy – Own work, CC BY 3.0, https://commons.wikimedia.org/w/index.php?curid=15853913.

- **Hazards:** Batteries that are misused or malfunctioning may burst. Gases can accumulate in a rechargeable battery and possibly burst if it is attempted to be overcharged or charged too quickly. An explosion can also result from a short circuit. When a battery is submerged in fire, the battery's internal steam can explode. Because the chemicals found in batteries have the potential to be harmful, leakage is also a concern. Leakage from the batteries is hazardous to handle and has the potential to destroy the device they are housed in. The extensive use of batteries raises a number of environmental concerns. Batteries require the handling of numerous hazardous chemicals and the consumption of large amounts of resources. Used batteries are frequently disposed of incorrectly, adding to the waste generated by electronics. Because batteries may contain hazardous materials, disposing of them incorrectly can be extremely risky. Toxic metals, including lead and mercury, are prevented from entering the environment through electronic recycling programs. Battery consumption is dangerous and can be fatal.
- **Homemade Batteries:** A battery can be created from any liquid or moist material that contains enough ions to be electrically conductive. Even little amounts of electricity can be produced by sticking electrodes made of various metals inside carbonated cola, potatoes, lemons, and bananas. Two coins and a piece of paper dipped in salt water can be used to create a voltaic pile. The current may rise when several coins are stacked in succession (Bitenc et al., 2020).

4.1.6. Understanding the Environmental Potential of Batteries

Batteries are essential in the field of sustainable technology and renewable energy. Batteries allow for a cleaner and more effective energy ecosystem by powering electric vehicles (EVs) and storing energy produced from renewable sources like solar energy and wind power. But like any technology, batteries have their own set of environmental issues, especially when it comes to how they are made, used, and disposed of after their useful life. First, let us look at how batteries affect the environment:

- **Production Phase:** Lithium, cobalt, and nickel are raw materials used in battery manufacturing that are often extracted using methods that harm the environment, such as mining and deforestation. Even though less conflicted materials are used in battery technologies like LFP, labor and environmental aspects of the value chain still need to be optimized.

- **Use Phase:** Batteries have an overall environmental impact that goes beyond their operational lifespan, even though they are a cleaner option than fossil fuels when used. For instance, the average degradation level of EV batteries nears the end of their vehicle life, ranging from 70% to 80%. Furthermore, improvements in battery technology are creating batteries that have a longer lifespan than the cars they power, opening up possibilities for prolonged use.
- **End-of-Life Management:** When batteries are disposed of improperly, harmful chemicals may seep into the ground and contaminate nearby areas, endangering both human health and ecosystems. Low battery recycling rates highlight the need for better recycling infrastructure since they worsen pollution and resource depletion.

4.1.7. The Importance of Recycling

Recycling helps lessen the negative effects of batteries on the environment by recovering valuable materials and lowering the demand for virgin resources. Nevertheless, obstacles including intricate battery chemistries, financial considerations, and legal concerns impede extensive recycling initiatives. One of the primary advantages of lithium-ion batteries, for instance, is that they don't contain valuable materials like nickel and cobalt, which means that the materials recovered during recycling aren't worth much, making the recycling of these batteries not very profitable. But more work is being done to develop recycling, so it's not just batteries that get more efficient. Society can enable this development and save money and batteries by extending the battery's lifespan.

4.1.7.1. Second-Life Applications: Extending Battery Lifespan

Especially in Battery Energy Storage Systems (BESS), the idea of second-life applications is an emerging strategy to address the environmental impact of batteries. Second-life applications extend the useful life of batteries by using them for stationary storage applications. This makes recycling batteries later and provides a more affordable alternative for energy storage. Second-life batteries are also essential for advancing grid stability, circular economy ideas, and lessening environmental impact as we move toward a sustainable energy future.

4.1.7.2. Future Innovations in Battery Technology

The demand for novel battery technologies that reduce their environmental impact is growing along with the significance of eco-friendly practices. Future developments in battery technology should see an increase in energy storage capacity, increased efficiency, and a decrease in the use of hazardous materials. A current research and development focus is investigating substitute materials for use in battery manufacturing. Researchers and engineers are working hard to find abundant and sustainable materials that can take the place of the currently used rare and costly elements. This lessens the effect on the environment while also increasing the affordability and accessibility of batteries. Furthermore, improvements in recycling methods and procedures might also be included in next battery technology advancements. In order to recover a larger percentage of valuable materials and minimize waste production, researchers are trying to develop more economical and efficient battery recycling techniques. All things considered, battery technology has a lot of potential to improve sustainability and lessen its negative effects on the environment when it comes to energy storage solutions (Bitenc et al., 2020).

4.2. BATTERY CHEMISTRY AND ENVIRONMENTAL IMPLICATIONS

The voltage and performance characteristics are reflected in the battery chemistry. The precise chemistry of the materials that make up each battery determines which category it belongs in. Typical battery chemistries consist of lead, nickel, and lithium in different combinations. Additionally, there are regulatory requirements specific to each battery chemistry for shipping, handling, and disposal. The majority of older batteries used rechargeable lead-acid or non-rechargeable alkaline chemistries. Each of these battery types had a nominal voltage in increments of 2–213 and 1–5 volts, respectively, and represented a single electrochemical cell. Older naming conventions have been strained by new special battery chemistries. NiMH (Nickel Metal Hydride) and NiCd (Nickel Cadmium) rechargeable batteries normally produce 1.25 V per cell. Given the 16% drop in voltage, some devices might not work properly with these cells, but the majority of modern devices work well with them. In contrast, a pair of alkaline cells (3 V), which lithium-ion rechargeable batteries are frequently intended to replace, have an output of 3.7 V per cell, which is 23% higher. Lithium-chemistry non-rechargeable batteries, which offer a very high energy density, generate approximately 0.5 V per cell, making them comparable to alkaline batteries. While older names did not indicate the batteries' chemistry or size, many new battery sizes do. Only battery size-related types are included in the summary (Tables 4.1 and 4.2).

Table 4.1. *Variety of Anode Materials for Li-ion Batteries*

Anode Materials	**Capacity (mA h g^{-1})**	**Merits**	**Remarks**
Hard carbon	200–600	Excellent working potential	Less columbic efficiency
Carbon Nanotubes	1115	Cost-effective	Greater hysteresis voltage
Graphene	770/1115	Safe to operate	More irreversible capacity
$LiTi_4O_5$	176	Safer, Low cost	Lower capacities
TiO_2	320	Greater power capability, good life cycle	Lower energy densities
Germanium	1623	Greater Specific Capacities, Greater Energy densities, Excellent safety	Lower irreversible capacity, More fading capacity, less life cycle
Silicon	4213		
Silicon oxide	1562		
Antimony	661		
Tin	992		
Tin oxide	793		
Metal Oxides (Co_3O_4, CoO, MoO_2/O_3, NiO, RuO_2, Cr_2O_3)	500–1200	Large capacity, energy, Minimal cost, ecofriendly	Lower columbic efficiency, Unstable SEI formation, not good life cycle, more hysteresis potential.
Metal phosphides/nitrides and sulfides	500–1800	Greater capacities, lower potential, the lower polarization	Weak rate capability, short life span, higher production cost.

Source: Kim H-J, Krishna T, Zeb K, Rajangam V, Gopi CVVM, Sambasivam S, Raghavendra KVG, Obaidat IM. A Comprehensive Review of Li-Ion Battery Materials and Their Recycling Techniques. Electronics. 2020; 9(7):1161. https://doi.org/10.3390/electronics9071161.

Table 4.2. *Summary of Recycling Technologies for Different Types of Batteries*

Battery Type	Recycling Technology	Advantages	Disadvantages
Alkaline, Ni-Cd, Ni-MH	Mechanical separation	Can handle diverse battery types and chemistries Separates metals from non-metals Scalable and flexible process	Does not recover pure metals Further refining steps needed Dismantling and size reduction required first
	Pyrometallurgy	Effective for large-scale metal recovery	High energy consumption Hazardous emissions Slag waste generation
	Hydrometallurgy	Selective metal recovery High purity products	Multi-stage, complex processes High reagent consumption Effluent treatment required
Lithium-Ion	Mechanical separation	Can handle diverse battery types and formats Separates metallic and non-metallic materials Facilitates further processing	Does not recover pure metals or compounds Further refining steps needed Pre-processing and size reduction required first
	Pyrometallurgy	High throughput capability	High energy consumption Hazardous emissions Material losses to slag
	Hydrometallurgy	Selective recovery of materials High purity products	Complex multi-stage processes High reagent consumption Effluent treatment requirements

Battery Type	Recycling Technology	Advantages	Disadvantages
All Battery Types	Biotechnological Methods	Environmentally friendly Energy efficient Selective leaching	Early development stage Process optimization required
All Battery Types	Direct Recycling	Reduces processing steps Maintains material value	Limited commercial viability currently

Source: Toro L, Moscardini E, Baldassari L, Forte F, Falcone I, Coletta J, Toro L. A Systematic Review of Battery Recycling Technologies: Advances, Challenges, and Future Prospects. Energies. 2023; 16(18):6571. https://doi.org/10.3390/en16186571.

The electrochemistry, or more accurately, the basic chemistry of batteries. This is the electrolyte, anode, and cathode. Which ones are the most recent, what are they, who makes them, where they fit in the roadmap, and what are the benefits and drawbacks of each? Power Density versus Energy Density is typically plotted. The .s in this plot are actual cell datasheet data.

The main chemistries are:

- **Lithium Ion:** Lithium ions travel from the negative electrode to the positive electrode in a rechargeable lithium-ion battery during discharge and vice versa during charging. The energy density of current production cells is about 280 Wh/kg. A lithium transition metal oxide, such as manganese or cobalt, or a combination of transitional metals, such as LCO, LMO, NCA, NMC, LFP, and LMFP, serves as the cathode. Typically, the anode is made of graphite, which has the ability to intercalate or release lithium. To boost capacity, the anode may also contain silicon. Lithium titanate (LTO) is an alternative anode.
- **Sodium Ion:** Like a lithium-ion battery, but with sodium ions (Na^+) acting as the charge carriers. With the exception of the substitution of sodium compounds for lithium compounds, the operation of the sodium-based chemistry and cell construction are nearly identical to those of the commercially available lithium-ion battery kinds.
- **Lead Acid:** The Lead Acid Battery is a battery with electrodes of lead oxide and metallic lead that are separated by an electrolyte of sulfuric acid. Energy density 40–60 Wh/kg.

- **Nickel Metal Hydride:** The Nickel Metal Hydride battery consists of an aqueous potassium hydroxide electrolyte, a metal hydride anode (many different metal alloys are used), and a nickel-hydroxide cathode. Toyota hybrids make extensive use of this rechargeable battery chemistry, which has been surpassed by lithium ion. 40–110 Wh/kg of energy density at the cell level.
- **Solid-State:** Each battery technology that makes use of solid electrolyte and electrodes. While there are considerable obstacles with cycling, manufacturing, and the solid sandwich's durability, this offers potential improvements in energy density and safety. Solid State is still very much in the early stages of technological development, but it is very important and has a long way to go before it is widely industrialized.

There are many other types of battery cell and have been listed below.

- **Aluminum Air:** Low cost and high energy density. Power is produced by the oxidation of aluminum by oxygen from the air, which occurs in the air cathode, anode, and aqueous electrolyte. Nonetheless, a significant problem that lowers capacity and jeopardizes calendar life is the corrosion of the aluminum anode.
- **Aluminum Ion:** Theoretically, the aluminum metal anode has a high volumetric and gravimetric capacity because it can exchange three electrons during the electrochemical process. However, before this kind of cell can take advantage of the high energy density with stability over a lifetime of cycles, anode, cathode, and electrolyte development are needed.
- **Dual-Ion:** A battery technology that provides grid-based storage at an affordable price. Using graphite as the cathode and anode materials, cations and anions take part in the intercalation and deintercalation processes.
- **Fluoride-Ion:** Seen as a replacement for lithium and possibly the post-lithium technology with "up to 7x the Wh/kg" of current Lithium technology.
- **Graphene Aluminum Ion:** Graphene cathode and aluminum anode, high charge rate battery technology.
- **Lithium Air:** Touted as the technology that will surpass lithium ion batteries in terms of energy density. But despite a great deal of scientific effort being put into understanding the chemistry and reactions, the material is still insufficiently developed to become

a practical battery at this time. If anything, solid state lithium ion batteries are now the main topic of discussion.

- **Lithium Rich Manganese:** A new generation of high-energy density lithium-ion batteries is anticipated to use these materials as cathode materials due to their high specific capacity, which is a result of their combined cationic and anionic redox activity.
- **Lithium Sulfur:** With a potential energy density of more than 600 Wh/kg, this battery chemistry is arguably the most developed among Li-ion batteries along with the possibility of significant cost savings and increased safety. However, cycle life has been subpar due to several difficulties. As a result, the potential sparks interest and much-needed basic chemistry research.
- **Magnesium-Ion:** Because there are twice as many electrons in each magnesium ion as there are in lithium-ion batteries, these batteries function extremely similarly to lithium-ion batteries and have a comparable energy density. Lithium is less common than magnesium. With magnesium-ion, however, there are a lot more potential adverse effects.
- **Nickel Cadmium:** Rechargeable battery that uses nickel oxide hydroxide and metallic cadmium as electrodes.
- **Nickel Hydrogen:** This has a hydrogen negative electrode from the hydrogen-oxygen fuel cell and a nickel oxide positive electrode from the nickel-cadmium cell. With a cycle life of about 200,000, the energy density is low at about 60 Wh/kg and the cost is high, making it suitable for a specialized use in spacecraft.
- **Nickel Iron:** The NiFe battery, nickel(III) oxide-hydroxide positive plates and iron negative plates, with an electrolyte of potassium hydroxide. A very low gravimetric energy density of 19–25Wh/kg.
- **Potassium Ion:** Possibility of a low-cost storage based chemistry, but cycling is limited by the large change of ~60% of graphite caused by the insertion of K ions, which places a high strain on the lattice. The development of lithium and sodium ion batteries is lagging behind.
- **Sodium-Nickel-Chloride Molten Salt Battery:** Also known as the Zebra Battery.
- **Zinc Air:** Possibly the most advanced technology for metal-air batteries. It usually consists of a separator, a caustic alkaline electrolyte, an oxygen-permeable cathode, and zinc anode. promising excellent stability, low cost, and high energy density.

- **Zinc Bromine:** Promising flow battery technology.
- **Zinc Carbon:** A primary battery chemistry, commonly used in batteries for radios, toys and household goods (Figure 4.6).

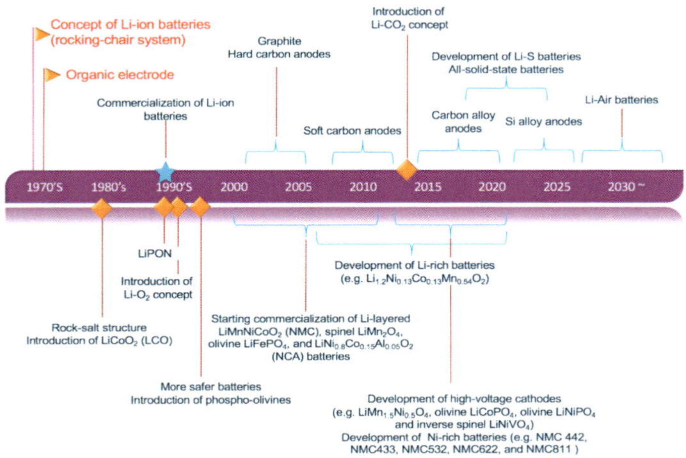

Figure 4.6. *Historical evolution and advances of Lithium-ion battery technologies.*

Source: https://pubs.rsc.org/image/article/2019/TA/c8ta10513h/c8ta10513h-f1_hi-res.gif.

4.2.1. Purpose

Every alkaline battery is powered by a specific mixture of potassium hydroxide, high-density manganese dioxide, and zinc. When the zinc anode is oxidized and the manganese dioxide cathode is reduced, an alkaline battery generates electricity. To get even more specific, the following is the equation for a basic alkaline cell reaction:

$$Zn + 2\ MnO_2 + H_2O \rightarrow ZnO + 2\ MnOOH$$

During this battery chemical reaction, water (H_2O) is consumed and hydroxyl ions (OH-) are produced by the MnO_2 cathode under the following reaction:

$$2\ MnO_2 + 2\ H_2O + 2\ e^- \rightarrow 2\ MnOOH + 2\ OH^-$$

At the same time, the anode is consuming hydroxyl ions and producing water:

$$Zn + 2\ OH^- \rightarrow ZnO + H_2O + 2\ e^-$$

In practical terms, this battery chemistry means that devices are powered by the electrons (e) produced during the reaction. The quality of the raw materials and the accessibility of water and hydroxyl ions during the reaction determine how efficient the reaction will be. To put it simply, every battery is made to keep the anode and cathode apart in order to stop a reaction. Only when the circuit is closed will the stored electrons begin to flow. This occurs when a device's battery is inserted and turned on; this works similarly to turning on and off a light switch in your home. The manganese dioxide's greater ability to attract electrons will cause the circuit's wire to carry electrons from the zinc anode electrode to the cathode electrode when the circuit is closed. Electricity is created by the chemical reaction of the battery and the flow of electrons through the wire. This electricity is ready to power anything, including your wireless mouse, portable clock, handheld game, digital camera, and wireless mouse.

4.2.2. Environmental Impact of Battery Production and Disposal

Many of the gadgets we use on a daily basis, including phones, toys, portable power tools, headphones, and other devices, depend on batteries. Batteries play a vital role in the transportation and energy industries on a larger scale. Because of the growing environmental concerns associated with the use of fossil fuels, many automakers have shifted to producing electric vehicles. Fossil fuel combustion is already known to increase pollution and contribute to climate change when it comes to products like cars (Bitenc et al., 2020).

4.2.3. Battery Usage in Today's World

The way we use and dispose of batteries is changing due to their widespread use in the power and automotive industries. The industry keeps looking for smaller, more durable batteries as volume rises, from tiny device batteries to lithium-ion battery packs found in electric cars. Because there are constant discussions about how to lower the world's carbon footprint, the electric vehicle market has been expanding. On the one hand, there's no getting around the worries regarding fossil fuel combustion.

This has made it more imperative to move to more recent systems that support electric vehicles. But more and more people are talking about how much electric cars actually cost the environment. More attention is being paid to electric car batteries in particular, as disposal techniques can be anything but ecologically friendly and battery manufacturers have an adverse effect on the environment. Numerous systems have been modified to transition away from the burning of fossil fuels and toward renewable energy sources. Examining the life cycle of products is now more crucial than ever due to the impact of greenhouse gas emissions on the environment.

4.2.3.1. Battery-Powered Electric Vehicles & Ubiquitous Electronic Devices

On the one hand, it is evident that major automakers are moving away from gasoline-powered cars and toward electric ones. The US Environmental Protection Agency estimates that fuels derived from petroleum account for 95% of the world's transportation energy. Lithium-ion battery-powered electric cars contribute to the decrease of road pollution. Therefore, to some extent, electric car batteries do assist us in lessening our impact on the environment. There has also been a sharp increase in demand for electric vehicles on the international market. This demand consequently causes a rise in the production of electric batteries as well as an increase in the quantity of spent lithium-ion batteries. In addition to focusing on electric vehicles, we also need to investigate the pervasive use of electronics. Numerous appliances both inside and outside the house are powered by batteries.

4.2.3.2. Growing Battery Demand Translates Into a Growing Environmental Impact

Researchers thus see rising concerns regarding spent batteries' effects on the environment and ecology. Research showed that in 2018, there was a potential 8% compound annual growth rate. By 2030, it's anticipated that the percentage will rise to between 18% and 30%. The rising demand for new goods and electronics necessitates an increase in output. Herein lies the relevance of worries regarding the carbon footprint of batteries and battery-operated devices. While it is easy to commend batteries designed with energy storage in mind, there are many other factors to take into account throughout their lifecycle in addition to emissions reductions when they power our electric vehicles. The production of batteries harms the environment in numerous ways when regulations regarding manufacturing practices and waste management are lax. Improper handling and disposal of batteries can contaminate the air, soil, and water, starting from the conversion process to the mining of materials like lithium. Furthermore, the health of humans and aquatic life are directly threatened by the toxic nature of batteries. Examining the effects of batteries on the environment also requires us to consider the demand for particular kinds of batteries. These days, the most popular kind of rechargeable battery is the lithium-ion battery, or Li-ion battery. Lithium-ion batteries are used by manufacturers in phones, computers, and of course, electric vehicles. This increases the demand for lithium as a result.

4.2.4. Lithium-Ion Batteries

One popular and widely used battery type in our daily lives is the lithium-ion battery, also known as the li-ion battery. Li-ion batteries are mostly used by manufacturers in computers and consumer electronics. Li-ion batteries are a kind of rechargeable batteries that are electric and can be used repeatedly. These batteries offer a high energy density as well. When compared to other electric batteries, their high power output per kilogram is what draws manufacturers to them. About 150 watt-hours are stored per kg in a Li-ion battery. In addition to being vital to the survival of our electronic devices, they play a significant role in the electric vehicle sector. This type of battery is used in many EV batteries. Lithium ions travel from the anode to the cathode in these batteries. Through this process, the device receives energy from the battery. During the recharge phase, it goes through this cycle backwards. Nonetheless, over time, the battery's capacity is gradually decreased by the cycle of charging and discharging. Naturally, compared to a car battery, manufacturers of small devices require a much different quantity of natural resources and battery materials. EV batteries have a greater need for energy storage. Consequently, producers must use raw materials such as graphite, nickel, and cobalt. These necessitate extraction techniques that produce lithium while also harming the environment.

4.2.5. Other Popular Battery Types

Apart from lithium batteries, the world uses a host of other battery types. Some other popular ones include:

- **Lithium Batteries.** Lithium batteries are a kind of non-rechargeable battery, not to be confused with li-ion batteries. High energy densities are provided by the primary cell construction of the lithium battery. There are three battery types available: AA, AAA, and 9V. Lithium batteries are used by manufacturers in both compact and large electronic devices. Their light weight makes them ideal for portable devices.
- **Lead Acid Batteries.** An older battery technology that was investigated for its affordability, efficiency, and longevity was the lead acid battery. Numerous battery power applications can benefit from this kind of battery. Because of their high surge and current capacity, they are frequently used to start internal combustion engines in conventional automobiles. The production of lead acid batteries has been supplanted by lithium batteries nowadays. Lead-acid batteries are still produced by manufacturers, nevertheless, for a variety of uses, including backup power systems and motorbikes (Bitenc et al., 2020).

4.2.6. Are Batteries Bad for the Environment?

These days, lithium, an alkali metal, is used in a lot of our electronics and electric vehicles. It is nearly hard to not own anything that uses lithium batteries. Exporting nations benefit economically from this raw material, on the one hand. But energy-intensive lithium production and extraction also present an environmental risk. The production of lithium-ion poses problems for both the environment and human population. Lithium mining uses a lot of energy and water during production. It also causes issues with soil and air pollution, which have an impact on global safety and the climate. Concerning the immediate effects on humans, there is the moral dilemma of hazardous conditions in mines. There are instances of child labor in developing nations where producers obtain these raw materials, putting both adults and children in dangerous situations. Examining the effects is vital given the exponential growth in lithium production and consumption. According to studies, the market for lithium-ion batteries is expected to grow at a rate of US$30 billion in 2017 to $100 billion in 2025. In addition, a study from the World Economic Forum-led Global Battery Alliance, a public-private alliance, claims that batteries could allow for a 30% decrease in greenhouse gas emissions from the transportation and electricity industries.

4.2.7. Environmental Effects Associated with the Production of Li-Ion Batteries

Like petroleum, in order to use materials like lithium, we must mine and extract them from the earth. Therefore, when analyzing the effects of battery manufacturing, the mining and extraction process comes first. In addition to lithium, cobalt and nickel are other materials used by battery makers to increase battery life. When trying to extract lithium, manufacturing companies frequently have to source lithium deposits in distant nations. In Mexico, East Asia, and South America are some of the world's largest deposits. Salt flats in Chile and Argentina provide the world with about one-third of its lithium. Huge amounts of water are used by miners in these areas to extract the resource. Water is a problem during extraction in the South American Andes Mountains because the process needs a lot of water and the mountains are relatively dry. This makes it possible for the element to surface as a brine with salt. Lithium carbonate is required to make a battery. Concentrated lithium carbonate is produced by refining salts containing lithium in evaporation pools. Before they can filter the solid salts, workers let pools of brine that has been salted evaporate. The duration of this evaporation phase can range from 12–18 months. This process, which is water-intensive, requires roughly 500,000 gallons of water for every ton of lithium.

Lithium refining involves not only the heavy use of water but also the use of hazardous chemicals like hydrochloric acid. As a result, these substances may seep into public water sources. These practices have an impact on local farmers, the general public, and aquatic environments. Lithium is extracted from rocks by miners using conventional techniques in places like Australia and North America. But even this needs the use of chemicals. From chemicals in the soil to canals full of tainted water, the environmental cost is evident. It should come as no surprise that producers would want to increase output given the high demand for lithium. Factories can heat the water to reduce the amount of time the brine evaporation period. But this negates the goal by including fossil fuels in the energy mix.

- **Cobalt Mining:** An essential component of a battery's electrode is cobalt. Consequently, producers of electronics and cars also take this raw material out. In addition to the harmful effects that lithium production and extraction have on the environment, cobalt mining is a destructive process. It's crucial to first understand that the Democratic Republic of the Congo (DRC) is home to about 70% of the world's cobalt reserves. Therefore, this region bears the brunt of the global cobalt demand. In addition to the effects on the environment, there are human repercussions for the local populace. There is a great deal of child and slave labor, hazardous working conditions, and a high rate of congenital disabilities from chemical exposure. Nickel and cobalt are mined underground, unlike lithium. This process has a variety of negative effects, such as the physical destruction of habitats and the chemical pollution of nearby areas. Because cobalt is necessary to make batteries, chemists are looking into greener substitutes (Gaines and Dunn, 2014; Aouf, 2020)

4.2.8. Battery Disposal

Another area of concern is the disposal of batteries. When used battery components are disposed of incorrectly, hazardous environmental problems can arise. These batteries have the potential to cause environmental harm because they contain elements that could be toxic. Erroneous disposal may occasionally result in explosions.

- **The Effects of Battery Waste:** Solid waste from homes that ends up in landfills is battery waste. Thus, your batteries wind up in a landfill when you throw them away. They deteriorate and leak here. The battery leaks its chemicals into the ground as it corrodes. In addition, they have the ability to contaminate groundwater and surface water sources by getting inside them. These battery chemicals also endanger aquatic life by getting into the ocean. Lithium ion batteries is volatile when exposed, aside from chemicals leaking. As such, it has the potential to start landfill fires, which would release additional toxic gases into the atmosphere. This makes it simple to understand how poor waste management, even on a small scale, can have long-term, significant effects. These substances have the potential to harm human safety as well as the environment. This also holds true for batteries used in electric vehicles, in addition to batteries used within homes.

- **Disposing of Electric Vehicles Batteries:** EV batteries are bigger and heavier than batteries for conventional cars. There are multiple li-ion cells in these batteries as well, which need to be disassembled. The hazardous materials have the potential to explode if they are disassembled improperly. Recycled materials have been commended as a potential solution. Li-ion battery recycling is not as common as it is for traditional lead-acid batteries, though. Global organizations understand the necessity of establishing standards, though. These guidelines aim to lower the quantity of batteries that brands and manufacturers discard or dump. This emphasizes the necessity of recycling and reusing materials in order to promote sustainable development. Proposals from the European Union, for example, emphasize that producers of electric vehicles must ensure that their products are managed properly.

4.2.9. Battery Recycling as a Solution

Recycling is a highly researched and financially supported field, so it makes sense that recycling would be recommended as a solution. Batteries can be recycled so they can have a new life instead of being thrown out with the trash. That being said, the type of battery plays a major role in the recycling process and rate. Let's examine lead-acid batteries first. This battery is recycled all over the world. Additionally, studies show a 99.3% battery recycling rate7. The fact that each battery component is recyclable and reusable is the best part. That is, components used in the production of new batteries include lead, plastic, and sulfuric acid. This is a success in terms of sustainability. Recycling lowers waste and eases the strain on new lead mining. It also lessens the quantity of lead that ends up in landfills. Li-ion recycling is a distinct process. This variant is more difficult to work with than lead-acid variants, which recycling facilities can handle with ease. Lithium-ion battery recycling is technically feasible but a more involved procedure. Because of its complex life cycle, lithium is hardly ever recycled. These batteries sustain irreversible damage during their lifetime.

This indicates that there is no easy way to repurpose them. Recyclers must therefore disassemble the battery in order to remove the lithium. They can then remanufacture them afterward.

Producers add a variety of additives to the battery's electrolyte liquid during the manufacturing process. In one way or another, these additives help to enhance the li-ion battery. An additive, for example, can increase the battery's resilience to changing weather conditions. However, using a different additive might help to expedite the manufacturing process. Repurposing the metals in the battery becomes challenging because of the potential for different combinations. Additionally, it renders the method costly.

Businesses must carefully and safely disassemble the parts of used batteries in order to try recycling. This is due, in part, to the electrolyte mixture's rapid explosive potential when exposed to high temperatures, which can happen quickly with improper handling. With all of these variables in place, it is simple to understand how and why the recycling rate of lithium battery material is so low.

Currently, the percentage of these batteries that are recycled globally is unstable. Estimates, however, place the value at only 5%. Large-scale recycling is justified because automobile Li-ion batteries have only been widely used in cars for about five years, during which time a significant portion of them have not yet reached the end of their useful lives (Gaines and Dunn, 2014; Aouf, 2020).

4.2.9.1. Possible Solutions

There's no doubting that electric cars are a more environmentally friendly option than vehicles that run on gasoline. But as we've seen, there are some environmental consequences associated with the battery-making process. This means that in order to produce batteries as environmentally friendly as possible, sector-wide improvements are required. The production and disposal of batteries need to be done in a way that is safer and more environmentally friendly.

This will safeguard people's health in addition to having a positive effect on the environment. Environmentally friendly improvements can be made through advancements in fields such as battery technology. Making the move to renewable energy sources is also a big step. Another option would be to use batteries for longer periods of time before recycling. Experts report that after the initial use of an electric car, the battery still has some capacity. They may have a second life even though automakers are no longer able to use them to power the vehicles. For example, they could be used to store extra solar energy.

4.3. ENVIRONMENTAL IMPACTS OF LITHIUM-ION BATTERY PRODUCTION

Lithium serves as the anode in lithium batteries. This kind of battery, also known as a lithium-ion battery, is mostly found in electric cars and electronic devices. The British chemist M. Stanley Whittingham invented the first kind of lithium battery in the early 1970s, using titanium and lithium as the electrode materials. The high cost of titanium and the disagreeable odor the reaction produced limited the battery's uses. First developed in 1985, the lithium-ion battery of today is based on Akira Yoshino's Whittingham attempt. Although switching to lithium-based batteries for all fossil fuel-powered devices may not be the best course of action for the planet, lithium-ion batteries can be utilized as a component of a sustainable solution. Although it is a natural resource that could run out, there isn't currently a shortage. There are approximately 14 million tons of lithium left, which is 165 times the volume produced in 2018, according to Volkswagen researchers.

4.3.1. Extraction

Commercial lithium extraction comes from three main sources: hard rock deposits, clay rich in lithium, and brines of salt. There will inevitably be some environmental disruptions with each method. Sites for extracting salt brine are the most widely used methods for obtaining lithium; approximately 66% of the world's lithium production comes from these sites. When compared to other extraction techniques, brine extraction has the biggest environmental advantage because very little machinery is required for the entire process. Hard rock deposits and clays rich in lithium, on the other hand, call for more conventional mining techniques that entail large machinery. Despite this advantage, the methods are all frequently employed because their recovery rates are all fairly close to one another. Hard-rock deposits recover 94% of their resources after extraction, compared to 97% after brine extraction.

4.3.2. Continental Brine Extraction

Open-air evaporation is used in brine extraction to gradually concentrate the brine. Large amounts of water are lost as a result of evaporation. It is important to note that the brine that is evaporating generally has a very high salinity, rendering the water unfit for human or agricultural use. In order to create Li_2CO_3 and $LiOH{\bullet}H_2O$, the concentrated brine is then transported to a nearby production facility. The majority of the air pollution from brine extraction sites is caused by these production facilities, which release airborne pollutants like sulfur dioxide.

Most brine extraction sites are found in South America, mainly in Chile and Argentina. This region, known as the lithium triangle, contains about half of the world's lithium reserves. The two operational lithium mines in Chile, the second-largest lithium producer in the world, are situated on the Salar de Atacama salt flat in the Atacama Desert and are operated by SQM and Albemarle. The brine from these mines has about 350 g/L of total dissolved solids, according to tests done on it. According to research on this mine and the local water tables, Salar de Atacama's overall water storage dropped by -1.16 mm annually between 2010 and 2017. Communities in the area are divided in a complicated way; some accept payments from mining companies and participate in their community development projects, while others are either ignored by these initiatives or turn down the companies' offers because of their previously mentioned environmental concerns. Records exist in Tagong, a small town in China's Garzê Tibetan Autonomous Prefecture, of hazardous substances like hydrochloric acid spilling into the Liqi River from adjacent lithium mining operations. Large animals and dead fish were consequently observed floating down the Liqi River and other adjacent rivers close to the Tibetan mines. Subsequent analysis revealed that the possible cause of this could have been evaporation pools that have been left unattended for months or even years.

4.3.2.1. Hard-Rock Deposits

Hard-rock deposits can also be used to extract lithium. Through spodumene ores, these deposits are most frequently found in Australia, the largest producer of lithium in the world. Throughout the world, continental brines are far more common than sporumene ores and other hard-rock deposits containing lithium. The operating costs are very similar to those of operating a brine extraction operation, Although the deposits are much less common and accessible for mining. The result is that even though salt brines are much more common and generally have less of an impact on the environment, hard-rock deposit extraction sites are still being developed and utilized (Gaines and Dunn, 2014; Aouf, 2020).

4.3.2.2. Lithium-Rich Clays

The process of extracting lithium from clays rich in lithium first requires mining the clays, which pollutes the atmosphere greatly. Lithium can be found in a number of minerals found in clay, including cookeite, jadarite, swinefordite, hectorite, masutomilite, and zinnwaldite. Following the removal of these minerals from the earth, the clays are treated in order to extract the lithium; usually, this is accomplished by chemical reactions such as acidification. If not managed appropriately, this chemical process may produce hazardous chemicals and gases as byproducts, which can quickly lead to pollution. The third main source of lithium is clays rich in lithium, though they are much less common than lithium-containing hard rock ores and salt brines. Specifically, less than 2% of the world's lithium products come from clays rich in lithium. In contrast, the extraction of brine accounts for 39% and the extraction of hard-rock ores for 59% of the production of lithium.

4.3.2.3. Disposal

Metals like cobalt, nickel, and manganese—which are hazardous and can contaminate water supplies and ecosystems if they leak out of landfills—are present in lithium-ion batteries. Furthermore, improper disposal of lithium-ion batteries has been linked to fires in landfills and battery recycling facilities. Because of this, recycling lithium-ion batteries is mandated by some legal systems. Even though improper disposal of lithium-ion batteries has negative environmental effects, recycling rates are still low due to expensive and undeveloped recycling processes. According to a 2014 study done in Australia, 98% of lithium-ion batteries were disposed of in landfills in 2012–2013.

4.3.3. Recycling

From the time they are made until they are recycled, lithium-ion batteries need to be treated with the utmost care. Reusing is essential to reducing the negative effects lithium-ion batteries have on the environment. Emissions and energy usage can be decreased by recycling the batteries because less lithium would need to be extracted and processed. Regarding recycling lithium batteries in the United States, the EPA has guidelines. Rechargeable and single-use batteries are processed differently, so it is advisable to bring batteries of all sizes to designated recycling facilities. This will enable the separation of the individual metals that can be recovered for additional use in a safer manner. Lithium-ion battery recycling presently employs three main techniques: hydrometallurgical metal reclamation, mechanical recycling, and pyrometallurgical recovery. According to a 2016 study done with multiple recycling facilities in Australia, mechanical recycling was able to recover the highest percentage of materials from lithium-ion batteries, averaging seven out of 10 possible materials. The same study also discovered that, on average, pyrometallurgical processes recovered only half of the potential materials, while hydrometallurgy recovered six out of 10 materials.

4.3.3.1. Pyrometallurgical Recovery

The pyrolysis, incineration, roasting, and smelting processes are all part of the pyrometallurgical recovery process. The majority of conventional industrial processes are currently unable to recover lithium. The primary procedure is the extraction of additional metals, such as copper, nickel, and cobalt. The efficiency with which materials and capital resources are recycled is extremely low. High energy needs are combined with gas treatment systems that reduce the amount of gas byproducts produced.

4.3.3.2. Hydrometallurgical Metals Reclamation

In hydrometallurgy, materials are dissolved using chemical reactions to create a solution, which is then precipitated to extract the desired raw material. All organic materials, including plastic, are destroyed during the recycling process. Nevertheless, hydrometallurgy produces recovered metals with a very high purity, which makes it a useful recycling technique. It's frequently employed in copper recovery. This technique has been applied to other metals in an attempt to mitigate the sulfur dioxide byproduct issue that arises from more traditional smelting.

4.3.3.3. Direct/Mechanical Recycling

Disassembling outdated lithium-ion batteries to remove valuable, useable parts and/or materials to be repurposed with new batteries is known as direct or mechanical recycling. Old batteries are crushed or shredded in this process, and the materials are then extracted. Cross-contamination may result from this, rendering some materials or components non-recyclable. Even though this type of recycling is an option, the cost is usually still higher than that of mining the ores.

The need for a more effective recycling program is detrimental due to the growing demand for lithium-ion batteries, as many businesses are competing to develop the most effective solution. One of the most urgent problems is that recycling is not given priority during the battery's manufacturing process. This recycling method has the advantage over the other two, which can both produce hazardous chemicals and gases, in that there is typically very little, if any, pollution during the process (Gaines and Dunn, 2014; Aouf, 2020).

4.3.4. Application

Lithium-ion batteries have a wide range of applications because they are small, lightweight, and rechargeable. Though their applications in the military and aerospace industry are growing, they are primarily found in electric vehicles and handheld electronics.

4.3.4.1. Electric Vehicles

EVs are the main industry and source of lithium-ion batteries. Sales of electric vehicles have skyrocketed in recent years, and as of 2019, EV incentives were in place in over 90% of all global auto markets. We can observe a considerable improvement in the environmental impacts from the decrease in reliance on fossil fuels as a result of the growth in EV sales and their ongoing sales. Recent research has looked into the various applications of recycled lithium-ion batteries, particularly those from electric cars. It has been demonstrated that grid companies can benefit from the secondary use of lithium-ion batteries recovered from electric vehicles in China for peak shaving power loads. Remanufacturing of lithium batteries is necessary because of the risks to the environment posed by spent lithium-ion batteries as well as the potential shortage of battery components for electric vehicle futures. A test carried out in China using the EverBatt model found that lithium-ion battery remanufacturing will only be profitable if the cost of purchasing spent batteries doesn't increase. Recycling will have a big positive impact on the environment as well. The application of remanufacturing results in a 6.62% reduction in overall greenhouse gas emissions (Gaines and Dunn, 2014; Aouf, 2020).

4.3.5. Environmental Impacts of Lithium-Ion Batteries

One benefit of storing energy in lithium-ion batteries is that it can help us reach sustainability targets related to energy consumption. For example, it can reduce our dependency on fossil fuels and increase our use of renewable energy sources and lithium-ion batteries. However, these benefits also have a number of disadvantages that affect the battery's supply chain as a whole. A list of some of these benefits and drawbacks is provided below.

Because electric vehicles are thought to produce fewer greenhouse gases than gasoline-powered vehicles, the green movement is pushing electric vehicles on the American public. But lithium-ion batteries—which are necessary for electric vehicles—have problems with greenhouse gas emissions both during the extraction and processing of the raw materials and during the batteries' eventual disposal. As the sales of electric vehicles rise, so do the problems associated with mining and disposal. Research illustrates the enormous rise in anticipated demand for EV batteries worldwide. The Inflation Reduction Act, popularly known as the "climate bill," includes purchase incentives and proposed vehicle standards that would force the public to drive electric vehicles (Figure 4.7).

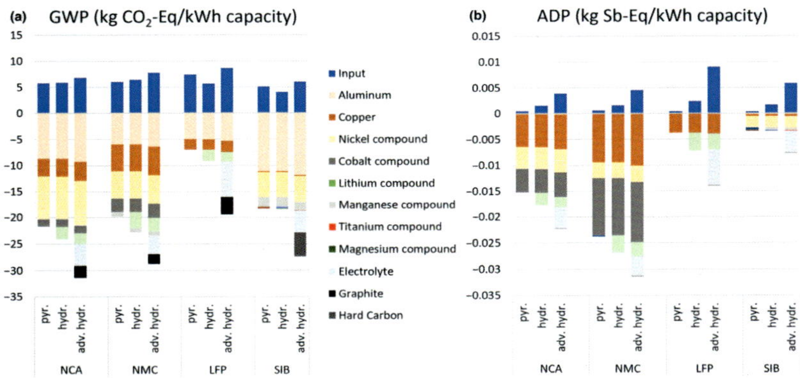

Figure 4.7. *Environmental benefits of battery cell recycling, broken down to the contribution of the different fractions recovered by the recycling processes: (**a**) global warming potential (GWP), (**b**) abiotic resource depletion potential (ADP). Negative values indicate net benefits (reduction of impacts due to recovered materials), positive values are environmental impacts (due to process inputs and emissions, e.g., energy and chemicals, off-gasses).*

Source: Reprinted from Journal of Industrial Ecology, Vol. 24, Mohr et al., Toward a cell-chemistry specific life cycle assessment of lithium-ion battery recycling processes, no. 6, pp. 1310–1322, Copyright (2022), distributed under Creative Commons Attribution License; Pražanová A, Knap V, Stroe D-I. Literature Review, Recycling of Lithium-Ion Batteries from Electric Vehicles, Part II: Environmental and Economic Perspective. Energies. *2022; 15(19):7356. https://doi.org/10.3390/en15197356.*

As more and more electric vehicles are seen on the road worldwide, there is growing environmental concern over the disposal of the lithium-ion batteries that power these vehicles, which produce more carbon dioxide emissions during production than gasoline-powered cars do. The mining and processing of the necessary minerals accounts for about 40% of the climate impact associated with the production of lithium-ion batteries. The production of battery cells, modules, and packs as well as the mining and refining of battery materials demand large energy inputs that contribute to greenhouse gas emissions. China, the global leader in the EV battery supply chain, generates nearly 60% of its electricity from coal, a fuel that emits a lot of greenhouse gases. The Wall Street Journal claims that the mining and manufacturing of lithium-ion batteries has a greater negative impact on the environment than that of fossil fuel-powered car batteries. When compared to a generic battery, the average lithium-ion battery requires three times as much cumulative energy demand (CED) during production (Figure 4.8).

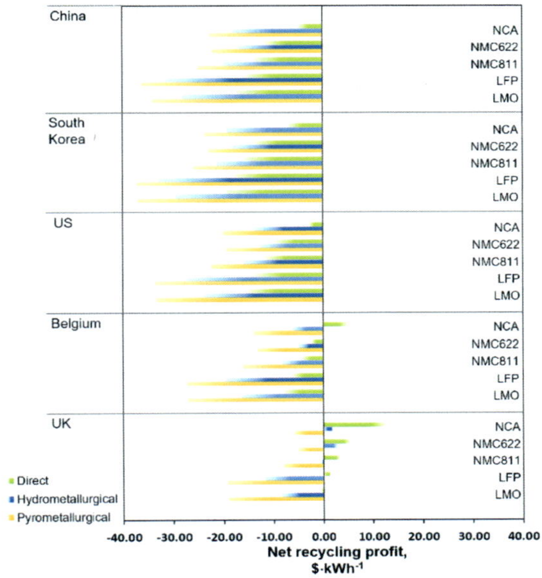

Figure 4.8. *Net recycling profits for commercial EV battery packs.*

Source: Reprinted from iScience, Vol. 24, Lander et al., Financial viability of electric vehicle lithium-ion battery recycling, no. 7, Copyright (2022), distributed under Creative Commons Attribution License; Pražanová A, Knap V, Stroe D-I. Literature Review, Recycling of Lithium-Ion Batteries from Electric Vehicles, Part II: Environmental and Economic Perspective. Energies. *2022; 15(19):7356. https://doi.org/10.3390/en15197356.*

Another risk to the climate is the battery disposal process. The battery's cells may leak heavy metals and other toxins into the groundwater and soil if it ends up in a landfill. According to an Australian study, 98.3% of lithium-ion batteries wind up in landfills, which raises the possibility of long-lasting landfill fires. Lithium-ion battery fires were blamed for 124 reported landfill fires in the Pacific Northwest between June 2017 and December 2020. With 21 fires reported on the site in 2018, that number is expected to rise to 47 by 2020, indicating an increasing frequency of fires. Governments are pushing for lithium-ion battery recycling because of the problems with environmental waste that come with them and because there is an increasing demand for batteries as more and more electric vehicles are sold. In contrast to the 99% of lead car batteries that are recycled in the US, only about 5% of the world's lithium batteries are recycled. On the other hand, recycling lithium batteries can be dangerous. Cutting a cell in the wrong spot or too deeply can cause it to short circuit, catch fire, and release poisonous fumes. The chemistry and construction of batteries vary greatly, making the development of effective recycling systems challenging. Furthermore, even with rapidly rising prices, it is frequently less expensive for battery manufacturers to purchase freshly mined metals than to use recycled materials because the cells are frequently held together with strong glues that make them difficult to disassemble.

But governments are starting to mandate recycling at least in part. In order to encourage the reuse of EV battery components, China, which has the largest EV market and produces the most lithium-ion batteries, implemented regulations in 2018. A certain proportion of recycled materials must be used in the production of new batteries, according to battery recycling regulations that the European Union passed last year.

4.3.5.1. Composition of Lithium Ion Batteries

A pack, which is the structure in which the cells are mounted, a battery management system, and cells, which contain the active materials, make up a lithium-ion battery. Because of its low weight, aluminum is a crucial component of the pack; however, its high energy content accounts for 17% of the battery's carbon footprint. The mining, conversion, and refinement of the active materials used in the cells—where nickel, manganese, cobalt, and lithium are turned into cathode powder—accounts for about 40% of the battery's carbon footprint. Since energy is required for the manufacturing process, the actual production of cells accounts for 20% of the carbon footprint and is the second most energy-demanding activity.

4.3.5.2. Issues with Mining of Battery Raw Materials

Brine extraction and open-pit mining are two common forms of mining needed to extract minerals for batteries. Pumping a lot of water into salt flats to bring mineral-rich saltwater to the surface is a key component of brine extraction, one of the main methods used to extract lithium. Lithium is filtered out of the mixture after the water evaporates. However, the water-intensive procedure runs the risk of tainting the water source. The region known as the Lithium Triangle, which includes parts of Argentina, Bolivia, and Chile's Andean Mountains, contains more than half of the world's lithium reserves. The region is among the driest on Earth, and up to 65% of the water used there is for lithium mining. Any increase in demand will exacerbate the problems with water. Cobalt is another component required for electric vehicles; most of it is mined in the DRC. Hazardous byproducts from cobalt mining can contaminate the environment. Sulfur is frequently found at cobalt mine sites; when exposed to air or water, it produces sulfuric acid, which seeps into rivers, streams, and aquatic life. Additionally, the DRC uses child labor for cobalt mining, with Chinese companies owning or funding about 80% of the country's industrial cobalt mines.

4.3.5.3. What are the Environmental Benefits?

- **Renewable Energy Sources:** Energy from renewable sources, including solar, wind, tidal currents, biofuels, and hydropower, can be stored in lithium-ion batteries. By using renewable energy, we can reduce our carbon footprint and obtain fuel for our homes and cities from naturally replenished sources.
- **Resilience:** Similar to battery systems, stored energy can be helpful in the event of emergencies that endanger the power supply. Reserving energy in transportable containers makes us more resilient, enabling us to bounce back from adversity like natural disasters.
- **Long Service Life:** Rechargeable thousands of times, lithium-ion batteries offer an extended lifespan. Long-term battery and product use can lead to lower consumption and less electronic waste, even though lithium-ion battery recycling is still not optimal.
- **Smarter Energy Use:** We can use devices powered by lithium-ion batteries to monitor and manage our energy usage and efficiency.

4.3.5.4. What are the Environmental Drawbacks?

- **Intensive Extraction:** Open-pit mining and brine extraction are the two types of mining that are frequently needed to extract minerals for batteries. Pollution and erosion may result from these extraction techniques.
- **Open-Pit Mining:** Vegetation has to be removed to create room for an open pit. A deep pit is then excavated. When combined, these elements foster erosion. Dust and poisonous soils with elevated heavy metal concentrations can result from mining. These dusts turn into pollutants that increase the risk of disease in both humans and animals.
- **Brine Extraction:** Natural subterranean drinking water reserves are drained by the process of brine extraction. More quickly than the water cycle can replenish it, this water is lost from the ecosystem. Lithium processing in brine extraction also involves hazardous chemicals. Water, soil, and air quality are all harmed by the release of these chemicals.

- **Electronic Waste:** Lithium-ion batteries, sometimes referred to as e-waste or electronic waste, are discarded after use. According to the United Nations, e-waste is one of the most urgent problems facing the environment and public health today. Many electronic devices are disposed of improperly; instead of being disassembled by experts, they end up in landfills or commercial waste, or they are dangerously disassembled or burned to extract valuable small parts. Because of the numerous combustible materials they are combined with, batteries disposed of in landfills and commercial waste can unintentionally short circuit and start a large fire in those locations that is very challenging to contain (Gaines and Dunn, 2014; Aouf, 2020).

4.3.6. The Future of Sustainable Battery Technology

Picture a future in which the maxim "less is more" becomes more than just a catchphrase and forms the basis of cutting-edge battery research. In front of you is a sight of the future, not just a work of art.

4.3.6.1. The Power of Less: Crafting Nano-crystals

With the ability to create atomic masterpieces in the form of nano-crystals, nano-structuring represents the pinnacle of precision. They are materials, tailored to Angstrom (é) precision at the unit cell level, the most fundamental atomic level, not just collections of particles. First, advanced materials with special qualities are created from nano-crystals through a chemically regulated process. The next step involves developing these synthesized nano-crystals into a specially prepared slurry that can be used in high-throughput precision manufacturing processes. Unmatched control over a material's properties is made possible by this atomic and molecular level engineering, opening the door to the development of high-performing and material-efficient batteries. Here, we observe the maximization of material intensity in the undulating waves created by electromagnetic fields—not by sheer bulk, but rather by the thoughtful placement of nanocrystals.

This is the ultimate in efficiency: more surface area, interaction, and overall performance with less material used. By fully utilizing each nanocrystal, waste is drastically reduced and the impact of each individual particle is maximized. This exact coordination of nano-crystals—as opposed to atoms—showcases the skill of accomplishing more with less. This is about redefining the properties of the material to unlock new capabilities that are unattainable at larger scales, not just about achieving smaller sizes.

4.3.6.2. The LFP Cathode Triumph: Blade Battery

An excellent illustration of how nanostructuring can transform material efficiency is the LFP cathode. LFP blade batteries use a great deal less raw material while achieving superior energy density and thermal stability thanks to the exact arrangement of nanoscale structures. This lowers the cost of producing batteries and lessens the environmental impact of material extraction and processing, increasing the accessibility of renewable energy storage and EVs.

4.3.6.3. Why Atom-Level Precision Matters: Anode Perspective

It is only natural that we would wonder about the potential of anodes given the success of the LFP cathode. The potential for atom-level precision to drastically change the performance properties of materials makes it significant. With regard to battery technology, and more especially anodes, this accuracy enables us to modify the mechanical, thermal, and electrical characteristics of materials to fulfill particular requirements. Anodes with precise arrangements, for example, can result in increased conductivity, quicker lithium-ion intercalation, and enhanced durability. Nanostructuring holds the potential to address the intrinsic problems with conventional anode materials, like silicon's volume expansion problems and graphite's restricted lithium-ion storage. Through nanoscale anode reengineering, we expect increased lithium-ion storage capacity, better charge rates, and longer battery life—all while leaving a negligible material footprint. We can attain previously unimaginable material efficiencies and performance improvements by building anodes from the ground up. This involves expanding the surface area available for reaction, which enhances the battery's capacity and power production. Furthermore, we can experiment with a range of materials and possibly find more abundant or sustainable substitutes for the scarce resources currently in use thanks to our ability to precisely control the composition and structure of nano-crystals. One of the most intriguing potential applications of bottom-up nanotechnology is the ability to optimize production by switching materials within the same R&D manufacturing line with little to no modifications. The time and expense involved in switching between various battery chemistries or making adjustments for new material science discoveries could be significantly decreased with this flexibility.

4.3.6.3.1. Exploring Nanostructuring's Broad Horizons

- **Enhanced Performance:** Nanostructured materials often showcase superior electrical, thermal, and mechanical properties, culminating in batteries that are both more powerful and durable.
- **Material Efficiency:** Nanostructuring optimizes the utility of every

atom, reducing the quantity of material required for exceptional performance and preserving valuable resources.

- **Cost Efficiency:** Diminished material reliance leads to notable cost reductions in raw materials, potentially lowering the financial barrier to sustainable technology adoption.
- **Environmental Impact:** By curtailing raw material usage, nanostructuring aligns with global sustainability objectives, mitigating the environmental toll of material extraction and processing.

Certain ternary and binary metal-oxides, like TiOx, provide a wealth of opportunities. These materials are ready to be used and have qualities that go against accepted notions. Certain metal-oxides represent a significant advancement with electrical conductivity comparable to graphite and lithium intercalation capacities that break previous records. The groundwork for innovation has already been laid, setting the stage for these cutting-edge materials to emerge. The foundation of electrode performance is made up of the two characteristics of conductivity and capacity. Nonstoichiometric titanium oxides have the potential to revolutionize energy storage due to their exceptional electron transport efficiency and capacity to accommodate lithium. Here, in the synthesis of materials that provide the best of both worlds, lies the real promise for a battery revolution. Imagine a future where batteries are not only more potent and long-lasting but also more sustainable and affordable. What if the same ideas that have made LFP cathodes more competitive could also be applied to anodes and other battery components? The future of nanostructuring holds the promise of a time when less really is more. As we continue to explore this nanostructured future, battery technology appears to have countless opportunities for innovation and sustainability.

4.3.6.3.2. Conductivity and Capacity: The Twin Pioneers

Performance of an electrode is based on two key factors: conductivity and capacity. Certain titanium oxides have been identified as dual pioneers because they have the high conductivity required for effective electron transport as well as a remarkable capacity to intercalate lithium, exceeding energy density limits (Gaines and Dunn, 2014; Aouf, 2020).

REFERENCES

1. Aouf, R. (2020). IBM proposes sustainable successor to lithium-ion battery. *Dezeen*. Retrieved from: https://www.dezeen.com/2020/01/06/ibm-research-sustainable-battery-sea-water/ (accessed on 9 November 2024).
2. Bitenc, J., Lindahl, N., Vizintin, A., Abdelhamid, M. E., Dominko, R., & Johansson, P. (2020). Concept and electrochemical mechanism of an Al metal anode-organic cathode battery. *Energy Storage Materials, 24*, 379–383. https://doi.org/10.1016/j.ensm.2019.07.033.
3. Gaines, L. L., & Dunn, J. B. (2014). Lithium-ion battery environmental impacts. In *Lithium-Ion Batteries* (pp. 483–508). Elsevier. https://doi.org/10.1016/B978-0-444-59513-3.00021-2.
4. Harper, G., Sommerville, R., Kendrick, E., Driscoll, L., Slater, P., Stolkin, R., & Anderson, P. (2019). Recycling lithium-ion batteries from electric vehicles. *Nature, 575*(7781), 75–86. https://doi.org/10.1038/s41586-019-1682-5.
5. Lu, Y., Zhang, S., Dai, S., Liu, D., Wang, X., Tang, W., & Huang, J. (2020). Ultrasensitive detection of electrolyte leakage from lithium-ion batteries by ionically conductive metal-organic frameworks. *Matter, 3*(3), 904–919. https://doi.org/10.1016/j.matt.2020.05.021.
6. Wang, Y., Dong, H., Katyal, N., Hao, H., Liu, P., Celio, H., & Mitlin, D. (2022). A sodium–antimony–telluride intermetallic allows sodium-metal cycling at 100% depth of discharge and as an anode-free metal battery. *Advanced Materials, 34*(1), 2106005. https://doi.org/10.1002/adma.202106005.
7. Wilson, J. (2022). Imperial graduate awarded €2.5 million European funding for saltwater battery. *Imperial College London*. Retrieved from: https://www.imperial.ac.uk/news/234235/imperial-graduate-awarded-25-million-european/ (accessed on 8 November 2024).

Chapter 5

Energy Consumption and Carbon Footprint

Contents

INTRODUCTION

Since they can have smaller carbon footprints than cars with internal combustion engines, electric vehicles (EVs) have become a major part of the solution to combat climate change and reduce greenhouse gas emissions. The energy source used for charging, the vehicle's efficiency, and the lifecycle emissions related to battery production are some of the factors that affect an electric vehicle's energy consumption and carbon footprint.

EVs use less energy overall than cars powered by gas or diesel. Internal combustion engines normally only convert about 20% of the energy from fuel into motion, with the remaining energy being lost as heat. In contrast, electric drivetrains convert over 90% of the electrical energy from the battery into vehicle movement. Because of their high efficiency, EVs can cover a greater distance with a given amount of energy.

However, the electricity source used for charging has a significant impact on the carbon footprint of EVs. An electric vehicle's total carbon footprint can be substantial if it is charged with electricity produced from fossil fuels like coal or natural gas, but it will still be less than that of an equivalent internal combustion engine vehicle. Conversely, EVs can operate with almost zero emissions when powered by renewable energy sources like solar, wind, or hydro power. Therefore, as the electrical grid gets greener, the advantages of EVs for the environment grow.

Moreover, EV batteries' overall carbon footprint is increased by the substantial energy and resource consumption during production. Lithium, cobalt, and nickel are examples of raw materials whose mining and processing require a lot of energy and can have negative effects on the environment. However, because of the vehicle's longer lifespan and lower operating emissions, these emissions are frequently offset. In an effort to lessen the negative effects of battery production on the environment and enhance recycling procedures, battery technology is also developing.

Overall, even though EVs do have a carbon footprint, especially when it comes to the production of batteries, it is generally less than that of conventional cars, especially as the electrical grid transitions to greener energy sources. The advantages of EVs are anticipated to rise along with the use of renewable energy sources and advancements in battery technology, making them an essential part of the movement to lessen greenhouse gas emissions and slow down global warming (Wiedmann, 2010; Huang and Huang, 2020).

5.1. ENERGY CONSUMPTION

Modern society is heavily influenced by energy consumption, which affects every facet of daily life, from home comfort to industrial productivity. The increasing need for energy due to the growing global population and economies has sparked worries about sustainability and its effects on the environment. The amount of energy used by people, organizations, and governments to carry out different tasks, such as heating and cooling buildings or powering automobiles and industrial operations, is commonly referred to as energy consumption.

Fossil fuels, including coal, oil, and natural gas, are the main sources of energy consumption and have historically been the foundation of industrial development. Because of their high energy density and relative abundance, these energy sources are preferred. Their environmental costs, however, are high and they contribute to greenhouse gas emissions, air pollution, and climate change. Fossil fuel combustion contributes to atmospheric pollution, the release of carbon dioxide, and other gases that intensify the effects of global warming, such as increased sea levels, extreme weather, and ecological disruption.

There has been a major push toward renewable energy sources, like solar, wind, hydro, and geothermal power, in response to these environmental concerns. Since renewable energy sources don't harm the environment as much and can help reduce greenhouse gas emissions, they offer a more sustainable option. Because of cost reductions and technological improvements, renewable energy is now more competitive than conventional fossil fuels. For example, the price of solar panels has dropped dramatically, increasing the accessibility and widespread use of solar energy. Similar to this, advancements in wind turbine technology have increased output and efficiency, which has helped to increase the proportion of renewable energy in the world's energy mix.

Reducing overall energy consumption also requires energy efficiency. By adopting energy-efficient appliances, upgrading building insulation, and streamlining industrial processes, one can increase energy efficiency and use less energy to accomplish the same tasks. Energy efficiency lowers energy costs and eases the burden on the energy infrastructure in addition to assisting in the reduction of energy consumption. Economic and social factors are also connected to the concept of energy consumption. As developing nations industrialize and raise their standards of living, they frequently see greater growth in energy consumption. Although this growth can spur economic expansion, it also presents difficulties in controlling energy consumption and reducing its negative effects on the environment. On the other hand, industrialized countries must balance energy use with sustainability objectives and make the switch to cleaner energy sources.

International agreements, national policies, and individual acts are all part of the global efforts to address energy consumption and its effects. The goals of programs like the Paris Agreement are to slow down the increase in global temperature and to advance sustainable energy use. In order to enhance energy management techniques and develop clean energy technologies, governments and organizations are also spending money on research and development.

Energy consumption is a crucial problem with broad effects on society, the economy, and the environment. Increasing energy efficiency and switching to renewable energy sources are crucial tactics for resolving the issues related to energy usage. Encouraging technological advancements and implementing sustainable practices can help meet energy needs while reducing negative environmental effects and building a more resilient and sustainable future. There are a lot of subtopics and perspectives you can look into when studying energy consumption.

Here are some specific topics that could be both intriguing and relevant:

- **The Impact of Renewable Energy Adoption on Global Energy Consumption Patterns**: Examining how the shift to renewable energy sources like solar and wind is changing overall energy consumption trends and its implications for sustainability.
- **Energy Consumption and Climate Change: A Critical Analysis**: Analyzing the relationship between energy consumption patterns, greenhouse gas emissions, and climate change, and exploring strategies to mitigate negative impacts.
- **The Role of Energy Efficiency in Reducing Overall Consumption**: Investigating how energy-efficient technologies and practices in residential, commercial, and industrial sectors contribute to reducing total energy consumption.
- **Energy Consumption in Developing vs. Developed Countries**: Comparing energy consumption patterns, challenges, and solutions between developing and developed nations, focusing on how economic development impacts energy use and sustainability.
- **The Future of Energy Consumption: Trends and Innovations**: Exploring emerging technologies and trends in energy consumption, including smart grids, energy storage solutions, and advancements in energy-efficient devices.
- **The Environmental and Economic Impacts of Fossil Fuels**: Analyzing the effects of continued reliance on fossil fuels for energy, including environmental degradation and economic costs, and discussing potential alternatives.

- **Behavioral Factors Influencing Energy Consumption**: Investigating how individual and societal behaviors affect energy use, and exploring strategies for promoting energy conservation and sustainable practices.
- **Energy Consumption and Urban Development: Challenges and Solutions**: Examining how urbanization and city planning affect energy consumption and exploring sustainable development practices that address these challenges.
- **The Role of Government Policies in Shaping Energy Consumption**: Analyzing how policies, regulations, and incentives influence energy consumption patterns and drive the transition to cleaner energy sources.
- **The Impact of Energy Consumption on Public Health**: Exploring the connections between energy consumption, air quality, and public health, and discussing ways to mitigate health risks associated with energy use.

Each of these topics offers a rich area for research and discussion, providing insights into the complex and multifaceted nature of energy consumption and its broader implications.

5.1.1. Energy Consumption, Conversion, Storage, and Efficiency

In the modern era, energy has become the unsaid life force that keeps our world going. Energy supports almost every aspect of our lives, including running our cars, heating our homes, and supplying fuel for industrial engines. The impact of energy on our economies, geopolitics, and environment is enormous, and its significance cannot be emphasized. Global energy demand is growing, and this is being correlated with improved living standards and rising prosperity. Our energy landscape is shaped by a number of interconnected components that make up the global energy system, including energy consumption, storage, conversion, and efficiency.

5.1.1.1. Energy Consumption

The use of energy in various forms to satisfy societal demands and human needs is referred to as energy consumption. This covers using electricity, renewable solar energy, wind, hydro, and geothermal energy, as well as natural gas, coal, oil, biomass, and other energy sources. Energy consumption lies at the core of our modern civilization and is directly correlated with global prosperity. However, there is a strong correlation between environmental issues and energy consumption. For instance, carbon dioxide, a byproduct of burning fossil fuels, is the main contributor to climate change and global warming.

Every year, the US Department of Energy (DOE) releases data on the country's energy sources and consumption. In 2021, the United States consumed 97.3 quadrillion BTU of energy, an increase of roughly 5% from 2020. The portion of renewable energy, which includes biomass, solar, and wind energy, increases significantly by more than 10%, which is a remarkable finding. This rise is in line with people's growing awareness of energy use's negative environmental effects. One of the main challenges is continuing to use more energy in a way that is both safe and responsible for the environment. The conversion of renewable energy into electricity, effective energy storage, and sustainable and efficient energy consumption must all be prioritized in order to meet this challenge.

5.1.1.2. Energy Conversion

Energy conversion is the process of changing one energy source into another to fulfill particular energy requirements and applications. The first law of thermodynamics states that although energy cannot be created or destroyed, it can change into other forms. Energy conversion processes include mechanical, chemical, thermal, optical, and electrical ones that occur constantly in our surroundings. Chemical energy is converted into mechanical energy by automobile engines. Solar panels directly generate electricity from optical energy. Electricity is transformed into thermal energy by space heaters. LED lights transform electrical power into light or optical energy.

The conversion of different energy sources into electricity is one of the most fascinating energy conversions because, in the modern world, electricity is just as essential to human survival as food and water. In our daily lives, we use electricity for appliances like cell phones and air conditioners, as well as for lighting, heating, and cooling. Electric generators come in a variety of forms. The principle behind almost all electrical generators is Faraday's law of induction, which was established in 1831. Thermoelectric generators, fuel cells, and photovoltaic cells are some other kinds of electrical generators. By using the photovoltaic effect, photovoltaic cells directly convert solar energy into electrical power. Fuel cells work by reacting a hydrocarbon fuel with oxygen to produce electricity through an electrochemical process. The Seebeck effect is used by p- and n-type semiconductor legs in thermoelectric generators to convert thermal energy directly into electrical power.

One of the energy sources that is producing electricity the fastest in the world today is renewable energy, such as solar, wind, and geothermal energy.

Improving energy conversion processes' sustainability and efficiency is essential for cutting greenhouse gas emissions, minimizing resource depletion, and lessening the effects on the environment.

5.1.1.3. Energy Storage

Extra energy is stored so it can be used at a later time in various forms. Energy storage systems come in a variety of forms, such as thermal and mechanical systems and batteries. Alessandro Volta, an Italian chemist, created the first electrochemical battery in 1800. This battery used a chemical reaction to store and release a charge. In the 1970s, modern lithium-ion batteries with high energy densities went on sale. These days, the batteries are widely used in everything from grid-scale energy storage projects to electric cars and consumer electronics. Systems for storing thermal energy can hold cold or heat. Sensible heat, latent heat, and thermo-chemical heat storage are the three primary categories of thermal energy systems. Sensible heat is typically stored in a water tank, whereas latent heat is stored in phase change materials. Reversible chemical reactions are used in thermo-chemical energy storage systems to store heat or cold. Three primary categories of mechanical energy storage systems are flywheel, compressed air, and pumped hydro. The most popular large-scale mechanical energy storage system in use today is pumped hydro storage, which uses two reservoirs, one higher than the other.

The effective use of different types of energy is made possible by energy storage, which is essential in closing the gap between supply and demand for energy. Applications for energy storage technologies can be found in many different fields, such as utility-scale energy systems, commercial, industrial, residential, and transportation. There are still many issues that need to be resolved before energy storage devices can be widely used. Cost competitiveness, technological constraints, legal obstacles, grid integration problems, and environmental factors are a few of these.

5.1.1.4. Energy Efficiency

Energy efficiency is the practice of using less energy to produce a given amount of work, which lowers costs, reduces waste, and lessens the impact on the environment. Energy efficiency gains are driven by technological developments in energy-efficient appliances and industrial processes. For instance, while offering the same amount of illumination, LED (light emitting diode) lighting uses a lot less electricity than conventional incandescent lighting. By adding thermal insulation material to walls and roofs, buildings can reduce heat loss and more effectively maintain the desired temperature. Fuel consumption and emissions are decreased by driving fuel-efficient vehicles, such as electric and hybrid cars. When compared to separate heat and power generation, combined heat and power systems may maximize energy efficiency because they produce both electricity and useful heat.

Given that electricity is essential to global industry, infrastructure, transportation, and communication, the efficiency of electric generation systems is especially fascinating. The achievable efficiency of heat engines, and consequently heat engine-based electric generators like steam turbine generators, gas turbine generators, and internal combustion engine generators, is limited by the second law of thermodynamics. Electric generator energy efficiency is influenced by a number of variables, such as generator type, operating environment, design, and technology. It is imperative to consistently strive for enhanced efficiency via technological and operational practice advancements in order to minimize environmental impact and reduce energy consumption.

5.1.2. Factors Affect Electric Vehicle Energy Consumption

An electric car's operational expenses are directly proportional to its energy consumption, but what does this actually mean? Measured in kilowatt hours per hundred kilometers, consumption is an important factor to consider when determining a vehicle's range based on its intended use. The way a driver operates an electric vehicle influences how much energy it uses, just like in conventional combustion-powered vehicles. Anticipating traffic and road conditions while driving can help you maximize your electricity consumption by lowering the energy needed to power your electric car. On the highway, energy consumption rises with speed. Air resistance and increased road surface friction force the motor to expend more watts in order to maintain speed. Last but not least, internal temperature control affects consumption as well. Heating and air conditioning considerably shorten the battery life of an electric vehicle.

5.1.3. Electric Car: Energy Consumption When Charging on a Domestic Socket

5.1.3.1. Different Types of Charging Methods

For safety reasons, charging electric cars on conventional outlets is not advised (the Renault Twizy being an exception, as it can be charged using a regular domestic socket). As with the rest of Renault's electric range, installing a dedicated charging station is advised for the majority of models. A "secure socket," which is the least expensive option in terms of installation and purchase prices, is made to guard against overloading and supply power that is appropriate for the home electrical circuit as well as the vehicle.

The Wallbox is a more sophisticated charging box that mounts to your wall. It allows you to control the vehicle's consumption from a power outlet by supplying it with varying intensities. Power levels for different models range from 3:07 kW to 22 kW. For shorter charging sessions, higher wattage is helpful and advised, for instance, if your electricity contract offers discounted rates during specific hours of the day. This kind of box shows the duration and associated cost of each charging session, giving you precise control over how much energy you use. Some Wallboxes can even modify the maximum charge intensity automatically, in real time, based on the current available on the property.

5.1.3.2. Grants for Installing a Charging Point

A number of subsidies are offered in France to incentivize the installation of an electric vehicle charging station at home. Recently, there was introduced a bonus for energy-efficient renovations. This bonus is applicable to all kinds of households, including single-family homes and shared apartment buildings. This bonus has a flat rate of 300 euros for installing an electric vehicle charging station. Other grants, such as the ADVENIR bonus, are also available for shared apartment buildings. This program aims to reimburse up to 50% of material and installation costs. Additionally, nearby nations provide a range of grants. While Germany supports owners of electric vehicles by covering between 10% and 30% of the cost of a Wallbox (plus additional bonuses provided by various local initiatives), the United Kingdom reimburses up to 75% of the costs associated with the installation of a home charging point.).

5.1.3.3. Contract-Dependent Costs

The price per kilowatt hour and the terms of your energy provider's contract are the two main factors that affect how much it costs to recharge your electric car at home. Simply multiply the price per kilowatt hour by the capacity of your battery in kilowatt hours (kWh) to determine the cost of a full charge. Based on the average domestic rate in France in 2019, which is approximately 0 point 15 euros per kWh, this works out to less than 3 euros for every 100 kilometers and about 8 euros for a 52 kWh battery (like the ones used on the New ZOE) full recharge. The substantial variations in the price of a full charge in various nations—just over 15 euros in Germany, compared to 11 euros in Italy and 9 euros in the UK—can be attributed to variations in the cost of electricity throughout Europe (Figure 5.1).

Figure 5.1. *The cost to recharge your electric vehicle.*

Source: By Michael Movchin / Felix Müller, CC BY-SA 3.0, https://commons.wikimedia.org/w/index.php?curid=27783265.

5.1.4. The Energy Consumption of an Electric Vehicle on A Public Charging Station

5.1.4.1. Chargers Designed for Use at Charging Stations

Apart from the cables that connect to your domestic power grid, electric cars also have adapters made specifically for charging at public stations. Like other models from the manufacturer, the Renault ZOE comes with a Caméléon® charger as standard equipment. It is special in that it can adjust to various AC (alternating current) stations, which are the most prevalent kind of infrastructure for charging, enabling the driver to maximize the use of the electrical network that is available. ZOE can therefore charge up to 22 kW at AC stations, which are the most prevalent in public areas, allowing it to cover an average distance of 125 kilometers in one hour*. You can increase your electric vehicle's range by doing this.

The Renault ZOE also has a CCS connector, which makes it possible for the car to refuel even faster at 50 kW fast-charge stations, regaining up to 150 km of range in 30 minutes. This connector uses direct current (DC), which is mainly available at highway rapid charging stations. The method of charging affects energy costs directly and varies in terms of power and charging time. The cost increases with the speed of charging (Wiedmann, 2010; Huang and Huang, 2020).

5.1.5. Calculate the Consumption of an Electric Vehicle: Variable Pricing on Public Station

How much does it cost to recharge an electric car on a public charging network? That amount can vary up to twofold, depending on the location, power source, and operator of the electric charging station. Free stations are frequently found near big-box retailers or in the parking lots of shopping centers. Additionally, some networks allow charging at night, or during off-peak hours. Without a subscription, the average cost of a roadside charging station in France is 1.30 euros every five minutes; however, with a subscription, this cost can be as low as 0.50 euros. A ZOE can gain an additional 100 kilometers of range in just 20 minutes at a cost of 2–5 euros using a 50 kW rapid charge station, which are located alongside major highways. Different countries have different costs per kilowatt hour for different types of infrastructure. For instance, in the Netherlands, a subscription is available for 0.59 euros per kilowatt hour, or 17.70 euros for a 30 kWh charge. Some operators in England charge 12 pounds (13.20 euros) for an hour of fast charging at a 50 kW station. In general, customers can take advantage of discounted rates on thousands of stations by subscribing to a network spanning all of Europe. In any case, electric charging is far less expensive than the equivalent fossil fuels (diesel and gasoline).

5.1.5.1. Ways to Pay to Recharge an Electric Car

For pragmatic reasons, the great majority of networks that run charging stations do not accept payments by bank card, and even fewer accept payments in cash. The easiest solution is to choose a badge, whether it is free or paid, until support for payment methods is expanded. When buying an electric vehicle, the manufacturer may provide this. For instance, Renault sells the Z in some nations: a Pass card that is compatible with the MY Renault app. The closest charging station is found using this app's directions, and charging is initiated by using the badge. The Z is also known as a "recharge card." Throughout Europe, Pass provides access to a network of over 190,000 charging stations. Similar services are provided by numerous online vendors, along with an application that makes card management easier overall. Payments can be made directly from these platforms using a number of different ways, including bank transfer, PayPal, bank card, and more.

5.1.6. Energy Consumption When Charging An Electric Car in Europe: Country Disparities

Public charging stations in Europe follow European regulations, particularly when located near motorways. The methods used by various nations to bill for public charging vary greatly. While most of the neighboring countries base their calculations on the number of kilowatt hours that are "filled up," stations in France use a time-based rate. Furthermore, the cost of energy consumption for an electric car varies depending on the country in which it is being charged, as it does for all electric devices due to the variation in electricity prices between nations (Figure 5.2).

Figure 5.2. *Energy consumption when charging an electric car.*

Source: By Tennen-Gas – Own work, CC BY-SA 3.0, https://commons.wikimedia.org/w/index.php?curid=8249799.

5.1.6.1. How Much do Renault ZOE and Twingo Electric Consume?

The battery capacity of a car determines how much energy it uses each time it is charged. For instance, on the Renault ZOE, the Z. E. When using a 50 kW fast-charge station, the battery and its 52 kWh capacity allow it to drive 100 kilometers (WLTP cycle*) for about 8 euros in France or about 10/12 euros in Germany. The amount may vary based on the suggested billing method for a charge. With regard to Twingo Electric, its 55 kWh battery allows it to zigzag throughout the town for a full week on a single charge. Furthermore, it should be mentioned that safe driving techniques—particularly ecodriving—should not be undervalued. They significantly reduce energy use, enabling drivers to save money right away.

5.2. CARBON FOOTPRINT

Carbon footprint is the total amount of carbon dioxide (CO_2) emissions from an individual's or other entity's activities (e.g., structure, company, nation, etc.). It comprises both the direct emissions that come from burning fossil fuels for transportation, heating, and manufacturing as well as the emissions needed to generate the electricity needed to run the appliances that power the products and services that are used. Furthermore, the idea of the carbon footprint frequently takes into account the emissions of additional greenhouse gases, like nitrous oxide, methane, and chlorofluorocarbons (CFCs).

The idea of ecological footprint, which was first introduced in the early 1990s by Canadian ecologist William Rees and Swiss-born regional planner Matthias Wackernagel at the University of British Columbia, is connected to and evolved from the carbon footprint concept. An ecological footprint is the total land area needed to support a population or an activity. It covers environmental effects like the amount of land used for food production and the use of water. On the other hand, a carbon footprint is typically measured in tons of CO_2 or CO_2 equivalent annually.

A carbon footprint, also known as a greenhouse gas footprint, is a numerical value or index that enables comparison of the overall quantity of greenhouse gases released into the atmosphere by a certain activity, product, nation, or business. The standard way to report carbon footprints is in tonnes of emissions (CO_2-equivalent) per unit of comparison. These units can be tonnes CO_2 equivalent annually, kilograms of protein consumed, kilometers traveled, articles of clothing, and so on. The carbon footprint of a product is the total amount of emissions over its life. These cover the entire supply chain, from the point of production to the point of disposal and final consumption.

In a similar vein, an organization's carbon footprint consists of both its direct and indirect emissions. Scope 1, 2, and 3 emissions are what the Greenhouse Gas Protocol (for carbon accounting of organizations) refers to. There are numerous online resources and approaches available for calculating carbon footprints. They vary according to whether a nation, group, item, or specific person is the subject of the attention. For instance, if a customer wants to be environmentally conscious, the carbon footprint of a product may help them choose what to buy. The carbon footprint can be used to differentiate between economic activities that have a high and low footprint in relation to climate change mitigation efforts. Therefore, the concept of carbon footprint enables comparisons between the climate impacts of products, companies, individuals, and nations. It also aids in the development of priorities and plans for lowering carbon footprints.

A good way to represent a carbon footprint is in terms of carbon dioxide equivalent (CO_2eq) emissions per unit of comparison. This represents the total emissions of greenhouse gases. Not just carbon dioxide is included; all greenhouse gases are included. Additionally, it examines emissions from businesses, events, associations, and services. Certain definitions solely consider emissions of carbon dioxide. Other greenhouse gases like methane and nitrous oxide are not included in this.

The carbon footprint can be calculated using a variety of techniques, some of which may be slightly different for various entities. It is standard procedure for organizations to employ the Greenhouse Gas Protocol. There are three carbon emission scopes included. Emissions of direct carbon are included in scope 1. Scopes 2 and 3 deal with emissions of indirect carbon. Indirect emissions resulting from an organization's operations that originate from non-owned or non-controlled sources are referred to as scope 3 emissions.

Consumption-based emissions accounting is a widely used method in countries to determine their annual carbon footprint. Analysis of global supply chains is made possible by consumption-based accounting that uses input-output analysis supported by supercomputers. For the UNFCCC, nations also compile national greenhouse gas inventories. Only activities within the nation itself are responsible for the GHG emissions reported in those national inventories. This method is referred to as production-based accounting or territorial-based accounting. It doesn't account for the production of goods and services that locals import. Emissions from imported goods and services are taken into account in consumption-based accounting. For this reason, consumption-based accounting is more thorough.

This thorough reporting of carbon footprint, which includes emissions from Scope 3, addresses weaknesses in the current framework. Global Highway GHG inventories submitted to the UNFCCC do not account for international travel. In-depth carbon footprint reporting examines the point at which goods and services are consumed, as well as the ultimate demand for emissions.

5.2.1. Definition of Carbon Footprint

A population, system, or activity's total emissions of carbon dioxide (CO_2) and methane (CH_4), taking into account all pertinent sources, sinks, and storage within the defined population, system, or activity's temporal and spatial boundaries, is referred to as the carbon footprint. computed as the carbon dioxide equivalent (GWP100) utilizing the applicable 100-year global warming potential. When reporting carbon footprints, scientists use CO_2-equivalent units, which are equivalent to tonnes of CO_2 emissions. They could report them by the year, by the individual, by the kilogram of protein, by the kilometers traveled, and so forth. All of the most significant greenhouse gases are included in the Greenhouse Gas Protocol. The Kyoto Protocol addresses seven greenhouse gases, which are the subject of this standard. These gases are carbon dioxide (CO_2), methane (CH_4), nitrogen dioxide (N_2O), hydrofluorocarbons (HFCs), perfluorocarbons (PCFs), sulfur hexafluoride (SF_6), and nitrogen trifluoride (NF_3).

In contrast, only carbon dioxide is included in the IPCC's definition of a carbon footprint as of 2022. The measurement of the exclusive total amount of carbon dioxide (CO_2) emissions that are either directly or indirectly caused by an activity or accumulate over the course of a product's lifecycle is known as the carbon footprint.

The definition that was suggested in the UK in 2007 was also adopted by the authors of the IPCC report. The definition of carbon footprint in that publication only included carbon dioxide. It used the justification that other greenhouse gases were harder to measure to support this. Their varying potentials for global warming account for this. They added that the carbon footprint indicator would become less useful if all greenhouse gases were included. However, there are drawbacks to this strategy. The fact that some industries or products with a high methane footprint, like livestock, appear less harmful to the climate than they actually are is one drawback of excluding methane.

5.2.2. Types of Greenhouse Gas Emissions

A set of guidelines for monitoring greenhouse gas emissions is known as the greenhouse gas protocol. Within the value chain, the standards separate emissions into three scopes (Scope 1, 2, and 3). Scope 1 refers to greenhouse gas emissions that are directly caused by the organization, such as when fossil fuels are burned. Scope 2 emissions are those that are produced indirectly by an organization, such as when it buys secondary energy sources like steam, heat, electricity, or cooling. Finally, Scope 3 refers to indirect emissions resulting from upstream or downstream processes (Figure 5.3).

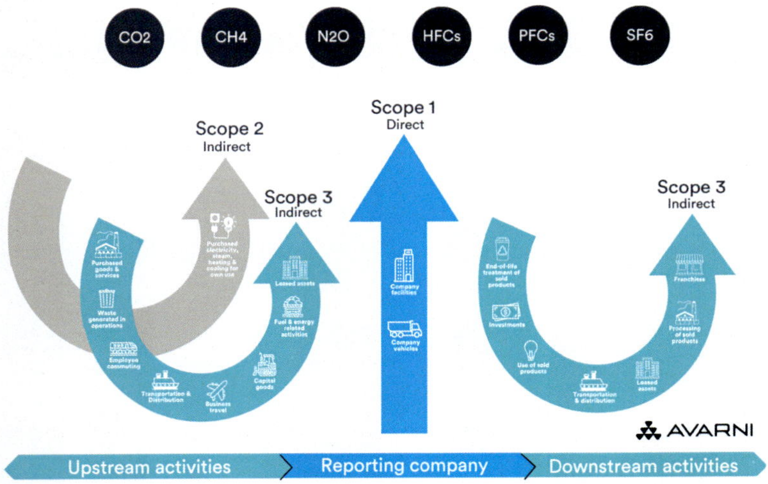

Figure 5.3. *Overview of greenhouse gas protocol scopes and emissions across the value chain, showing upstream activities, reporting company and downstream activities.*

Source: https://uploads-ssl.webflow.com/620f1a8dcbe491415cb12e7c/64210ef422b66759cfbb3840_avarni_infographic_scope-1-2-3-min.png.

5.2.2.1. Direct Carbon Emissions (Scope 1)

Direct or *Scope 1* carbon emissions come from sources on the site that is producing a product or delivering a service. An example for industry would be the emissions from burning a fuel on site. On the individual level, emissions from personal vehicles or gas-burning stoves are Scope 1.

5.2.2.2. Indirect Carbon Emissions (Scope 2)

Emissions from sources either upstream or downstream of the process under study are referred to as indirect carbon emissions. They go by the names Scope 2 or Scope 3 emissions as well. The indirect emissions associated with buying heat, steam, or electricity for on-site use are referred to as scope 2 emissions. Transportation of fuels and materials, energy used outside the production facility, and waste generated outside the facility are a few examples of upstream carbon emissions. Transportation of products and waste, end-of-life processes and treatments, and emissions related to product sales are a few examples of downstream carbon emissions. The GHG Protocol states that determining emissions both upstream and downstream is crucial. There might be more than one count. This is due to the possibility that downstream emissions from one person's consumption habits could originate from upstream emissions.

5.2.2.3. Other Indirect Carbon Emissions (Scope 3)

All other indirect emissions resulting from an organization's operations are included in scope 3. However, they come from non-owned and non-controlling sources. Businesses are able to evaluate the emissions impact of their entire value chain and determine where to concentrate their reduction efforts by using the GHG Protocol's Corporate Value Chain (Scope 3) Accounting and Reporting Standard. Suppliers and product users are examples of scope 3 emission sources.

They go by the name of the value chain as well. This also includes other indirect emissions and the transportation of goods. Roughly 30% of US businesses reported Scope 3 emissions in 2022. The International Sustainability Standards Board is working on a proposal to incorporate emissions from Scope 3 into all greenhouse gas reports.

5.2.3. Purpose and Strengths

The average global temperature is rising faster now than it has in the past. The main cause of it is the burning of fossil fuels by humans. Deforestation, along with industrial and agricultural practices, is another factor contributing to the increase in greenhouse gases in the atmosphere. Production of cement is one of these. The two most prominent greenhouse gases are methane and carbon dioxide. In the 21st century, greenhouse gas emissions—and consequently, the carbon footprint of humanity—have been rising. The goal of the Paris Agreement is to cut greenhouse gas emissions to a level where the increase in global temperature is kept to 1.5 °C over pre-industrial levels.

The idea of carbon footprint compares the effects on the climate of various entities, including people, goods, businesses, and nations. If consumers wish to contribute to slowing down climate change, they may be able to select products with a lower carbon footprint thanks to a label on the product. Such a label could indicate, for meat products, that beef has a higher carbon footprint than chicken.

Developing a plan to lessen an organization's carbon footprint requires an understanding of its size. For most businesses, neither scope 1 (on-site activities) nor scope 2 (energy supplied to the organization) accounts for the great majority of emissions. Instead, Scope 3 emissions are the source of them. They originate from the supply chain's extended upstream and downstream segments. It is therefore impossible to identify all significant emissions if Scope 3 emissions are disregarded. This will reduce available options for relief. Big businesses in industries like apparel or autos would have to look at over 100,000 supply chain routes to accurately report their carbon footprints. The necessity of reducing carbon emissions has long been recognized. This is also known as carbon leakage by scientists. The concept of carbon footprint tackles issues related to carbon leakage that are not covered by the Paris Agreement. When importing nations contract out production to exporting nations, carbon leakage happens. Rich nations are frequently the ones that outsource, and low-income nations are typically the ones that export. Impacts are usually shifted from developed to developing nations. Countries can move their dirty industries abroad to give the impression that their greenhouse gas emissions are declining. However, if you consider their emissions concerning their consumption, they might be rising.

Numerous negative effects on the environment are caused by carbon leakage and the associated global trade. Increasing air pollution, water scarcity, biodiversity loss, raw material consumption, and energy depletion are a few of these. Both production- and consumption-based accounting have been supported by academics. This promotes shared accountability between producers and consumers. At the moment, nations submit annual GHG inventories to the UNFCCC based on the emissions from their territories. This is referred to as the production-based approach or the territorial-based approach. By tackling the issue of carbon leakage, including consumption-based calculations in the UNFCCC reporting requirements would aid in closing loopholes.

As of right now, countries are not required by the Paris Agreement to include the greenhouse gas emissions linked to international transportation in their national totals. Each of these emissions is disclosed separately. The Annex 1 Parties' commitments to limitation and reduction under the Kyoto Protocol and the Climate Convention do not apply to them. International travel-related greenhouse gas emissions are included in the carbon footprint methodology. In other words, it places the burden of emissions resulting from global trade on the importing nation.

5.2.4. Carbon Footprint Calculation

A product, service, or sector's carbon footprint must be calculated, which calls for in-depth knowledge and careful consideration of all relevant factors. Different scales can be used to calculate carbon footprints. They can be applied to entire nations, towns, and neighborhoods as well as to industries, businesses, and goods. To determine one's personal carbon footprint, there are numerous free online carbon footprint calculators available. Utilizing tools like the Scope 3 Evaluator, businesses can report emissions at every stage of the value chain. Consultants and researchers can model global sustainability footprints with the aid of software tools. There are several questions that need to be addressed in every circumstance. These include the activities that are associated with which emissions and the percentage that belongs to which company. Software is necessary for managing a business. Therefore, innovative approaches to enterprise resource planning are required to raise the performance of corporate sustainability.

A nation's reported per capita emissions (such as those reported under the United Nations Framework Convention on Climate Change) are not the same as its carbon footprint. Carbon footprints concentrate on the greenhouse gas emissions related to consumption rather than those related to production. They typically account for emissions related to international transport and shipping, which are not included in standard national inventories, and they include emissions associated with goods that are imported into a country but are produced elsewhere. Because of this, a nation's carbon footprint may grow even while domestic carbon emissions decline.

The United States has the highest carbon footprint per person. As per the Carbon Dioxide Information Analysis Center and the UNDP, the average American citizen had a per capita carbon footprint of 20.6 metric tons (22.7 short tons) of CO_2 equivalent in 2004, which was approximately five to seven times higher than the global average. Global averages differ significantly, with people living in developed nations typically having larger footprints. In the same year, for instance, Tanzania and Brazil had carbon footprints of 0 point1 metric ton (0 .1 short ton) and 1 .8 metric tons (roughly 2 short tons) of CO_2 equivalent, respectively, while France had a per capita carbon footprint of 6 .0 metric tons (6 .6 short tons).

Transportation and residential energy use account for the majority of a person's carbon footprint in developed nations. For instance, during the first 10 years of the 21st century, those sources accounted for about 40% of all emissions in the United States. These emissions constitute the emissions that a person has direct control over and are included in their "primary" carbon footprint. The "secondary" carbon footprint refers to the remaining portion of a person's carbon footprint, which is made up of carbon emissions from the consumption of goods and services. The carbon emissions from the production of food are included in the secondary footprint. It can be applied to explain diets with higher percentages of meat, which needs more energy and nutrients to produce than grains and vegetables, as well as foods that have traveled long distances. The secondary carbon footprint is also influenced by the production and delivery of consumer goods. A bottle of water, for instance, has a carbon footprint that comprises the CO_2 or CO_2 equivalent released during the bottle's production as well as the amount released during delivery to the customer. There are numerous resources available to determine the carbon footprints of people, companies, and other entities. The World Resources Institute's and the World Business Council for Sustainable Development's Greenhouse Gas Protocol, as well as ISO 14064, an international standard that addresses greenhouse gas emissions specifically, are frequently utilized techniques for determining an organization's carbon footprint. Numerous institutions, including the U.S. S. Carbon calculators are available for use on the Internet for individuals by the Environmental Protection Agency, the Nature Conservancy, and British Petroleum. Using these calculators, individuals can assess how their estimated carbon footprints compare to global and national averages.

5.2.5. Carbon Footprint Reduction

Numerous actions can be taken by both individuals and businesses to lessen their carbon footprints and thereby aid in the mitigation of global warming. To offset all or part of their carbon footprint, they can buy carbon offsets, which are, in general, investments made in technology or activities that reduce carbon emissions. They practically become carbon neutral if they make enough purchases to offset their carbon footprint. Reducing carbon footprints requires changing lifestyles and purchasing patterns in addition to increasing energy efficiency. Changing one's energy and transportation habits can affect one's overall carbon footprint. For instance, compared to driving, taking public transit like buses and trains lowers a person's carbon footprint. By adding insulation to buildings, installing energy-efficient lighting, or generating the electricity they need from renewable energy sources, both individuals and businesses can lower their respective carbon footprints. For instance, there are no direct carbon emissions when electricity is generated using wind power. Reducing meat consumption and changing one's shopping habits to include items that require less carbon emissions to produce and transport are two more lifestyle decisions that can minimize a person's secondary carbon footprint.

5.2.6. Ecological Footprint

An individual's or a group's ecological footprint (EF) is a measurement of the demands they place on the world's natural resources. It has grown to be one of the most popular metrics for assessing how humans are affecting the environment. It has been used to draw attention to resource consumption disparities between and within nations as well as the apparent unviability of current practices. Instead of attempting to determine how many people a given land area or the entire planet can support, the EF estimates the biologically productive land and sea area needed to provide the renewable resources that a population consumes and to absorb the wastes it generates. It does this by using current technology and resource-management practices. It quantifies the needs for productive areas, which include built-up land for infrastructure and housing, grazing lands for animal products, forests to produce wood products, marine areas for fisheries, and forests required to absorb carbon dioxide emissions from energy consumption. The EF, which is expressed in "global hectares" (GHA), can be calculated at different scales for people, places, nations, and the entire human race. (A hectare is equivalent to 2,047 acres. The resulting numbers can also be compared to the available biocapacity, or productive area.

The EF concept originated with Canadian ecologist William Rees, who supervised Swiss urban planner Matthias Wackernagel as he developed it for his dissertation. The idea is explained in Wackernagel and Rees' 1996 book, Our Ecological Footprint. The sustainability and equity of the present production and consumption patterns have been called into question by EF calculations. The per capita global footprint is determined by the nonprofit organization The (GFN), which collaborated with hundreds of cities, companies, and other organizations to promote the EF as a sustainability metric. The global footprint per person in 2023 was 2.6 GHA. The EF of humanity exceeded Earth's biocapacity by 1.1 gha that year, as the global biocapacity was 1.5 gha per person. Put differently, 1.7 "Earths" would be required to meet the demands on resources today, or it would take more than a year and a half for Earth to replenish the resources used in a single year. The result of this "ecological overshoot," which started in the middle of the 1970s, is the depletion of biological resources necessary for life, including rangelands, fisheries, forests, and agricultural land.

5.2.6.1. Using the EF

EF analysis can determine whether a nation is using the ecological "capital" of other regions of the world to meet its needs or if it is living within the biocapacity of its own territory and is therefore a "ecological debtor.." The demands that people in different societies place on the environment are reflected in per capita EFs, which range from the United States (7.5), Germany (4.5), China (3.6), and Qatar (13.1 gha/person) at the high end to Rwanda (0.55) at the low end (2022 data). These numbers serve as the foundation for claims that roughly five Earths would be required if all of humanity consumed as much as the typical American does. The degree of wealth within a country also has a significant impact on EFs.

To determine whether a country is headed toward sustainable development, which is defined as having a per capita EF lower than the available per capita biocapacity and a high rating (above 0.8) on the United Nations Human Development Index (HDI), researchers have combined footprint analysis with measures of human development. (The HDI is a measure of economic and social progress that combines average life expectancy, income, and educational attainment for a given nation. Environmental activists and educators have used the EF to draw attention to unsustainable consumption patterns. Less frequently, they have used it to spread knowledge of the larger structural factors that underlie these patterns of consumption. Such objectives have led to the appearance of numerous online footprint calculators on the websites of nongovernmental organizations. With the aid of these calculators, individuals can compute their individual EF and compare it to estimates of available biocapacity or to the average footprints of other people both locally and worldwide. In the meantime, to test widely divergent social theories about the forces underlying the ecological impacts of humans on the planet, social scientists have utilized the EF as a comprehensive indicator of those impacts.

EF analysis is becoming more widely accepted by governments and corporations, Although it has the potential to provoke a radical critique of the existing economic and social order. Global firms like BP and Wal-Mart launched initiatives to reduce their EFs, and the World Business Council for Sustainable Development started looking into ways to get to a "one-planet economy" by 2050. Many national governments have also given the idea some thought or support, including those of Wales, Switzerland, the United Kingdom, Japan, the United Arab Emirates, and the European Union as a whole.

5.2.6.2. Critiques of the EF

The EF has drawn a variety of criticism in spite of its quick rise and extensive use. The fact that the EF offers a solitary aggregate indicator of ecological impacts is one of its draws. However, such aggregation necessitates the simplification of a complex reality; for instance, the EF is predicated on the idea that technology is constant both globally and historically. Additionally, detractors contend that the EF methodology encourages more intensive farming practices that, while they may boost yields per unit of land in the short term, may not be as sustainable in the long run—for example, by speeding up land degradation. Similarly, despite other ecological advantages, organic farming practices that produce lower yields than conventional agriculture may seem to have a larger environmental impact. Some contend that EF analysis is unduly anthropocentric, failing to allocate space for the needs of other species and concentrating solely on land and marine areas that are beneficial to the human economy. In fact, other indicators are required to measure changes in biodiversity, as the EF does not. Advocates of the EF recognize that it cannot capture all major environmental impacts due to data gaps for certain issues and challenges in translating some ecological demands—for which there is no regenerative capacity—into a land area measurement. The EF fails to account for several significant effects, including those about hazardous materials, greenhouse gases other than carbon dioxide, and water usage.

The idea that a nation's population should live within the biocapacity of its own territory and not rely on resources from outside sources through trade is another point of contention for some critics. Large, sparsely populated nations like Canada and Australia appear to be living well within their territories' ecological means even though their per capita footprints (at 7.42 and approximately 5.78, respectively, for 2022) are among the highest in the world. These countries may be very ecologically efficient, but they will almost inevitably be ecological debtors drawing on outside resources. The EF's supporters have worked to improve the idea while also acknowledging some of its drawbacks. The first set of guidelines for accurately calculating and communicating the EF was created in 2006, and updates are still being made. According to a 2008 review done for the European Commission, the EF is a helpful and simple-to-understand indicator of sustainable resource use, even though complementary sustainability indicators and additional advancements in data quality and methodology are required (Wiedmann, 2010; Huang and Huang, 2020).

5.2.7. Reducing the Carbon Footprint

5.2.7.1. Climate Change Mitigation

Reducing the carbon footprint of goods, services, and institutions contributes to slowing down global warming. We refer to these actions as mitigating climate change. Reducing the amount of greenhouse gases in the atmosphere that contribute to climate change is known as decarbonization or mitigation of climate change. Energy conservation and switching to clean energy sources from fossil fuels are two examples of climate change mitigation measures. Alterations to land use and the extraction of carbon dioxide (CO_2) from the atmosphere are examples of secondary mitigation techniques. The 2015 Paris Agreement's target of keeping global warming to below 2°C would still be greatly exceeded by the current climate change mitigation measures, which is why they are insufficient. By 2100, global warming would still be approximately 2 °C.

5.2.7.2. Reducing the Industry's Carbon Footprint

By giving a business a carbon credit, carbon offsetting helps lower its overall carbon footprint. Recognizing an equivalent decrease in carbon dioxide in the atmosphere makes up for the company's carbon dioxide emissions. One method of offsetting carbon emissions is reforestation, which involves replenishing depleted forests. Particularly important areas for improvement can be found in a carbon footprint study. Through the use of input-output analysis, the entire supply chain is examined. Using this kind of analysis, the supply chains with the highest emissions of greenhouse gases could be eliminated (Wiedmann, 2010; Huang and Huang, 2020).

REFERENCES

1. Allerhand, A. (2018). Who invented the earliest capacitor bank ("battery" of Leyden jars)? It's complicated. *Proceedings of the IEEE, 106*(3), 498–500. https://doi.org/10.1109/JPROC.2018.2795846.
2. Corbett, J. (2008). Carbon footprint. In B. W. Lerner & K. L. Lerner (Eds.), *Climate Change: In context* (Vol. 1, pp. 162–164). Gale.
3. Costantini, V., & Martini, C. (2010). The causality between energy consumption and economic growth: A multi-sectoral analysis using non-stationary cointegrated panel data. *Energy Economics, 32*, 591–603.
4. Faraday, M. (1936). *Faraday's Notebooks: Electromagnetic Induction.* London: G. Bell & Sons, LTD, England for The Royal Institution of Great Britain.
5. Hou, Q. (2009). The relationship between energy consumption growths and economic growth in China. *International Journal of Economics and Finance, 1*, 232–237.
6. Huang, Z., & Huang, L. (2020). Individual new energy consumption and economic growth in China. *North American Journal of Economics and Finance, 54*, 101010.
7. Moran, M. J., Shapiro, H. N., Boettner, D. D., & Bailey, M. B. (2018). *Fundamentals of Engineering Thermodynamics*. WileyPlus. ISBN: 978-1-119-39138-8.
8. Rehman, S., Al-Hadhrami, L., & Alam, M. (2015). Pumped hydro energy storage system: A technological review. *Renewable and Sustainable Energy Reviews, 44*, 586–598.
9. Scrosati, B. (2011). History of lithium batteries. *Journal of Solid-State Electrochemistry, 15*(7, 8), 1623–1630. https://doi.org/10.1007/s10008-011-1386-8.
10. Shaari, M. S., Hussain, N. E., & Ismail, M. S. (2013). Relationship between energy consumption and economic growth: Empirical evidence for Malaysia. *Business Systems Review, 2*.
11. Wiedmann, T., & Barrett, J. (2010). A review of the ecological footprint indicator—Perceptions and methods. *Sustainability, 2*(6), 1645–1693.
12. Wiedmann, T., & Minx, J. (2008). A definition of 'carbon footprint.' In C. C. Pertsova (Ed.), *Ecological Economics Research Trends* (pp. 1–11). Nova Science Publishers.

Chapter 6

End-of-Life and Recycling of EV Batteries

Contents

INTRODUCTION

The end-of-life (EOL) management and recycling of electric vehicle (EV) batteries have become critical topics as the adoption of EVs accelerates worldwide. EV batteries, predominantly lithium-ion, contain valuable materials such as lithium, cobalt, nickel, and manganese. These materials not only represent a significant economic value but also pose environmental risks if not properly managed at the end of their life cycle.

As EV batteries reach the end of their useful life, usually after 8–15 years depending on use, they can no longer provide the required performance for vehicle propulsion. However, they often retain a substantial capacity that can be repurposed for less demanding applications, such as energy storage systems in renewable energy projects. This "second life" application extends the utility of the batteries, delaying the need for recycling and reducing the demand for new resources.

Recycling of EV batteries is essential for recovering valuable materials and mitigating environmental impacts. The recycling process typically involves disassembling the batteries, extracting and purifying the metals, and safely disposing of hazardous materials. However, the recycling industry faces challenges such as the diversity of battery chemistries, the complexity of battery designs, and the economic viability of recycling processes.

Governments and industry stakeholders are increasingly focusing on developing efficient recycling technologies and establishing regulations to manage the EOL of EV batteries. These efforts are crucial for creating a circular economy where materials are reused, reducing the need for raw material extraction and minimizing environmental harm.

The EOL and recycling of EV batteries are integral to the sustainable growth of the electric vehicle industry. By advancing recycling technologies and policies, we can ensure that the environmental benefits of EVs are fully realized, contributing to a greener and more sustainable future.

6.1. END-OF-LIFE STRATEGIES FOR ELECTRIC VEHICLE LITHIUM-ION BATTERIES

The deployment of electric vehicles (EVs) has increased significantly over the past 10 years; in 2019, there were 21.1 million EVs sold worldwide. COVID-19 is predicted to cause sales to decline to 11.7 million in 2020, but they are predicted to increase further, with an estimated 80.5 million EVs sold by 2025 (BNEF, 2020). These days, policy and regulatory developments as well as growing consumer demand are the main drivers of this growth, even though actual EV sales do not yet reflect this. However, what is facilitating and speeding up the adoption of EVs are developments in battery technology, infrastructure for charging, and declining battery prices (McKinsey and Company, 2019a).

The most popular battery type for EVs nowadays is lithium-ion batteries (LIBs), which have a higher energy density than other significant commercialized battery types (Iclodean et al., 2017). The performance and range of EVs have significantly improved thanks to the technological developments in the last 10 years. Research illustrates the geopolitical risks associated with sourcing critical raw materials (CRMs) like cobalt and lithium, which have led to criticism of the widespread use of LIBs due to their impact on the environment and ethics. Because of their non-reversible bonding techniques, the mixture of materials, and lack of standardization, LIBs are also challenging to recycle at the end of their useful lives (Harper et al., 2019).

Batteries are classified as hazardous waste, so retailers who sell batteries are responsible for managing and disposing of them properly. Additionally, because batteries present a risk to public safety, it is against the law to send batteries to landfills or be burned. These regulations are enforced in the UK by the Waste Batteries and Accumulators 2009 Regulation and the End of Life Vehicles (ELVs) Regulation (Figure 6.1) (European Parliament, 2018; HM UK Government, 2019, 2017).

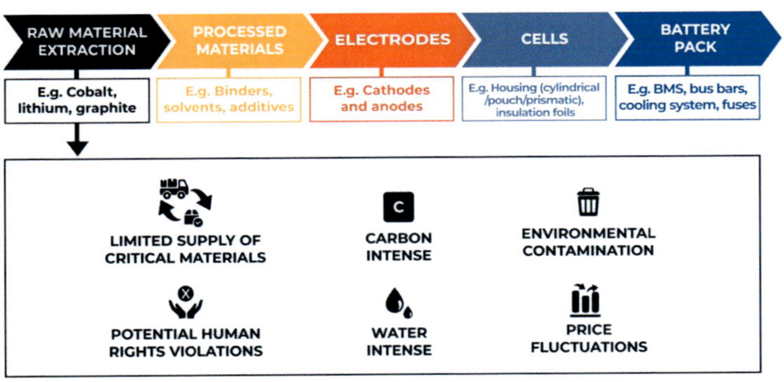

Figure 6.1. *Ethical and environmental issues with raw material extraction for LIBs.*

Source: Harper et al. (2019); https://autorecyclingworld.com/wp-content/uploads/2020/07/EoL-Strategies-for-EV-LIBs-2020-HSSMI.pdf.

Because existing recycling technology is costly, difficult, and unavailable, OEMs are left with limited options by the regulation (Harper et al., 2019). After December 2020, the EU Directive will no longer apply to the UK, but the UK regulations enforcing the Directive are expected to stay in effect. We can use current EV sales trends to estimate the end-of-life date of these battery packs, as noncommercial EVs are expected to have an 8–10-year first-life period (Figure 6.2). With about 6 million battery packs retiring from electric vehicles globally by 2030, post-consumer batteries are expected to become more and more important (BNEF, 2020; Jiao, 2019). However, the first-life period may end as soon as five years after purchase, based on HSSMI's experience with commercial EV fleets. As a result, OEMs in both the commercial and non-commercial sectors need to be cautious when anticipating battery returns before 2030.

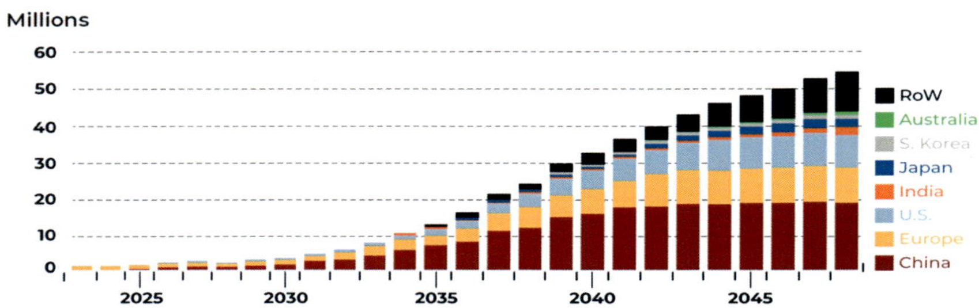

Figure 6.2. *Forecast of EVs reaching the end of life.*

Source: Based on sales trends from BNEF (2020); https://autorecyclingworld.com/wp-content/uploads/2020/07/EoL-Strategies-for-EV-LIBs-2020-HSSMI.pdf.

Furthermore, a large number of LIBs from manufacturing rejects and accidents will arrive, reaching their EoL far sooner than anticipated.

Currently, recycling and reuse of LIBs are considered low due to:

- Very small battery volumes reaching EoL; battery EVs (BEVs) have only been sold over the past 5–10 years, thus very few vehicles have reached the end-of-life stage;
- Design has not yet been optimized for disassembly; and
- A lack of standard pack and cell marking, leading to ambiguity regarding specific cell chemistry.

Up to 250 new EV models with LIBs from over 15 manufacturers will be available by 2025 (McKinsey and Company, 2019b). Lack of standardization will result in more complicated recovery techniques since each battery is made for a particular model of an electric vehicle. Due to the limited supply of post-consumer EV batteries, this is currently somewhat manageable. Nevertheless,

OEMs will need to come up with a workable longer-term solution given the anticipated rise in the number of EVs and the corresponding batteries that will be nearing the end of their first lives in the next 10 years.

End-of-life battery disposal does not have to be seen as something that is done merely to comply with regulations; rather, it can be viewed as a resource and financial opportunity. A LIB has a lot of valuable materials that can be recovered at the end of life. Figure 6.3 shows their value per kilogram (European Commission, 2019a). Based on the anticipated volumes of lithium-ion batteries and this residual value, the CEPS (2018) projects that the EU could recover between €408 and €555 million from cobalt, nickel, aluminum, and lithium by 2030. By 2030, the global market for EoL batteries is expected to be valued at $30 billion, according to McKinsey and Company (2019b). Because the recovered materials can be sold further along the value chain or used to make new batteries, end-of-life batteries therefore offer a significant financial opportunity.

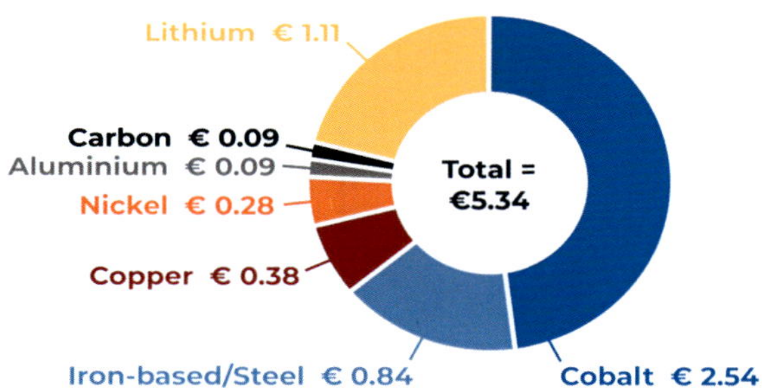

Figure 6.3. *Recycled material value in average LIB per kg.*

Source: Adapted from European Commission (2019a); https://autorecyclingworld.com/wp-content/uploads/2020/07/EoL-Strategies-for-EV-LIBs-2020-HSSMI.pdf

Additionally, the cost of batteries is rising due to shortages of critical raw materials (CRMs) like cobalt and lithium. It is therefore critical that manufacturers devise ways to facilitate easier and, ideally, less expensive access to these high-value materials. This is in addition to the fact that Europe lacks readily accessible mineral deposits or reserves.

OEMs will face a challenge as a result of the growing volumes and the current difficulties in treating batteries that are nearing the end of their useful lives. It also offers them a calculated opportunity to turn their duty into a business venture. A new strategy that OEMs can use to manage electric vehicle batteries nearing the end of their useful lives is summed up in the idea of a "circular economy," as seen in Figure 6.4 (Ellen MacArthur Foundation, 2019). The goal of this model is to shift supply chains from the conventional linear model of "take, make, consume, and dispose" to circular models in which materials and products are recycled back into the supply chain, given an extended lifespan, and their value is retrieved.

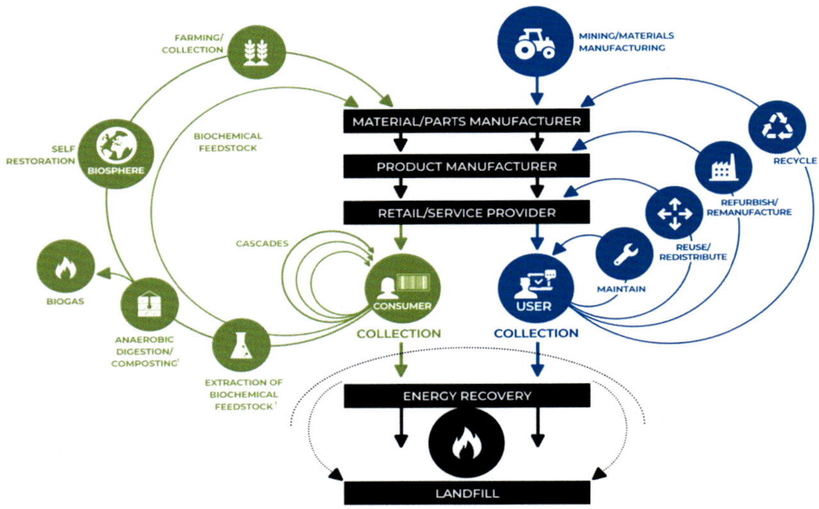

Figure 6.4. *Butterfly diagram.*

Source: Adapted from Ellen MacArthur Foundation (2019); https://autorecycling-orld.com/wp-content/uploads/2020/07/EoL-Strategies-for-EV-LIBs-2020-HSSMI.pdf.

The reasoning for and against each circular strategy, a summary of the pertinent factors to consider, and a decision tree to help determine the strategic direction will all be included in the upcoming chapters to aid in the decision-making process. Although the sections in this white paper can be used in any order, Figure 6.5 aims to illustrate how they can fit together:

- Consider the business context to inform which EoL strategies are relevant;
- Assess the relevant EoL strategies;
- Use the decision tree to determine at which SOH levels each strategy is most valuable.

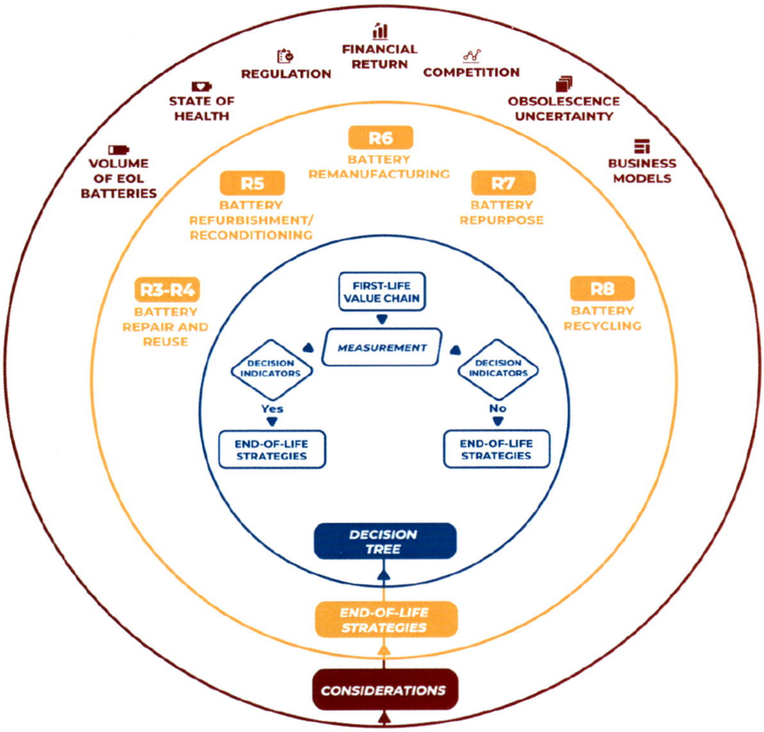

Figure 6.5. *Overview of the decision tree, EOL strategies, and considerations.*

Source: https://autorecyclingworld.com/wp-content/uploads/2020/07/EoL-Strategies-for-EV-LIBs-2020-HSSMI.pdf.

Several frameworks can be used to assess "circular" strategies:

- The waste hierarchy ranks waste management options according to what is best for the environment (DEFRA, n.d.).
- The Butterfly Diagram (Figure 6.4) visualizes the circular loops that products, components and materials can take at the end of their life as opposed to disposal. The model distinguishes between biological and technical materials, and although batteries are mostly related to the technical hemisphere, a recycling method called bio-hydrometallurgy involving bacteria to extract valuable metals relates it to the bio-hemisphere as well.
- Figure 6.6 by PBL further defines the looping strategies and additionally ranks them according to circularity. PBL's framework

has an additional three strategies – refuse, rethink and reduce – that are ranked even more circular and are important to evaluate. The choice of product recovery strategy will usually depend on the product type and its quality. These frameworks are not battery specific, but they can act as guidelines.

The principles of the circular economy are not wholly novel and share values with other approaches and ideologies. As a result, many businesses that embrace lean manufacturing, for instance, might also be unknowingly applying circular economy techniques and concepts.

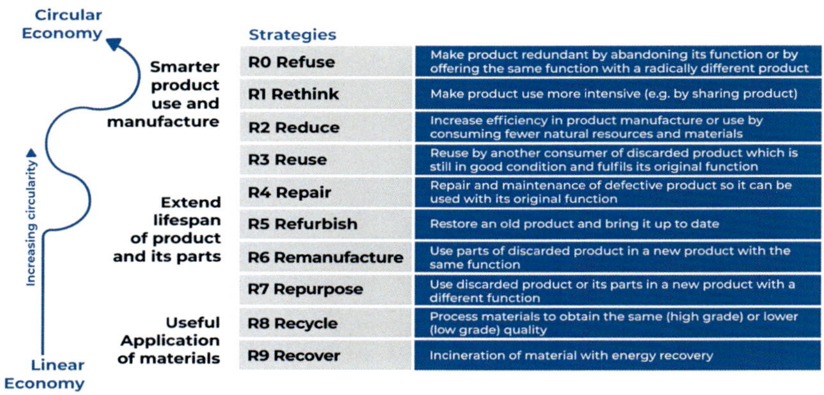

Figure 6.6. *The 9R framework.*

Source: Adapted from PBL (2017); https://autorecyclingworld.com/wp-content/uploads/2020/07/EoL-Strategies-for-EV-LIBs-2020-HSSMI.pdf.

Maximizing the recovery of a product's inherent value in terms of raw materials (such as platinum, gold, copper, steel, plastic, lithium, etc.) is the goal of the circular economy, embedded energy, and financial value with the least amount of waste and expense. The end-of-life care, product use, and business model are all included in the principles. EVs emit no emissions from their tailpipes, but the process of producing batteries and extracting raw materials is very carbon-intensive, particularly in certain nations. To counteract these initial emissions, it is crucial to maximize the raw material's lifetime and recovery. In contrast, the average emissions from the production of an EV are currently 1.3–2 times higher than those from the production of an internal combustion engine (ICE) car (European Environment Agency, 2018).

As modifications to a product and its process are very difficult to implement once decided, it is important to thoroughly evaluate circularity at the design stage. Below are some principles of circular product design:

- Long-lasting and slow degradation;
- Standardized and compatible products and components;
- Ease of maintenance and repair;
- Upgradeability, modularity, and adaptability;
- Design for disassembly;
- Design for test; and
- Easily recycled materials.

Based on available data, manufacturers who implement circular strategies at an early stage of the design process can reap significant benefits for users in terms of economic and operational aspects, as well as better alignment with customer demands and environmental benefits.

The Faraday Institution (2019) projects that by 2040, there could be 60,000 more full-time equivalent (FTE) jobs in the automotive sector and its battery supply chain in the UK alone. In addition, the production of conventional ICE vehicles will be producing fewer than half of the 83,000 full-time equivalent jobs in the EV and battery manufacturing sectors. These positions will be in the supply chain, R&D, EV manufacturing, and battery manufacturing (The Faraday Institution, 2019). In addition, the European Commission (2019b) projects that by 2025, the circular economy design will propel the entire value chain to reach €250 billion in Europe. This expansion shows that OEMs have a quickly developing chance to acquire a competitive edge by implementing circular economy strategies. The particular EoL tactics for LIBs will be discussed in the following section.

6.1.1. End-of-Life Strategies

The procedures shown in Figure 6.9 are part of the processes for recovering EV batteries; each step has expenses and difficulties of its own. The advantages and difficulties of circular EoL strategies for EV batteries are discussed in this section, along with the OEM application opportunities. PBL (Figure 6.6) will address them in the same sequence as the 9R framework.

6.1.1.1. R3 – R4: Battery Repair & Reuse

Reuse—not to be confused with repurposing (R7)—is the process of using a battery, either fully or partially, for the original application for which it was intended, possibly following maintenance. Repurposing is the use of something for a different purpose than it was intended. The battery would still be covered by warranty if it were reused, but it wouldn't last as long as it would have (Figure 6.7) (European Commission, 2019a).

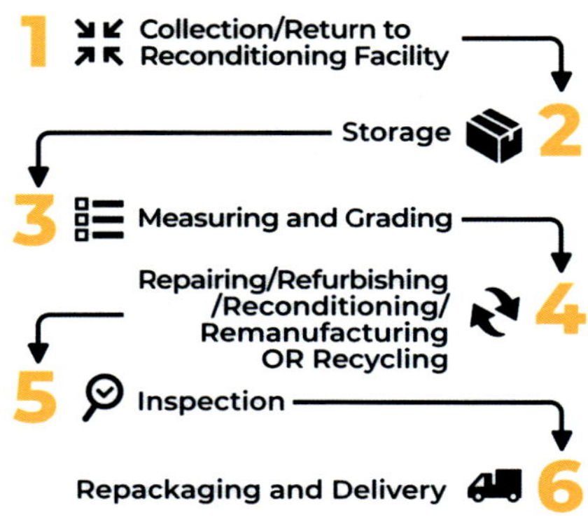

Figure 6.7. *Common steps in second life application.*

Source: https://autorecyclingworld.com/wp-content/uploads/2020/07/EoL-Strategies-for-EV-LIBs-2020-HSSMI.pdf.

6.1.1.1.1. Benefits

- The least energy and resource-intensive;
- The most economically and environmentally effective option for the life-cycle extension / end-of-life treatment of batteries; and
- Costs are well below 25% of the cost of producing a new battery.

6.1.1.1.2. Challenges

- For battery reuse, the costs are mainly related to the battery collection and sorting/grading steps; and
- Limited cases, as not many EV batteries have been through their first life cycles yet.

6.1.1.2. R5: Battery Refurbishment/ Reconditioning

The terms "refurbishment" and "reconditioning" are synonymous. LIBs can be refurbished, though little is known about these procedures at this time. Refurbishment is the process of restoring a used battery to a functional state, even in cases when there are no visible or reported defects in the cells or modules that are obviously older or used.

The first step in refurbishing an EV battery is to partially disassemble the battery pack. The next stage is to determine which cells are dead and replace them with new ones that can hold a sufficient charge. The battery pack is then reassembled, either in its original configuration or in a modified one that works better for the intended use. To properly identify the EV battery chemistries and designs, this process entails diagnostic and screening tests.

Reconfigurable battery packs that are specifically made for modular disassembly or cell replacement have been the subject of a few studies in the literature, but not many have investigated these concepts on an industrial scale. The reconditioned battery will have a shorter warranty than a new or remanufactured product, but it will still cover the entire item. The product's performance will not be as good as a new one.

6.1.1.2.1. Benefits

- Preserves most of the battery as a whole and only individual cells or modules are switched out.

6.1.1.2.2. Challenges

- Limited information on the practical implications of refurbishment;
- Current battery designs make this very difficult to achieve; and
- Costs may outweigh the value.

6.1.1.2.3. Examples

- Nissan refurbishes LIBs from the Leaf model in Japan (Kelleher Environmental, 2019);
- Spiers New Technologies specializes in among other things refurbishment; and
- Informal businesses refurbish batteries (Kelleher Environmental, 2019).

6.1.1.3. R6: Battery Remanufacturing

Remanufacturing is putting a used product back to work as good as new, if not better, and providing a warranty that matches or exceeds that of a newly manufactured product. Testing, disassembling, repairing, and replacing parts are all part of the remanufacturing process. The finished product is also tested to make sure it adheres to the original design specifications. Because of the use of newer components, performance following remanufacturing is anticipated to be at least as good as the original performance specification, if not better.

6.1.1.3.1. Benefits

- Can be returned to "new" or "better than new" standard/warranty (European Commission, 2019a); and
- The cost is 10% less than the cost of a new pack (European Commission, 2019a).

6.1.1.3.2. Challenges

- Non-standard remanufacturing processes make remanufacturing complex;
- Legislation and warranty are unclear. E.g.: lack of distinction between "new" and "better than new" battery packs if older battery cells degrade faster; and
- If third parties are remanufacturing the battery, at what stages do OEMs and third parties take responsibility? Who will take the liability if a remanufactured battery causes a serious risk to users?

6.1.1.3.3. Examples

- Nissan remanufactures LIBs from the Leaf model (Reuters, 2018);
- Spiers New Technologies specializes in among other things, remanufacturing; and
- Research project: Reclamation, Remanufacture of Lithium-ion Batteries (R2LiB) (UKRI, 2020).

6.1.1.4. R7: Battery Repurpose

Repurposing a battery, which is sometimes mistaken for reuse, is the act of using a battery entirely or in part for a purpose other than its intended one. Repurposing batteries that no longer meet the operating cycles and performance requirements of automobile use (80%) but are still ideal for other applications, like stationary energy storage, has been the main focus of existing second-life applications. Second-life batteries are expected to last an additional 12 years in energy storage, but their lifespan can potentially reach 30 years based on the application (Figure 6.8) (Casals et al., 2019).

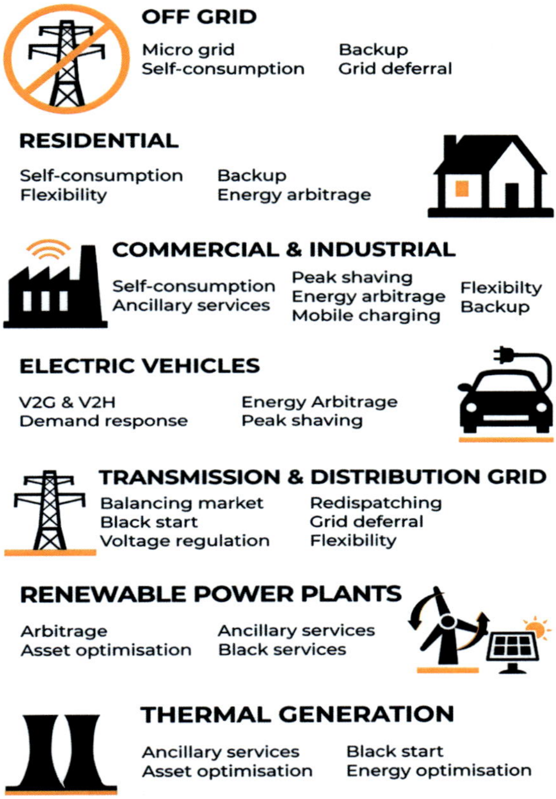

Figure 6.8. *Overview of potential battery repurpose use cases.*

Source: Adapted from Reid and Julve (2016); https://autorecyclingworld.com/wp-content/uploads/2020/07/EoL-Strategies-for-EV-LIBs-2020-HSSMI.pdf.

6.1.1.4.1. Benefits

- Bringing down the cost of energy storage can help integrate more renewable energy into the grid (European Commission, 2019a); and
- 25% of the cost of a new battery (European Commission, 2019a);

6.1.1.4.2. Challenges

- The state of health of the returned battery is uncertain and will dictate which repurposing applications are suitable (Jiao, 2019); and

- Some repurposing applications for lower capacity functions require further disassembly, adding complexity (Harper et al., 2019).

6.1.1.4.3. Examples

- Audi uses in factory vehicles (Audi, 2019a);
- Audi uses for factory energy storage (Audi, 2019b);
- Nissan and Volvo used for home energy storage (Volvo Buses, 2018; EATON, 2019); and
- Volvo uses mobile charging units (Volvo Cars, 2018).

6.1.1.5. R8: Battery Recycling

Since it is illegal to dispose of or incinerate any EV battery, they will eventually be processed for metal and material recycling. The LIB must be sent to a recycling facility in order to recover its basic materials when it can no longer be repaired or used. Lithium, nickel, cobalt, manganese, copper, aluminum, and iron are among the metals that were found. Both pyrometallurgy and multistage hydrometallurgical processes are used to recover these elements; biohydrometallurgical processes are also being studied (Velázquez Martínez et al., 2019). Figure 6.9 provides the detailed steps for the aforementioned processes (Swain, 2017).

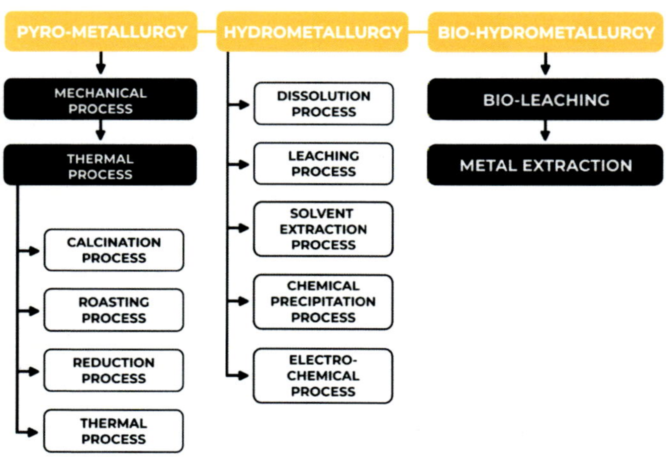

Figure 6.9. *Steps of pyro-metallurgy hydrometallurgy, and bio-hydrometallurgy.*

Source: Adapted from Swain (2017); https://autorecyclingworld.com/wp-content/uploads/2020/07/EoL-Strategies-for-EV-LIBs-2020-HSSMI.pdf.

6.1.1.5.1. Benefits

- Generally, recycling methods can recover up to 70%–90% of metals (Velázquez-Martínez et al., 2019);
- Lab tested methods combining mechanical, hydrometallurgical and pyrometallurgical processes can recover up to 99% of metals (Velázquez-Martínez et al., 2019); and
- From a material recovery point of view, cobalt, lithium, copper and nickel are the most valuable materials embedded in the LIB and can offer a financial opportunity (Lebedeva et al., 2016; European Commission, 2019a).

6.1.1.5.2. Challenges

- Due to the complexity of materials contained within batteries, their multiple chemistries and very low concentrations, many important metals (e.g., lithium, cadmium, manganese, rare-earth elements) are not recovered during mainstream recycling processes (Velázquez-Martínez et al., 2019); and
- The processes to recover valuable materials can be expensive to operate and highly polluting (Velázquez-Martínez et al., 2019).

6.1.1.5.3. Examples

- Umicore and SNAM are two of the larger recycling companies in Europe, receiving the majority of batteries from across the 28 member states.

6.2. ELECTRIC VEHICLE BATTERY END-OF-LIFE MANAGEMENT

6.2.1. Battery End-of-Life Definition

When a product no longer functions as intended, the battery must be retired owing to damage, or it no longer runs at a sufficient capacity, it has reached its End-of-Life (EOL) status. It's crucial to understand that the term EOL in this context refers to the point at which an electric vehicle battery no longer functions as intended because of performance decline, physical damage, or other safety-related issues.

In order to assess a battery's condition and determine whether it has reached the end of its intended first use, two commonly used metrics are Battery State of

Health (SOH) and State of Charge (SOC). Battery SOH is the total maximum capacity that a battery can have at any given time over its rated capacity, which is its initial life capacity. Conversely, SOC is the capacity that remains after accounting for total capacity (also known as rated capacity). SOH deteriorates with time; when the SOH falls to 70–80%, it is generally accepted that an EV battery has reached the end of its initial useful life. However, because batteries are intricate electrochemical systems with nonlinear degradation behaviors, a standard concept to specify a battery at EOL has not been agreed upon or defined. Determining the EOL of a battery can be difficult because it depends on application requirements and design specifications.

A battery that has reached the end of its intended use is taken out of the electric vehicle by the responsible party, who then evaluates it and decides whether to recycle it or put it to another use by reusing or repurposing it.

6.2.2. The Need for EV Battery End-of-Life Management

Globally, EV sales are increasing as a result of incentives linked to decarbonization and shifting consumer demand. In 2021, there were 16.5 million electric vehicles on the road globally. By 2050, it is anticipated that there will be 700 million EVs worldwide. "By 2050, all automobile sales in Europe (86%), China (81%), and North America (78%) will predominantly be electric vehicles," states the same report. While switching to EVs will lower net carbon emissions, if used EV batteries are not properly managed after their initial use, they can have a negative environmental impact.

Specifically designed to provide electric power for the traction to hybrid or electric vehicles of L category as provided for in Regulation (EU) No 168/2013, and with a weight above 25 kg, or designed to provide electric power for the traction to hybrid or electric vehicles of M, N, or O categories as provided for in Regulation (EU) 2018/858, is what is referred to as an "EV battery" for this project. Lithium-ion batteries are specifically the focus of this document's discussion on batteries.

The phases of an EV battery's life cycle that produce the greatest waste and emissions are those related to sourcing materials, manufacturing, and end-of-life operations. As per the CalEPA, it is imperative to establish a circular battery supply chain and mitigate its environmental impact by augmenting battery recycling, thereby enabling the extraction of raw materials for the production of new electric vehicle batteries. Reusing and repurposing batteries is crucial for achieving circularity, cutting down on waste, and replenishing available resources.

There are currently many different EV battery recycling technologies available, and more LiB-focused recycling facilities are being developed. But only roughly 5% of batteries are recycled at this point. A few of the many barriers to recycling include a lack of incentives, unclear who is responsible for managing EOLs, high recycling costs, and geographic considerations like close proximity to recyclers.

6.2.2.1. Material Sourcing

Lithium, nickel, and cobalt mining, which are necessary for producing lithium-ion batteries (LiBs), is bad for the environment and frequently raises ethical questions.

6.2.2.2. Recycling

While the majority of other battery types, like lead batteries, are recycled while 99% of lead batteries can be recycled, recycling lithium-ion batteries (LiBs) is more complicated and expensive because they require disassembling hundreds of Li-ion cells, which can be hazardous due to the materials they contain. The majority of EV battery systems' lack of recycling-related design considerations is another factor lowering LiBs' potential for recycling. These batteries frequently wind up in landfills when they are not properly recycled, reused, or repurposed. This method of disposal not only increases the risk of contaminating water and soil, but it also increases the risk of fires because the batteries contain toxic and flammable materials.

6.2.3. Current Regulatory Environment

Stakeholders in the battery market are being encouraged by recent proposals for regulations and policies in the US, EU, and Asia to identify accountable parties for managing EOL, boost recycling of electric vehicle batteries, create more effective recycling techniques with higher recovery rates, and design and manufacture batteries more sustainably.

6.2.3.1. United States

Although there are federal and state regulations in the US that govern activities related to batteries' EOL, there aren't any comprehensive policies in place for batteries at that point, and it's unclear how used lithium-ion batteries are defined legally as waste.

Used batteries are typically categorized as hazardous or universal waste in the US, and as such, they must comply with current waste regulations.

Jurisdictions also differ in their regulations. The US federal hazardous waste laws and regulations, which deal with the creation, handling, storage, treatment, transportation of hazardous wastes domestically or internationally, and disposal of hazardous wastes, are among the strictest regulations. Regulations governing the exchange of battery traceability data are likewise absent from the US.

6.2.3.2. European Union

A new rule from the European Union aims to improve EV battery EOL management and include provisions for digital battery records.

6.2.3.3. China

In 2018, the Chinese government set up a platform for traceability management so that electric car batteries could be tracked over time.

A set of regulations that place this obligation on EV and battery manufacturers or importers were released by the government. Vehicle manufacturers are in charge of selling batteries to a certified handler for recycling or reuse, as well as providing technical support to these businesses. Every battery made in China or imported into the country for use in electric vehicles has a unique code attached to it so that it can be tracked and properly processed when its initial life is coming to an end.

6.3. RECYCLING OF EV BATTERIES

Recycling EV batteries is essential to an electrified transportation system that is sustainable. By 2050, recycled batteries may provide a significant amount of essential minerals for electrification, negating the need for new mining significantly.

However, recycling those batteries can have a significant impact on the environment, so we need to use recycling techniques that have a low environmental impact and high rates of mineral recovery. Several battery recycling techniques in this section, along with why doing it correctly is crucial are discussed.

The three forms of recycling, which are outlined here, are covered in more detail later on, along with the prerequisite pre-processing. We've included a terms section at the end of this chapter for the italicized words because this technology can be complicated and confusing:

- **Hydrometallurgical Recycling:** It is the best available technology because it has high mineral recovery rates and results in low environmental impacts. It uses liquid solutions to separate minerals.

- **Direct Recycling:** It is still in development but has low environmental impacts and recovers the positive electrode intact, meaning this product skips a step in the battery manufacturing process. Direct recycling has lower lithium recovery rates than hydrometallurgical recycling but is ideal for manufacturing scrap and lithium-iron-phosphate (LFP) batteries.
- **Pyrometallurgical Recycling:** (Smelting) It is the least ideal technology because it does not recover lithium, aluminum, or manganese and results in the highest environmental impact. Additionally, the output must go through an extra step of hydrometallurgical refinement before it is ready for battery manufacturing (Figure 6.10).

Recycling Processes

	Recovery rates	Low environmental impact	Recovers highest quality product	Ideal for:
Hydrometallurgical recycling	90-99% nickel, cobalt, lithium, and manganese	✓	✗	Batteries containing nickel and cobalt
Direct recycling	90% nickel and cobalt, 50% lithium	✓	✓	Lower value batteries Manufacturing scrap and unused/aged batteries
Pyrometallurgical recycling	Nickel and cobalt recovered in alloy, 0% lithium recovery	✗	✗	

Figure 6.10. *Recycling process.*

Source: https://blog.ucsusa.org/wp-content/uploads/2023/10/Battery-Recycling-Table-1500x844.png.

6.3.1. Pre-processing

An automobile dismantler takes the EV's battery out and sends it to a battery recycler for processing before recycling. The battery pack is divided into smaller modules by the recycler. Usually rectangular in shape, these modules stack together like dominoes in a box. Rectangular battery cells are contained within each module. Tesla employs a different design, with cylindrical cells in place of rectangular ones. Since pyrometallurgical recycling burns the entire battery, the material is suitable for recycling following module or cell level deconstruction. To obtain a higher-value product, hydrometallurgical and direct recycling techniques need to separate the numerous cell layers from one another, necessitating further pre-processing.

In these further stages, a mechanical shredder is used to break the module or cell into tiny pieces. After that, the output is divided using shaker tables, magnets, and sifters to separate out a powder known as "black mass" that is composed of nickel, cobalt, and lithium.

6.3.2. Hydrometallurgical Recycling

- High recovery rates: 90%–99% nickel, cobalt, lithium;
- Low environmental impact; and
- Ideal for batteries containing nickel and cobalt.

In hydrometallurgy, minerals are extracted from the powder containing valuable metals (black mass) that was recovered during the aforementioned pre-processing using liquid and chemical-based solutions. The glue (binder) that adheres the positive electrode (cathode) and negative electrode (anode) to copper foil and aluminum (current collector) is present in the black mass powder.

Electrodes are separated from the foil by recyclers using a variety of techniques, which fall into two main categories: chemical pre-treatment and thermal pre-treatment. Chemical treatments dissolve the glue with the use of an acid; residues must be burned off by additional procedures like thermal treatment. By heating and burning off the glue and negative electrode, thermal pre-treatment alone can also remove the positive electrodes from the foil. Scrubbers and other additional purification equipment are required because burning produces gases.

Though it differs greatly from the smelting described in the pyrometallurgical section below, this thermal treatment is technically regarded as a pyrometallurgical recycling process. It operates at lower temperatures, so the cobalt, lithium, and nickel that we are recovering are not altered.

The mineral-containing powder is then what the recycler is left with. This is first dissolved in a liquid (leaching), usually a mixture of sulfuric acid and hydrogen peroxide. The next step is to extract the precious metals from the mixture. Solvent extraction is one strategy. This method naturally separates the minerals by using two liquids with different solubilities (think water and oil). As a result, the minerals separate from the liquids to make them easier to collect.

Water treatment is required for the wastewater produced by this process. Some operations within the facility reuse this water, but not all of them do.

Furthermore, a significant amount of sodium sulfate, a low-value and safe byproduct used to make household products like detergent, is produced during the process.

The chemicals and procedures employed in hydrometallurgical recycling determine the material output. Some companies recover a mixture of minerals, such as nickel-manganese-cobalt (NMC), and their final product is a positive electrode precursor. Since the battery manufacturer that purchases the recycler's product need not produce the positive electrode precursor, recovering these minerals jointly is optimal. Alternative techniques retrieve the separate elements (carbonate salts, nickel sulfate, cobalt, and lithium sulfate) needed to make a fresh electrode. Although recovery rates differ depending on specific processes, one business claims to have a 99% cathode material recovery rate (Figure 6.11).

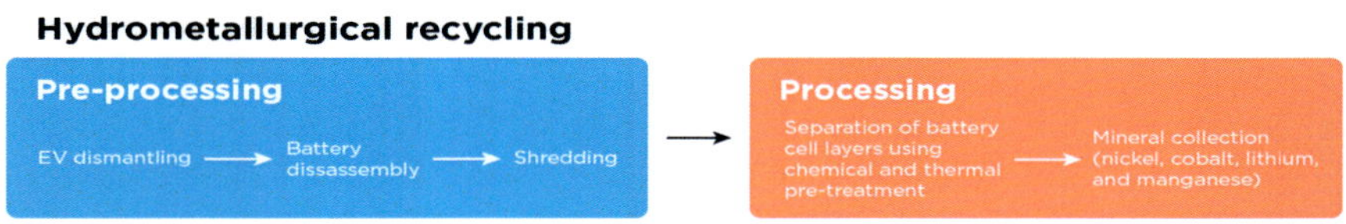

Figure 6.11. *Hydrometallurgical recycling.*

Source: https://blog.ucsusa.org/wp-content/uploads/2023/10/Hydro-Recycling-Flowchart.png.

6.3.3. Direct Recycling

- High recovery rates: 90% nickel and cobalt, 50% lithium;
- Lowest environmental impact;
- Recovers the highest quality product; and
- Ideal for:
  - Lower value batteries such as lithium-iron-phosphate; and
  - Manufacturing scrap and unused/aged batteries.

A revitalized positive electrode produced by direct recycling is prepared for use in the production of new batteries. Although this method is still being developed, batteries without lithium-iron-phosphate (LFP) or other high-value cobalt and nickel minerals will probably benefit the most from it. Hydrometallurgical recycling is not a cost-effective method of recovering individual minerals from lithium-ion batteries due in part to the lower input costs. Since LFP recovers the positive electrode and is a far more valuable product than just recovering the individual minerals, recycling it using direct recycling is profitable.

Direct recycling is best suited for waste material straight from manufacturing in higher-value lithium-ion batteries, like NMC, because battery use damages the electrode's crystalline structure.

Batteries are shred before being recycled. The recovered black mass is handled in a manner akin to hydrometallurgical recycling, with minor modifications made to the glue removal and positive/negative electrode separation processes. For instance, instead of burning the negative electrode, which is usually graphite, recyclers use the heated graphite's hydrophobic properties to draw it to the liquid solution's surface (the positive electrode is hydrophilic—repelled from the bubbles!). The other minerals remain below the foam that contains the graphite. We refer to this procedure as froth flotation.

Since the recovered positive electrode degraded and lost lithium during its initial use in a battery, it does not contain the required amount of lithium. As a result, additional lithium must be added (relithiation). In addition, the recovered positive electrode can be altered or upcycled by adding extra minerals (e.g., quartz) to guarantee recycled materials are still usable as lithium-ion battery chemistries continue to change. e.g., increasing the cobalt) (Figure 6.12).

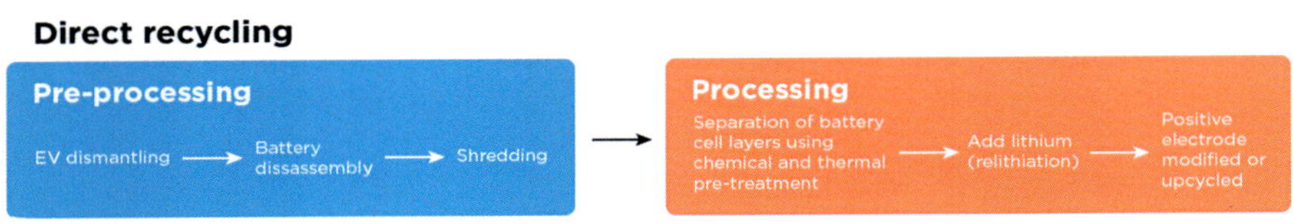

Figure 6.12. *Direct recycling.*

Source: https://blog.ucsusa.org/wp-content/uploads/2023/10/Direct-Recycling-Flowchart.png.

6.3.4. Pyrometallurgical Recycling

- Low recovery rates: nickel and cobalt recovered in the alloy, 0% lithium recovery;
- Highest energy use and greenhouse gas emissions.

Pyrometallurgical recycling is a broad term that encompasses all technologies that extract and purify metals using high temperatures. Technically, a number of the direct recycling and hydrometallurgical processes covered fall into this group. They still employ lower temperatures than the recycling through pyrometallurgy discussed in this section. Smelting is the term for this kind of high-temperature pyrometallurgy. Smelting uses a lot of energy to reach the required temperatures and does not recover all minerals.

In order to separate metals while they are still liquid, battery materials must be heated above their melting point. Since the battery module or cell is placed directly into the furnace, there is no need to shred the battery. To burn off the electrolyte, the battery is heated to 350°C–600°C. The metals are then melted into an alloy at a temperature of 1200°C –1450°C. Because of the harmful gases produced during this process, the recycler incorporates off-gas treatment. A metal alloy consisting of iron, nickel, copper, and cobalt is the end product. Lithium, aluminum, and manganese in the slag are lost during smelting. To recover the cobalt and nickel from the metal alloy, a hydrometallurgical process must also be performed. Much of the lithium is lost in the process as dust. Although the existing recyclers aren't pursuing it, lithium can be recovered through an additional hydrometallurgical process (Figure 6.13).

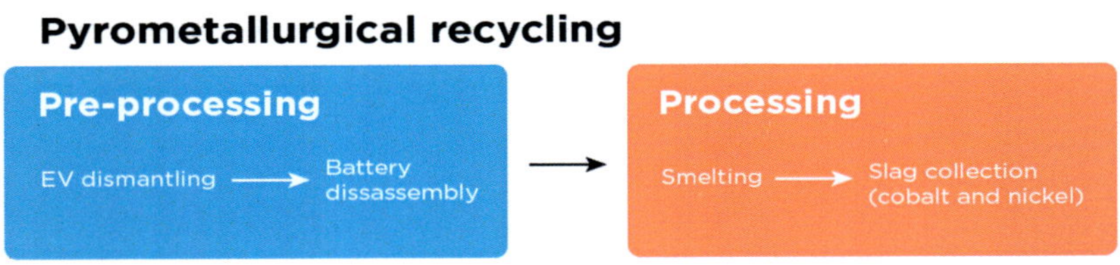

Figure 6.13. *Pyrometallurgical recycling.*

Source: https://blog.ucsusa.org/wp-content/uploads/2023/10/Pyro-Recycling-Flowchart-1500x274.png.

6.3.5. Battery Recycling Policy

Ensuring the safe recycling of all EV batteries would be possible with an effective recycling policy. Ideally, the US would implement extended producer responsibility (EPR) in line with our international allies. Automakers are obligated by EPR to recycle every battery. Automakers are well-positioned to establish a network for collecting retired batteries and design batteries that are easier to recycle, both of which are essential components of a sustainable supply chain and transportation system. However, as this section makes clear, recycling methods also count.

Recyclers need to recover as much of the minerals as possible with the lowest potential impact, which means using a hydrometallurgical or direct recycling method instead of the pyrometallurgical alternative.

Thanks to research centers like ReCell, the recycling sector is developing swiftly. Thus, rather than specifying the technology pathway, we propose that policy establish baseline recovery rates. The European Union (EU), for instance, has adopted this strategy. According to the EU Battery Law, recycling procedures must achieve a recovery rate of 50% for lithium and 90% for cobalt, nickel, and copper by 2025; by 2030, these percentages must rise to 80% and 95%, respectively. California is working on a recycling policy, and we hope the state will take a similar tack.

6.3.6. How to Recycle an EV Battery

The number of EVs sold is still rising. In the first three months of 2023 alone, 2.3 million EVs were sold worldwide, a 25% increase over the same period in 2022, according to the Paris-based International Energy Agency. At this rate, there could be more than 250 million EVs on the road worldwide by the end of the decade, accounting for more than 60% of newly sold automobiles (Figure 6.14).

Figure 6.14. *Discarded LIBs can overheat and catch on fire, while also posing an environmental threat due to toxic metals. So, what to do with them? Researchers generally talk about either recycling or reusing as part of less-demanding applications. But the right path to technical and economic viability is still unclear.*

Source: https://www.pnas.org/cms/10.1073/pnas.2400520121/asset/e7a71f01-c61b-4a11-8dcc-2d024adadcb3/assets/images/large/pnas.2400520121unfig01.jpg.

However, an issue is about to arise: EV batteries typically have a 10–20 year lifespan, and businesses are unprepared for the impending glut of tens of millions of used batteries. Furthermore, experts estimate that complete decarbonization will need more than 100 terawatt-hours of battery storage, or billions of tons of new batteries, or roughly 100 times the world's current production.

LIBs, which power EVs, are not suitable for disposal in landfills due to the possibility of overheating and fire. Hundreds of fires in the US have been started by smaller LIBs, sometimes in the back of garbage trucks. Another danger to the environment is EV batteries that are disposed of. Hazardous metals may seep into the environment if they are allowed to decompose in a landfill.

When it comes to what to do with them, researchers usually discuss two options: recycling and reusing in less demanding applications like golf carts or grid storage. Because the latter could, in theory, immediately result in a new supply of batteries and a source of income for businesses, scientists are working hard to create sustainable and affordable battery recycling technology. Head of the Electrification Section at Oak Ridge National Laboratory in Tennessee and chemical engineer Ilias Belharouak, who has been working on high-density LIBs for more than 20 years, says that every car company is doing research and development on recycling.

He says the fundamentals are covered. While some of the required procedures are currently in place, they are insufficiently effective to deal with the impending flood. "We're attempting to examine these current procedures, but we're also investigating approaches that don't currently exist in the industry," says Belharouak. Using parts from old batteries instead of freshly mined materials could be beneficial for all parties involved because it reduces the risk to the environment, offers a consistent supply of materials, and is profitable.

6.3.6.1. Lead and Lithium

LIBs, which were created in the 1970s, have a long runtime between charges due to their high energy density. When Sony started selling the first commercial LIB in 1991, they became available on the market. "However, recycling was never really discussed," Belharouak claims. "They never imagined the market would blow up and reach these heights of growth. Even then, he claims, recycling was primarily the responsibility of business."

When focusing on the best recycling techniques, scientists and engineers would be well to learn from history about what to pursue and what not to. The 1850s saw the development of the rechargeable lead-acid battery, and the 1920s saw the beginning of techniques for the large-scale recovery of lead. Since then, it has continued to be a workhorse; modern cars still use it for lights and ignitions. Lead-acid batteries can be recycled almost entirely. Most estimates place the recycling rate at higher than 95%. Lead and plastic recovered from old batteries can be readily reused in new ones. According to estimates from the

US Environmental Protection Agency, 99% of battery lead in the country was recycled in 2018, the most recent year for which data was available. However, it's not a clear-cut win: unsafe battery recycling plants, particularly in low- and middle-income nations, have resulted in widespread lead poisoning and what some experts describe as a serious environmental threat to children's health.

Recovering the valuable metals, like cobalt and lithium, without having to pay significant costs to the environment or human health is a big challenge for LIBs. With billions of tons of batteries expected in the future, experts must also consider recycling the less valuable (but still progressively more expensive) components, like copper and aluminum. In addition to being a much younger industry, recycling LIBs is more difficult than recycling lead-acid batteries. According to a widely discussed and cited 2013 report, only roughly 5% of LIBs were recycled. Though the number may have increased since then, Belharouak notes that it is difficult to estimate global rates because different companies define recycling of EV batteries differently and because the majority of recycling occurs in China, where information on recycling rates is not readily available.

Figure 6.15. *Researchers are working on the best ways to recycle EV battery components, including those seen here postshredding: (left to right) black mass (a mix of anode and cathode materials), plastic, and metals.*

Source: https://www.pnas.org/cms/10.1073/pnas.2400520121/asset/512b686d-abd2-451d-b43c-d7356de413c1/assets/images/large/pnas.2400520121unfig02.jpg.

But he issues a warning: the current approaches are ineffective, costly, and harmful to the environment. Disassembling LIBs is more difficult due to their greater complexity and quantity of components compared to traditional lead-acid batteries. According to him, industries haven't fully considered the other ingredients, like copper, aluminum, and the electrolyte, but the most popular recycling techniques concentrate on cobalt, nickel, and lithium (Figure 6.15). "Trying to expand these sectors while continuing to deal with the same problems will amplify the issues that are related to them," he claims."

6.3.6.2. Fire and Acid

The first step in recycling is material separation. Commercial approaches usually employ two main techniques, which are occasionally combined: hydrometallurgy, which extracts the recoverable metals from a soup of liquids (typically acids), and pyrometallurgy, which removes metal by intense heat.

Used batteries are disassembled, crushed, and heated—usually above 500°C—in order to perform pyrometallurgy. As a result, the lithium salts are left behind when the liquid portion of the electrolyte evaporates. They get even hotter. Metals like cobalt, nickel, copper, and iron melt at temperatures above 1,400°C and can be recovered and purified for further use. Some metal oxides decompose into beneficial substances.

Lithium is hard to extract and frequently stays in the slag. However, in recent years, scientists have discovered that by including specific salts, they can improve the recovery of manganese, which is also typically lost at this point, as well as lithium. It is reported that 90–100% of the lithium was extracted through experimental testing, which is a positive step. Nevertheless, the use of chemicals in this procedure creates more waste streams.

Expanded applications of pyrometallurgy include nickel-metal hydride batteries and other types of rechargeable batteries. However, it uses a lot of energy and generates hazardous byproducts for the environment, including greenhouse gases.

The other process, hydrometallurgy, can be done well below 100°C, so it doesn't require as much energy. According to Belharouak, the process essentially involves breaking down the lithium ion cells into acids, primarily sulfuric acid. The batteries become a caustic liquid as a result, which allows the metals in the cathode, including lithium, to precipitate out.

Nevertheless, hydrometallurgy has waste issues of its own. "This process generates a lot of pollutants and requires a lot of solvent," says Australian chemical engineer Zaiping Guo of the University of Adelaide. Researchers in Sweden, for instance, have demonstrated how oxalic acid can be used to recover nearly all of the lithium from shredded batteries with a significantly lower environmental risk. This suggests that there may be ways to lessen the environmental impact.

Some recyclers combine the two processes, adding costly steps to the process by using hydrometallurgy to extract lithium and other cathode materials from the residue left over from pyrometallurgy.

6.3.6.3. Full Circle

Furthermore, it's likely that neither pyrometallurgy nor traditional hydrometallurgy will be able to resolve the recycling problem. "If you focus solely on those two procedures, you will only see minor improvements—no revolutionary breakthroughs," claims chemical engineer Jeffrey Spangenberger of Chicago's Argonne National Laboratory. According to Guo, what's required is a closed-loop, eco-friendly system—a set of procedures that can mine almost all of the metal, of all kinds, from old batteries without creating new, problematic waste streams. Closed-loop recycling has the potential to improve air quality and the climate because it emits no nitrogen oxides or carbon dioxide.

Ascend Elements was founded in Westborough, Massachusetts, and materials scientist Eric Gratz says his company's strategy accomplishes those goals. He refers to it as "direct precursor synthesis," and it's a modified hydrometallurgical procedure. The same method of tearing apart the batteries is used. Ascend Elements, on the other hand, employs chemicals to leach out lithium as well as impurities, leaving behind nickel, cobalt, and manganese in an aqueous solution, in contrast to conventional hydrometallurgy, which uses chemicals to remove the metals. According to Gratz, the solution's precise ratios can be changed by gradually incorporating recycled or virgin materials. He claims that the solution can be used to create new cathode precursor material in the correct amounts.

"We retrieve our water; we retrieve every major component of the lithium-ion battery," explains Gratz. "Water is gathered and utilized again. Chemical waste does not exist. In addition, the company gives recovering all the metals—not just lithium—priority. According to Gratz, "all of the materials—copper, aluminum, plastics, graphite, and cathode material—are recycled.." New batteries made with the recycled material frequently had longer cycle lives than batteries made with virgin materials extracted from ore, according to results from lab tests of the approach published in Joule in November 2021. The researchers, led by Yan Wang, a co-founder of Ascend Elements, observed that the recycling process altered the cathode material's crystal structure by adding bigger voids and more surface area, which probably helped to increase efficiency.

A third technique, also known as direct recycling, uses used batteries to replenish the lithium ions in spent electrodes in order to revitalize them without the need for any caustic acids at all. Belharouak claims that people "become a little aged at the end of life.." It is possible for cathode materials to deteriorate and lose some of their lithium content, but if the damage is not too great, it may be possible to stop the process. He claims that they can be heated to a high temperature and combined with lithium hydroxide to restore them. The procedure can fix minor surface damage in addition to restoring the lithium content. However, he points out that it's difficult to standardize because different batteries require different amounts of lithium in order to restore the cathode to full function. A comparable method of direct recycling is being pursued by Princeton NuEnergy in New Jersey, which involves treating spent cathodes and anodes with low-temperature plasma to get rid of impurities and make the material suitable for use in new batteries.

However, Belharouak says it's too early to tell if it will become popular. To persuade battery manufacturers to use recovered materials, there are issues that need to be resolved, such as ensuring the materials' quality (Figure 6.16). "Ultimately, it all comes down to performance," he states, adding that a manufacturer might decide not to use recycled material if it doesn't provide the same level of performance."

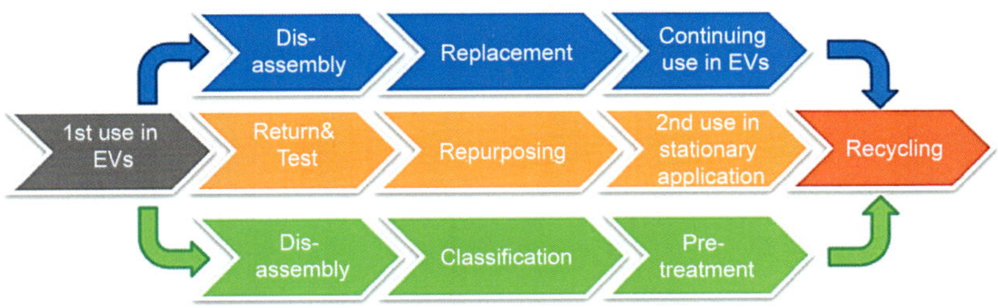

Figure 6.16. *Researchers are finding ways to reduce the environmental impact of using lithium in new electrode materials.*

Source: Kim H-J, Krishna T, Zeb K, Rajangam V, Gopi CVVM, Sambasivam S, Raghavendra KVG, Obaidat IM. A Comprehensive Review of Li-Ion Battery Materials and Their Recycling Techniques. Electronics. *2020; 9(7):1161. https://doi.org/10.3390/electronics9071161.*

6.3.6.4. Recharging Ahead

In an effort to address these issues, the US Department of Energy established the ReCell Center in February 2019 at Argonne National Laboratory. It centers on studies and advancements that might result in the kind of closed-loop system that Spangenberger and associates have in mind. Its primary goal is direct recycling. Better methods to recover and repurpose leftover electrode material are another area of focus for the DOE program. An aluminum foil or copper roll is used as the thin current collector, and the manufacturer coats it with either graphite for the anode or cathode material for the cathode to create the electrodes for a lithium-ion battery. Subsequently, they cut out designs based on the dimensions and form of the battery, discarding the remaining material.

If those scraps can be used to make new batteries, Belharouak claims they present an unrealized potential. Usually, they are thrown away as waste. He and his associates published the results of a successful test in April 2021 of an inexpensive, recyclable, and environmentally friendly solvent that effectively separates the coating from the foil that collects current. Graphite or cathode material can be recycled, and foil can be repurposed to create a fresh roll that can be cut again to create new electrodes. Three things need to happen, according to researchers, for EV battery recycling to be effective and desirable to manufacturers. According to Spangenberger, a successful procedure must first attract EV companies by providing them with incentives to recycle their old batteries in addition to producing batteries that they will actually use. "A fantastic way to increase recyclability in the future is to collaborate with manufacturers to create batteries that are simpler to recycle," he says.

Second, technology must scale up, according to Belharouak. Ascend Elements launched a sizable recycling plant in Covington, Georgia, in March 2023. The company estimates that when it reaches full capacity, it will be able to process batteries from 70,000 electric vehicles annually. However, he claims that each state would require a similar factory in order to meet demand. In 2021 alone, four million EVs were sold.

Lastly, experts concur that future battery designs must take recycling into account. Take the LIB in the 2017 Toyota Prius Prime, for instance, which has a battery pack made up of five modules with 19 connected cells apiece. The smallest unit of measurement is a cell. As of right now, replacing the entire battery is typically required for repairs when one cell in a bulky Prius battery pack fails. The damaged cell, or at least the module, could be removed with a more recyclable design without destroying the pack as a whole.

When EV warranties expire, batteries will start to accumulate; initially slowly, then probably quickly, similar to how EVs have been rolled out over the past 10 or so years. According to Gratz of Ascend Elements, chemists and companies will have ample time—possibly decades—to devise strategies for accelerating the technology due to the extended lifespan of contemporary EV batteries. He predicts that as battery technology advances, most of the materials used to make batteries will eventually be recycled rather than extracted from the earth. He points out that Ascend Elements primarily supplies cathode materials and that the company discovered early on that obtaining metals from recycled batteries would result in lower production costs than sourcing new ones.

Even though other recycling businesses are growing and reaching new markets in a similar way, they won't be able to satisfy the entire demand in the future. The Canadian recycling company Li-Cycle opened a facility in Germany in August 2023 that can shred LIBs of any kind and create "black mass," which can then be utilized to create new cathodes. In October 2022, a pilot recycling facility in McKinney, Texas, was established to evaluate novel direct cathode recycling technology created by Princeton NuEnergy researchers.

However, how batteries themselves develop in the coming years will have an impact on the recycling problem. For instance, Toyota has stated that it will release an EV in the coming years that utilizes a solid-state battery, which uses a solid electrolyte rather than a liquid one as the name implies. Because they won't need to go through as many cycles of charging and discharging, which weaken batteries, these batteries are expected to have an energy density more than twice that of today's LIBs. The investigation into recycling them and utilizing lithium from today in tomorrow's solid-state batteries is still in its early stages. It will be a significant task in any case, according to Belharouak. The anode in conventional LIBs is made of graphite. Instead, he claims that solid-state batteries use lithium metal, which becomes extremely reactive in the presence of air and moisture, complicating the recycling process. He asserts that any procedure must adjust to these difficulties. Belharouak also highlights the rising interest in EV battery "second-life" applications. For instance, the Natural Resources Defense Council outlines potential uses for used EV batteries in a report that was released in July 2023. These uses include grid energy storage and lower-demand vehicles like golf carts. Although it's not yet profitable or common, doing so shouldn't be too difficult. Used EV batteries are easily found online, and a few new businesses are creating gadgets that connect empty EV batteries to store extra energy, like solar power. In Leipzig, the German automaker BMW has installed used batteries from its cars in a battery storage farm that is grid-connected and stores energy from surrounding wind turbines. Using 148 new and used Leaf car batteries, the automaker Nissan created a power system for an Amsterdam stadium in 2018. The batteries gathered energy from solar panels mounted on the arena's roof. Reusing batteries extends the life of current batteries but won't eliminate the need for EV batteries or improved recycling technology, according to Belharouak. Better recycling and battery designs will help EVs grow no matter how the field develops, according to Spangenberger. "More people being able to drive electric vehicles is a good thing, and recycling should make it cheaper to make new batteries," he says. It could contribute to global decarbonization."

REFERENCES

1. Acis Energy. (2020). Solar PV systems–Generate your own free electricity. Retrieved from: http://acisenergy.com/ (accessed on 9 November 2024).
2. Bai, Y., Essehli, R., Jafta, C. J., Livingston, K. M., & Belharouak, I. (2021). Recovery of cathode materials and aluminum foil using a green solvent. *ACS Sustainable Chemistry & Engineering, 9*, 6048–6055.
3. Ballantyne, A. D., Hallett, J. P., Riley, D. J., Shah, N., & Payne, D. J. (2018). Lead acid battery recycling for the twenty-first century. *Royal Society Open Science, 5*, 171368.
4. Baum, Z. J., Bird, R. E., Yu, X., & Ma, J. (2022). Lithium-ion battery recycling─Overview of techniques and trends. *ACS Energy Letters, 7*, 712–719.
5. BloombergNEF (BNEF). (2020). *Electric Vehicle Outlook 2020*. Retrieved from: https://about.bnef.com/electric-vehicle-outlook/ (accessed on 9 November 2024).
6. Brinn, J. (2023). *Building Batteries Better: Doing the Best with Less* (Report No. 23-08-9). National Resources Defense Council.
7. Casals, L. C., Amante García, B., & Canal, C. (2019). Second life batteries lifespan: Rest of useful life and environmental analysis. *Journal of Environmental Management, 232*, 354–363. https://doi.org/10.1016/j.jenvman.2018.11.046.
8. Casals, L. C., Amante García, B., & Cremades, L. V. (2017). Electric vehicle battery reuse: Preparing for a second life. *Journal of Industrial Engineering and Management (JIEM), 10*, 266. https://doi.org/10.3926/jiem.2009.
9. Centre for European Policy Studies (CEPS). (2018). *Prospects for Electric Vehicle Batteries in a Circular Economy*.
10. Denchi Power. (2020). Denchi Power Ltd–Home. Retrieved from: https://www.denchipower.com (accessed on 9 November 2024).
11. Eaton. (2019). xStorage Home | Energy Storage. Retrieved from: https://www.eaton.com/gb/engb/catalog/energy-storage/xstorage-home.html (accessed on 9 November 2024).
12. Ellen MacArthur Foundation. (2019). Circular economy system diagram. Retrieved from: https://www.ellenmacarthurfoundation.org/circular-economy/concept/infographic (accessed on 9 November 2024).
13. European Commission. (2019a). *Circular Economy Perspectives for the Management of Batteries Used in Electric Vehicles*. Ricardo Energy & Environment.

14. European Commission. (2019b). *Implementation of the Strategic Action Plan on Batteries: Building a Strategic Battery Value Chain in Europe.* Retrieved from: https://eur-lex.europa.eu/legalcontent/EN/TXT/?qid=1554816272501&uri=COM:2019:176:FIN (accessed on 9 November 2024).
15. European Environment Agency. (2018). *Electric Vehicles from Life Cycle and Circular Economy Perspectives: TERM 2018–Transport and Environment Reporting Mechanism (TERM) report*. European Environment Agency.
16. EVgo. (2018). EVgo announces grid-tied public fast charging system with second-life batteries. Retrieved from: https://www.evgo.com/about/news/evgo-announces-grid-tied-public-fast-charging-systemsecond-life-batteries/ (accessed on 9 November 2024).
17. Gao, H., Tran, D., & Chen, Z. (2022). Seeking direct cathode regeneration for more efficient lithium-ion battery recycling. *Current Opinion in Electrochemistry, 31*, 100875.
18. Global Battery Alliance. (2018). *The Lithium-Ion Battery End-of-Life Market–A Baseline Study*.
19. Groupe Renault. (2017). *Renault Optimizes the Lifecycle of its Electric Vehicle Batteries–Groupe Renault* [Webpage]. https://group.renault.com/en/news-on-air/news/renault-optimizes-the-lifecycle-of-its-electric-vehicle-batteries/ (accessed on 9 November 2024).
20. Harper, G., Sommerville, R., Kendrick, E., Driscoll, L., Slater, P., Stolkin, R., Walton, A., Christensen, P., Heidrich, O., Lambert, S., Abbott, A., Ryder, K., Gaines, L., & Anderson, P. (2019). Recycling lithium-ion batteries from electric vehicles. *Nature, 575*, 75–86. https://doi.org/10.1038/s41586-019-1682-5.
21. HM UK Government. (2017). *Regulations: Batteries and Waste Batteries* [Webpage]. GOV.UK. https://www.gov.uk/guidance/regulations-batteries-and-waste-batteries (accessed on 9 November 2024).
22. HM UK Government. (2019). *Regulations: End-of-Life Vehicles (ELVs)* [Webpage]. GOV.UK. https://www.gov.uk/guidance/elv (accessed on 9 November 2024).
23. Iclodean, C., Varga, B., Burnete, N., Cimerdean, D., & Jurchiş, B. (2017). Comparison of different battery types for electric vehicles. *IOP Conference Series: Materials Science and Engineering, 252*, 012058. https://doi.org/10.1088/1757-899X/252/1/012058.
24. IEA. (2020). *Global EV Outlook 2020–Analysis* [Webpage]. IEA. https://www.iea.org/reports/global-evoutlook-2020 (accessed on 9 November 2024).

25. International Energy Agency. *Electric vehicles* [Webpage]. https://www.iea.org/energy-system/transport/electric-vehicles (accessed on 9 November 2024).

26. Jiao, N. (2019). *Second-Life Electric Vehicle Batteries 2020–2030.*

27. Jiao, N. (n.d.). *Post-Vehicle Value of Retired Electric Vehicle Batteries* [Webpage]. https://www.idtechex.com/en/research-article/post-vehicle-value-of-retired-electric-vehicle-batteries/18058 (accessed on 9 November 2024).

28. Kelleher Environmental. (2019). *Research Study on Reuse and Recycling of Batteries Employed in Electric Vehicles*. Energy API.

29. Kilgo, M. K., Anctil, A., Kennedy, M. S., & Powell, B. A. (2022). Metal leaching from lithium-ion and nickel-metal hydride batteries and photovoltaic modules in simulated landfill leachates and municipal solid waste materials. *Chemical Engineering Journal, 431*, 133825.

30. Lebedeva, N., Persio, F. D., & Boon-Brett, L. (2016). *Lithium-Ion Battery Value Chain and Related Opportunities for Europe*. Joint Research Centre.

31. Ma, X., et al. (2021). Recycled cathode materials enabled superior performance for lithium-ion batteries. *Joule, 5*, 2955–2970.

32. Mahmood, K., & Gutteridge, F. (2019). *EV Batteries Remanufacturing.* University of Strathclyde.

33. Mayyas, A., Steward, D., & Mann, M. (2019). The case for recycling: Overview and challenges in the material supply chain for automotive li-ion batteries. *Sustainable Materials and Technologies, 19*, e00087. https://doi.org/10.1016/j.susmat.2018.e00087.

34. McKinsey & Company. (2019a). *Breathing New Life into Used Electric Vehicle Batteries* [Webpage]. McKinsey & Company. https://www.mckinsey.com/business-functions/sustainability/our-insights/sustainability-blog/breathing-new-life-into-used-electric-vehicle-batteries (accessed on 9 November 2024).

35. McKinsey & Company. (2019b). *Expanding Electric-Vehicle Adoption Despite Early Growing Pains.*

36. McKinsey & Company. (2019c). *Second-Life EV Batteries: The Newest Value Pool in Energy Storage.*

37. Nissan. (2019). *Empowering Progress* [Webpage]. https://europe.nissanstories.com/en/releases/empowering-progress (accessed on 9 November 2024).

38. O'Neill, N. (2018). *Building Better Batteries for Your Future Electric Car | Imperial News* [Webpage]. Imperial College London. https://www.

imperial.ac.uk/news/186707/building-betterbatteries-your-future-electric/ (accessed on 9 November 2024).

39. Okorie, O., Salonitis, K., Charnley, F., Moreno, M., Turner, C., & Tiwari, A. (2018). Digitisation and the circular economy: A review of current research and future trends. *Energies, 11*(11), 3009. https://doi.org/10.3390/en11113009.

40. Pearce, F. (2020). Getting the lead out: Why battery recycling is a global health hazard. *YaleEnvironment360*. Retrieved from: https://e360.yale.edu/features/getting-the-lead-out-why-battery-recycling-is-a-global-health-hazard (accessed on 9 November 2024).

41. Pistoia, G., & Liaw, B. (2018). *Behaviour of Lithium-Ion Batteries in Electric Vehicles: Battery Health, Performance, Safety, and Cost*. Springer.

42. Powervault. (2020). *Powervault: Home Electricity Storage* [Webpage]. https://www.powervault.co.uk/ (accessed on 9 November 2024).

43. Reid, G., & Julve, J. (2016). *Second Life-Batteries as Flexible Storage for Renewables. Energies, 46*.

44. Reuters. (2018). *Nissan Spins Up New Plant to Give Second Life to EV Batteries*. Reuters.

45. Richa, K., Babbitt, C. W., Gaustad, G., & Wang, X. (2014). A future perspective on lithium-ion battery waste flows from electric vehicles. *Resources, Conservation and Recycling, 83*, 63–76. https://doi.org/10.1016/j.resconrec.2013.11.008.

46. Rouquette, L. M. J., Petranikova, M., & Vieceli, N. (2023). Complete and selective recovery of lithium from EV lithium-ion batteries: Modeling and optimization using oxalic acid as a leaching agent. *Separation and Purification Technology, 320*, 124143.

47. Sanderson, H. (2020). *Lithium Volatility Sends Shockwaves Through Supply Chains* [Webpage]. https://www.ft.com/content/78274a5c-33c2-11ea-a329-0bcf87a328f2 (accessed on 9 November 2024).

48. Spiers New Technologies. (n.d.). *Spiers New Technologies* [Webpage]. http://www.spiersnewtechnologies.com (accessed on 9 November 2024).

49. Swain, B. (2017). Recovery and recycling of lithium: A review. *Separation and Purification Technology, C*, 388–403. https://doi.org/10.1016/j.seppur.2016.08.031.

50. The Faraday Institution. (2019). *The Gigafactory Boom: The Demand for Battery Manufacturing in the UK*. Faraday Insights.

51. The Faraday Institution. (n.d.). *ReLiB–Reuse & Recycling of Lithium-Ion Batteries* [Webpage]. https://relib.org.uk/ (accessed on 9 November 2024).

52. Turner, J. M. (2015). Following the Pb: An envirotechnical approach to lead-acid batteries in the United States. *Environmental History, 20*(1), 29–56.
53. UKRI. (2019). *Faraday Battery Challenge: Funded Projects to Date.*
54. UKRI. (2020). *R2LIB (Reclamation, Remanufacture of Li Ion Batteries)* [Webpage]. https://gtr.ukri.org/projects?ref=104425 (accessed on 9 November 2024).
55. UNECE. (2017). *ADR*. United Nations.
56. Velázquez-Martínez, O., Valio, J., Santasalo-Aarnio, A., Reuter, M., & Serna-Guerrero, R. (2019). A critical review of lithium-ion battery recycling processes from a circular economy perspective. *Batteries, 5*(4), 68. https://doi.org/10.3390/batteries5040068.
57. Veolia. (2020). *Energy Storage Systems* [Webpage]. Veolia UK. https://www.veolia.co.uk/services/energy-storage-systems (accessed on 9 November 2024).
58. Volvo Buses. (2018). *Electric Bus Batteries Used to Store Solar Energy* [Webpage]. https://www.volvobuses.com/en-en/news/2018/dec/electric-bus-batteries-used-to-store-solarenergy.html (accessed on 9 November 2024).
59. Volvo Cars. (2018). *Volvo Cars Tech Fund Invests in Electric Car Charging Company FreeWire* [Webpage]. https://www.media.volvocars.com/global/en-gb/media/pressreleases/239948/volvo-cars-tech-fund-invests-in-electric-car-charging-company-freewire (accessed on 9 November 2024).
60. Waidha, A. I., et al. (2023). Recycling of all-solid-state Li-ion batteries: A case study of the separation of individual components within a system composed of LTO, LLZTO, and NMC. *ChemSusChem, 16*, e20220236.
61. Zhou, L. F., Yang, D., Du, T., Gong, H., & Luo, W. B. (2020). The current process for the recycling of spent lithium-ion batteries. *Frontiers in Chemistry, 8*, 578044.

Chapter 7

Impacts on Air Quality and Human Health

Contents

INTRODUCTION

Society depends heavily on transportation, but it is also primarily to blame for problems with air quality, public health, climate change, and other grave environmental threats. More people are driving cars, which emit a lot of pollutants like particulate matter (PM), nitrogen oxides, and volatile organic compounds (VOCs). This causes ozone to form in the atmosphere, which is harmful to human health and causes catastrophic global conditions. As of late, the majority of nations have concentrations of $PM_{2.5}$ and NO_2 above the WHO's recommended threshold for human health safety. Using EVs is an effective way to reduce air pollution because enhancing ambient air quality is important for safeguarding public health worldwide.

Generating electricity without emitting any emissions into the atmosphere, electric vehicles (EVs) are a growing solution to preserve public health and environmental sustainability. Reducing greenhouse gas emissions (GHGs) through the use of cleaner electricity in electric vehicles and charging is a global demand to improve air quality, create healthier communities, and manage traffic-related pollution where it is high. The impact of EVs will leverage a promising future, as evidenced by emission factors. More developments in practical emission control technologies and analysis are needed to support the development of regulations to control electric vehicles for effective social applications and to lessen the risks to the environment and public health. Electric cars therefore have the potential to change transportation.

7.1. AIR POLLUTION

The modern civilized world is very concerned about air pollution because it poses a serious toxicological risk to both the environment and human health. Although there are many different sources of emissions, the majority of air pollution is caused by motor vehicles and industrial processes. The World Health Organization lists lead, sulfur oxides, nitrogen oxides, carbon monoxide, particle pollution, and ground-level ozone as the six main air pollutants. Humans exposed to air suspended toxicants for short or long periods of time may experience varying toxicological effects, such as respiratory and cardiovascular disorders, neuropsychiatric issues, ocular irritation, skin conditions, and long-term chronic diseases like cancer. Numerous studies have demonstrated a clear correlation between exposure to low-quality air and rising rates of morbidity and mortality, primarily from respiratory and cardiovascular disorders. The incidence and progression of various diseases, including asthma, lung cancer, ventricular hypertrophy, Alzheimer's and Parkinson's disease, psychological complications, autism, retinopathy, fetal growth, and low birth weight, are thought to be primarily influenced by air pollution.

Currently, all vehicles obtain the energy required for propulsion through the combustion of hydrocarbon (HC) fuels. Heat and combustion products are released during combustion, which is a reaction between fuel and air. An engine transforms the heat into mechanical power and releases the combustion products into the atmosphere. A chemical compound called an HC is composed of hydrogen and carbon atoms in molecules. The only environmentally benign byproducts of an HC's combustion should be carbon dioxide and water. In fact, carbon dioxide is "digested" by green plants through photosynthesis. For plants to survive, carbon dioxide is an essential component. Animals that breathe in carbon dioxide do not suffer unless the concentration of the gas in the air is so high that oxygen is nearly absent.

In combustion engines, HC fuel never burns perfectly. In addition to water and carbon dioxide, the combustion products also include trace amounts of unburned hydrocarbons (HCs), carbon monoxides (CO), and nitrogen oxides (NO_x), all of which are harmful to human health.

7.1.1. Definitions

The term "air pollution" refers to any harmful consequences resulting from any source that contaminates the atmosphere or undermines the ecosystem. Both human activity and/or natural occurrences can contribute to air pollution. It is composed of various types of pollutants, encompassing substances in solid, liquid, and gaseous states. (Vallero, 2007)

Typically used to aid in risk assessment, the Pollutant Standard Index (PSI) is a numerical value and indicator of pollutants. It is a number with a range of zero to 500. Thom and Ott first proposed the PSI standard for reporting air quality in 1974. As a result, it would offer a way to assess how much each pollutant contributes relative to overall risk. The concentration of five primary air pollutants—particulate matter (PMs), sulfur dioxide (SO_2), nitrogen dioxide (NO_2), carbon monoxide (CO), and ozone (O_3)—in the atmosphere is what determines the PSI.

In line with Johnson et al. "A measure of the state of the air in relation to the requirements of one or more biotic species or to any human need is known as the air quality index, or AQI. "AQI is separated into numbered ranges, with color codes denoting each range.." It indicates the degree of health risk associated with air quality with a number ranging from zero, the healthy standard level, to above 300, the very hazardous level. The PSI is used to categorize air quality into six major indices, each of which is represented by a color code that indicates a different level of health concern. Basically, green is a color indicator for healthy air quality; in contrast, the colors yellow, orange, red, purple, and maroon represent moderate, unhealthy, very unhealthy, unhealthy, and hazardous air quality, respectively. The various classification systems used in various nations may cause these ranges and codes to vary (Johnson et al., 1997)

7.1.2. Toxicology of Air Pollution

Air pollution affects all living things, not just humans and animals. It also affects the environment as a whole. The environment, including animal life, and human health are impacted by regional variations, changes in the global climate, and environmental shifts.

7.1.2.1. Environmental Damages

In terms of ecology, air pollution can seriously harm soil, groundwater, and the air. The diversity of life is also gravely threatened by it. The detrimental effects of environmental contaminants on the extinction of animal and plant species are amply demonstrated by studies on the connection between air pollution and declining species diversity. Animal reproduction may also be impacted by airborne suspended pollutants. The following are other significant ecological effects of air pollution: acid rain, temperature inversions, and changes in the global climate brought on by greenhouse gas emissions. (Schneider SH, 1989)

7.1.2.2. Air Pollutants and Their Toxicities

Air pollutants are defined as any material in the air that has the potential to harm human health or the environment. The six main air pollutants that are harmful to both human health and the ecosystem are lead (Pb), nitrogen oxides, sulfur oxides, ground-level O_3, and particle pollution, according to the World Health Organization (WHO). Airborne contaminants that are suspended in the atmosphere include dust, fumes, smokes, mists, gaseous pollutants, hydrocarbons, volatile organic compounds (VOCs), polycyclic aromatic hydrocarbons (PAHs), and halogen derivatives. These pollutants, when present in high enough concentrations, can make a person more susceptible to a variety of illnesses, including cancer.

7.1.2.3. Particle Pollutants

One of the main categories of air pollutants is particle pollution. They are, to put it simply, a combination of airborne particles. Most pulmonary and cardiac-related morbidity and mortality are associated with PM, or particulate matter pollution. Their size range has been observed to be mostly between 2.5 and 10 μm ($PM_{2.5}$toPM_{10}).

The development and course of heart and lung disorders are closely linked to the size of particle pollution. Smaller particles are more likely to enter the lower respiratory tract and cause heart and lung problems because of this. Further evidence from multiple scientific studies has shown that fine particle pollution causes cardiac dysrhythmias, nonfatal heart attacks, worsened asthma, and reduced lung function in individuals with heart and/or lung diseases, all of which contribute to premature death. Particulate pollution can cause minor to serious illnesses, depending on the exposure level. The most common clinical symptoms of respiratory diseases brought on by air pollution are wheezing, coughing, dry mouth, and limitations on activities due to breathing difficulties.

Life expectancy may be significantly shortened by prolonged exposure to the ambient PM concentrations found today. Reduced life expectancy is primarily caused by an increase in lung cancer and cardiopulmonary deaths. Asthmatic bronchitis and chronic obstructive pulmonary disease (COPD), which cause reduced lung functions in both children and adults, are serious conditions that shorten life expectancy and lower quality of life. Cohort studies provide compelling evidence about how long-term PM exposure affects cardiovascular and cardiopulmonary mortality.

7.1.2.4. Ground-level Ozone

The main component of the atmosphere is the colorless gas O_3, which has the chemical formula O_3. It is present in both the troposphere, the uppermost part of the atmosphere, and the ground. The chemical reaction between nitrogen oxides and VOCs released by human activity or natural sources results in the production of ground-level ozone (GLO). There is a reasonable belief that GLO is linked to a higher risk of respiratory conditions, especially asthma.

O_3 takes up electrons from other molecules because it is a potent oxidant. The respiratory tract's surface fluid lining and the cell membranes that support it both contain high concentrations of polyunsaturated fatty acids. In these fatty acids, there are unstable double bonds. O_3 proceeds through an unstable zwitterion or trioxolane (depending on the presence of water) by attacking unpaired electrons to form ozonides. Aldehydes, hydrogen peroxide, and lipohydroperoxides are the end products of these reactions. It is believed that these pathways cause cell membranes and macromolecules to auto-oxidize and to start distributing lipid radicals. Additionally, it raises the possibility that epidermal keratinocytes may sustain DNA damage, which compromises cellular function. At concentrations found in many urban areas, O_3 causes a wide range of toxic effects in humans and experimental animals. These effects include changes in morphology, function, immunology, and biochemistry. A significant amount of inhaled O_3 reaches deep into the lungs due to its low water solubility, but the nasopharynxes of resting rats and humans scrub away approximately 17% and 40% of its reactivity, respectively. From an ecological perspective, O_3 can slow down the absorption of carbon by trees, resulting in deforestation that could eventually impact the world's food supply.

7.1.2.5. Carbon Monoxide

When combustion is not suitable, as occurs when burning coal and wood, fossil fuels produce carbon dioxide (CO), an odorless and colorless gas. About 250 times more strongly than oxygen, carbon dioxide has an affinity for hemoglobin, the body's oxygen carrier. A person may experience mild to severe poisoning, depending on the CO concentration and duration of exposure. Headache, weakness, nausea, vomiting, and finally loss of consciousness are some of the signs and symptoms of carbon monoxide poisoning. The signs and symptoms are strikingly similar to those of other ailments, like viral infections or food poisoning.

Levels of carboxyhemoglobin (COHb) above 40% may be lethal, while levels below 2% have not been shown to have any effects on human health. The known mechanisms behind CO toxicity include ischemia, hypoxia, and apoptosis. The loss of oxygen brought on by CO's competitive binding to the heme groups in hemoglobin is the mechanism underlying this toxicity. CO exposures that produce COHb levels greater than 5% can also cause cardiovascular alterations. A number of cardiovascular disease-related studies were conducted in the early 1990s by the Health Effects Institute in order to assess the possibility of angina pectoris in patients with COHb levels between 2% and 6%. It was shown by the results that although ventricular arrhythmias may occur in these circumstances, premature angina can still occur. For those who are predisposed, lowering the ambient CO level can therefore lower their risk of myocardial infarction.

When HCs burn too slowly because there is not enough oxygen in the air, carbon monoxide is produced. When animals or humans breathe it in, they become poisoned. The amount of oxygen that reaches the organs is decreased and the affected physical and mental capacities are lowered when carbon monoxide attaches to hemoglobin instead of oxygen once it enters blood cells. When carbon monoxide poisoning occurs, dizziness is the initial symptom and can quickly result in death. Compared to oxygen, carbon monoxide binds to hemoglobin more firmly. The bonds are so strong that they cannot be broken by the body's normal processes. Carbon monoxide intoxication patients need to be treated in pressurized chambers because the pressure facilitates the breaking of the carbon monoxide–hemoglobin bonds.

7.1.2.6. Sulfur Dioxide

Colorless and extremely reactive, SO_2 is one of the major air pollutants. It is primarily released by the burning of fossil fuels, spontaneous volcanic eruptions, and industrial operations. The health of humans, animals, and plants is all severely harmed by SO_2. Individuals who have lung conditions, kids, the elderly, and those who are exposed to higher levels of SO_2 are more susceptible to lung and skin diseases.

The two main health risks linked to exposure to high SO_2 concentrations are the exacerbation of pre-existing cardiovascular disease and respiratory irritation and dysfunction. The upper respiratory tract is where SO2 is primarily absorbed. It is a sensory irritant that can make people bronchospasm and secrete mucus. People living in developed areas who are exposed to SO_2 in the contaminated ambient air, even at lower concentrations (<1 ppm), may develop severe bronchitis.

When breathing through the mouth as opposed to the nose, more SO_2 enters the lungs. Deep, fast breathing increases airflow, which helps the gas enter the deeper lung more effectively. Exercisers in the contaminated air would therefore breathe in more SO_2 and probably experience more irritation. SO_2 is readily dispersed throughout the body when it deposits along the airway and dissolves into the surface lining fluid as sulfite or bisulfite. It appears that both locally and centrally mediated bronchoconstriction are caused by the sulfite's interaction with sensory receptors in the airways.

The USA's Environmental Protection Agency (EPA) states that 0.03 parts per million is the annual standard for SO_2. Because sulfur dioxide (SO_2) is soluble in water, it causes acid rain and soil acidification. Marine life, including both plants and animals, perishes when SO_2 lowers the oxygen content of the water. Exposure to SO_2 can harm mucous membranes, skin (redness, blisters), eyes (lacrimation, corneal opacity), and respiratory tracts. The most frequent clinical manifestations linked to SO_2 exposure are bronchospasm, pulmonary edema, pneumonitis, and acute airway obstruction. (Chen TM, et al., 2007)

7.1.2.7. Nitrogen Oxide

One significant ambient air pollutant that may raise the risk of respiratory infections is nitrogen oxides. They are air pollutants linked to traffic because they are mostly released by motor engines. These are deep lung irritants that, at high inhalation levels, can cause pulmonary edema. Although NO_2 can clearly cause toxicological issues, they are generally less toxic than O_3. It has been demonstrated that exposures at 2.0–5.0 ppm impact T-lymphocytes, in particular $CD8^+$ cells and natural killer cells, which are crucial to the host's defenses against viruses. Epidemiologic studies show that NO_2 has an impact on children's rates of respiratory infections, despite the possibly high levels.

The most frequent side effects of nitrogen oxide poisoning are coughing and wheezing, but other symptoms include headaches, diaphoresis, fever, bronchospasm, pulmonary edema, eyes, nose, or throat irritations, dyspnea, chest pain, and diaphoresis. According to a different study, humans can safely breathe nitrogen oxide at levels between 0.2 and 0.6 parts per million.

Oxygen and nitrogen in the air react to produce nitrogen oxides (NO_x). In theory, nitrogen is a harmless gas. On the other hand, nitrogen oxide formation is made possible by the high pressures and temperatures found in engines. The most crucial factor influencing the formation of nitrogen oxide is temperature. Although trace amounts of nitrous oxide (N_2O) and small amounts of nitric dioxide (NO_2) are present, nitric oxide (NO) is the most frequently found nitrogen oxide. When NO is released into the atmosphere, it combines with oxygen to create NO_2. Later, the UV rays from the Sun break this down into NO and extremely reactive oxygen atoms, which attack the membranes of living cells. Smog is partially caused by nitrogen dioxide, which also gives the air a brownish tint. It also produces nitric acid (HNO_3) when it reacts with atmospheric water, which becomes diluted with rain. In industrialized nations, the phenomenon known as "acid rain" is to blame for the devastation of forests. Marble historical monuments are also susceptible to deterioration from acid rain.

7.1.2.8. Unburned HCs

Unburned hydrocarbons (HCs) arise from incomplete combustion of hydrocarbons. HCs may be toxic to living things depending on their nature. Certain HCs could be direct carcinogens or poisons, like benzene or particulates. Another cause of smog is HCs. When the Sun's UV rays interact with unburned HCs and NO in the atmosphere, they produce ozone and other byproducts. Two oxygen atoms combine to form the molecule known as ozone. It attacks the membranes of living cells, causing them to age prematurely or die. Although colorless, it is extremely dangerous and poisonous. Exposure to high concentrations of ozone is particularly harmful to toddlers, the elderly, and those with asthma. Deaths from high ozone peaks in urban pollution have been documented on an annual basis.

7.1.2.9. Lead

Pb, also known as plumb, is a hazardous heavy metal that is extensively utilized in numerous industries. Pb pollution can originate from both external and internal sources. It is released by motor engines, especially those that run on gasoline that contains lead tetraethyl. Other sources of Pb emissions into the environment include wastewater treatment facilities, smelters, and battery plants. Environmental pollution may be taken into consideration as a potential source of lead exposure, according to an assessment of the blood lead level in traffic police officers. Children and fetuses are particularly vulnerable to even minute amounts of lead. Pb builds up in soft tissue, bone, and blood within the body. Pb can also have an impact on the liver, kidneys, nervous system, and other organs because it is not easily eliminated from the body.

Lead absorption by the lungs is contingent on particle size and concentration. Nearly 90% of lead particles inhaled from the atmosphere, small enough to be retained, are indeed retained. These retained particles are then absorbed through the alveoli, leading to toxicity. Lead is a potent neurotoxin, posing a substantial risk to infants and children, classified as high-risk groups. Adverse effects of lead exposure in childhood include mental retardation, learning disabilities, impairment of memory, hyperactivity, and antisocial behaviors. Therefore, it is imperative to decrease the lead concentration in ambient air.

Pb exposure is frequently chronic and shows no overt signs. The nervous system is the primary target of lead toxicity, though it can also affect the cardiovascular, renal, and reproductive systems. Through the inhibition of N-methyl-D-aspartate receptors, Pb impedes the regular operation of intracellular second messenger systems. In addition, Pb has the ability to substitute calcium as a second messenger, which can alter proteins via a variety of cellular mechanisms such as the activation or deactivation of protein kinase.

The most typical signs of lead poisoning are reduced sensations, anemia, aggression, constipation, headaches, irritability, loss of memory and concentration, and abdominal pain. Pb exposure can cause a variety of issues, including high blood pressure, infertility, renal and digestive dysfunctions, as well as pain in the muscles and joints.

7.1.2.10. Other Air Pollutants

Other significant air pollutants include VOCs like benzene, toluene, ethylbenzene, and xylene; PAHs like acenaphthene, anthracene, and benzopyrene; and other organic pollutants like dioxins, which are undesirable chemical pollutants almost entirely produced by human activity and industrial processes. These pollutants are also classified as carcinogens and mutagen compounds and are thought to be accountable for the incidence and progression of cancer in humans. Pollutant emissions are a result of fuel impurities. The main impurity is sulfur, which is primarily present in jet fuel and diesel but is also present in gasoline and natural gas. Sulfur oxides (SO_x) are released when sulfur (or sulfur compounds, such as hydrogen sulfide) is burned with oxygen. Upon combustion, the primary byproduct is sulfur dioxide (SO_2). Sulfur trioxide is created when it comes into contact with air and then combines with water to form sulfuric acid, which is a primary ingredient in acid rain. It should be mentioned that emissions of sulfur oxide are mostly caused by the burning of coal in power plants and steel mills, in addition to transportation-related sources. Furthermore, there is disagreement regarding the precise role played by natural sources like volcanoes.

To extend the life or performance of engines, petroleum companies incorporate chemical compounds into their fuels. The chemical known as tetraethyl lead, or just "lead," was added to gasoline to increase its resistance to knock and enhance engine performance. However, the release of lead metal during the combustion of this chemical causes saturnism, a neurological condition. In the majority of developed nations, its use is currently prohibited, and other chemicals have taken its place.

7.2. GLOBAL WARMING

The greenhouse effect brought on by the atmosphere's concentration of gases like methane and carbon dioxide causes global warming. The infrared radiation reflected from the Earth by the Sun is trapped by these gases, keeping the energy in the atmosphere and raising its temperature. A higher Earth temperature causes significant ecological harm to ecosystems and increases the frequency of natural disasters that have an impact on human populations.

Given the harm that global warming has done to the environment, it is concerning when certain endangered species go extinct because it throws off the natural resources that sustain certain populations. Concerns have also been raised regarding certain species' migration from warm to historically colder northern seas, where they may wipe out native species and the economies that depend on them. Barracudas from the Red Sea have been seen in the Mediterranean Sea, where this might be taking place.

Because of the extent of the harm they cause, natural disasters receive more attention than ecological disasters. El Niño, a meteorological phenomenon that frequently causes tornadoes, floods, and droughts in the South Pacific, is thought to have been induced by global warming. Another significant effect of global warming is the melting of the polar icecaps, which raises the sea level and sometimes permanently submerges coastal areas and even entire nations. When coal and HCs burn, carbon dioxide is produced. A significant portion of carbon dioxide emissions (32% between 1980 and 1999) are related to transportation. Figure 7.1 displays the distribution of carbon dioxide emissions.

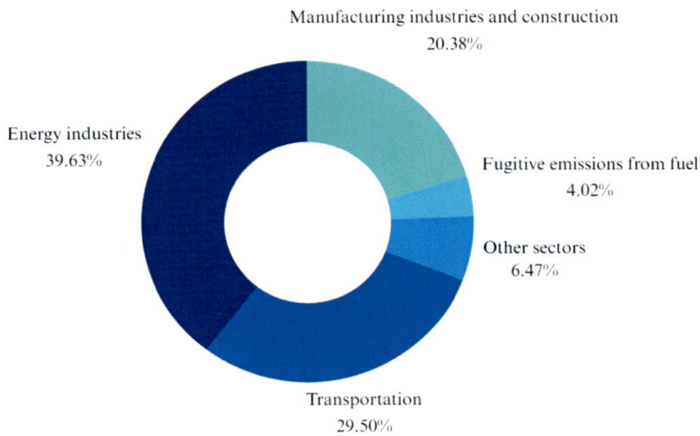

Figure 7.1. *Thailand's GHG emissions in the energy sector in 2019.*

Source: Janhuaton T, Ratanavaraha V, Jomnonkwao S. Forecasting Thailand's Transportation CO_2 Emissions: A Comparison among Artificial Intelligent Models. Forecasting. *2024; 6(2):462-484. https://doi.org/10.3390/forecast6020026.*

The carbon dioxide emissions trend is depicted in Figure 7.2. It is evident that the transportation industry is currently the main source of carbon dioxide emissions. It should be highlighted that developing nations, which account for a sizable portion of the global population, are expanding their transportation sectors quickly.

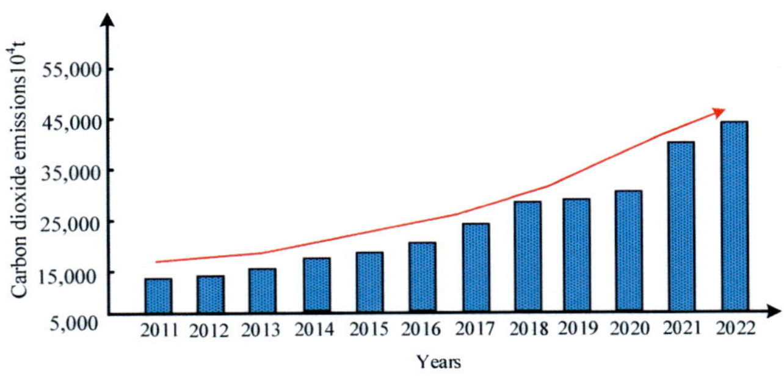

Figure 7.2. *Evolution of CO_2 emission.*

Source: Zhang H, Li S. Research on the Factors Influencing CO_2 Emission Reduction in High-Energy-Consumption Industries under Carbon Peak. Sustainability. *2023; 15(18):13437. https://doi.org/10.3390/su151813437.*

The substantial rise in Earth's temperature that has been observed in recent decades is thought to be mostly caused by human activity's release of carbon dioxide into the atmosphere (Figure 7.3). It is significant to remember that carbon dioxide is actually absorbed by plants and stored in the ocean as carbonates. Carbon dioxide builds up in the atmosphere as a result of these natural assimilation processes' limitations and inability to absorb all carbon dioxide emissions.

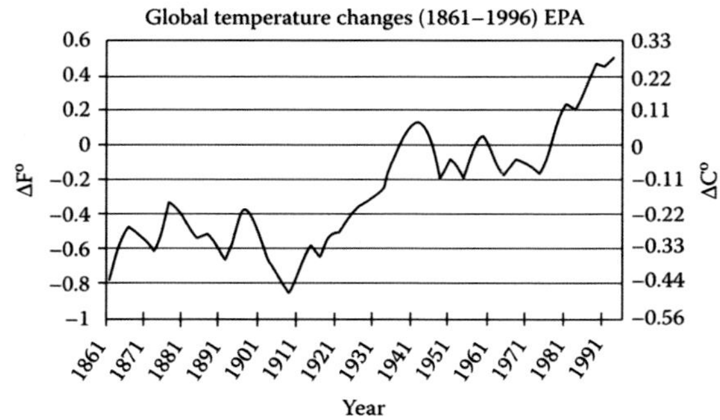

Figure 7.3. *Global greenhouse gas (GHG) emissions.*

Source: Intergovernmental Panel on Climate Change; Ameyaw B, Yao L. Analyzing the Impact of GDP on CO_2 Emissions and Forecasting Africa's Total CO_2 Emissions with Non-Assumption Driven Bidirectional Long Short-Term Memory. Sustainability. *2018; 10(9):3110. https://doi.org/10.3390/su10093110.*

7.3. EVOLUTION OF ROADWAY EMISSIONS

7.3.1. Primary and Secondary Pollutants

Road traffic emissions are made up of gaseous and PM, both of which include active carbonaceous products. Road traffic pollutants are classified into two categories: (1) primary pollutants, which are relatively non-reactive in the atmosphere and include CO, black carbon (BC), and ultrafine particles (UFPs); and (2) secondary pollutants, which are produced by the chemical processing of NO_x, VOCs, and SO_2 in roadways and result in the production of O_3 and secondary aerosol (i.e., SO_4^{2-} and NO_3). Evaluate individual concentration gradients and chemical transformations in such air pollutants, as both primary and secondary pollutants show heterogeneous distributions close to roads and change significantly in abundance over time.

7.3.2. Changes in Pollutant Concentrations Downwind of Roadways

Traffic emissions are thought to have a significant negative impact on between 30% and 45% of residents in large North American cities who live close to a road. Moreover, most urban residents' outdoor activities take place near roads, which exposes pedestrians and other road users to high levels of pollution. The surroundings of the roadway may also contain pollutants that affect the occupants of nearby buildings. Because of the close proximity of vehicle emissions, it is anticipated that the pedestrian level, or breathing height, of streets will experience particularly high levels of pollutants. The concentrations of pollutants in roads are often higher than in surrounding wider urban areas. Targeting air quality mitigation for long-term averages (hours, days, or years) may not fully account for exposure to real nonlinear changes in pollutant concentrations near roadways, which exhibit recurrent peaks in short intervals of time. To effectively assess the risk to one's own health and assist governments in creating related policies, it is therefore crucial to understand the dynamic and chemical processes governing the spatial air pollution gradients near roadways (e.g., the use of green spaces and road design) to lessen these negative effects on health.

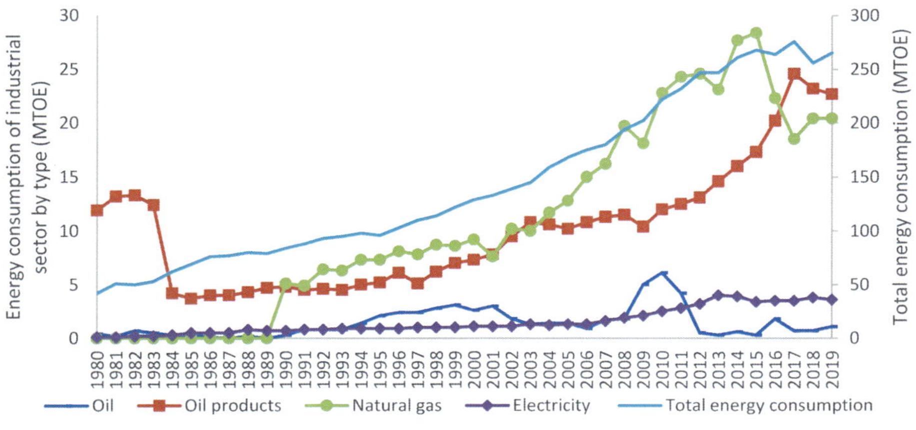

Figure 7.4. *Total energy consumption of Saudi Arabia's industrial sector, by fuel (1980–2019).*

Source: Alajmi RG. Energy Consumption and Carbon Emissions: An Empirical Study of Saudi Arabia. Sustainability. *2024; 16(13):5496. https://doi.org/10.3390/su16135496.*

In North America, the number of air pollution measurements close to roads has been rising recently. Pollutant gradients have been observed traveling downwind from roads in general, but there has been significant variation in the trends over time. Zhu et al. discovered that the levels of UFPs, PM, BC, and CO were two to five times higher than background concentrations close to the freeways in southern California, and that the distance from the freeways caused the pollution levels to exponentially drop to nearly background (background) levels. In California, USA, variations were found in both the seasons and the hours around Interstates 405 and 710. Clements et al. Investigated the spatial gradients of vehicular-emitted air pollutants of three highways (i.e., State Highway 71, Inter state 35 and Farm to Market Road 973) in Austin, TX, and found that concentrations of CO, NO, and NO_x returned to background levels within a few hundred meters of the roadway (Figure 7.4).

The concentrations of acetaldehyde and acrolein increased further downwind of State Highway 71 (SH-71), indicating chemical generation from the oxidation of primary vehicular emissions, while the concentrations of most carbonyl species decreased with distance downwind of SH-71 (Table 7.1). Regardless of wind speed, the concentrations of fine PM mass, UFPs, PAHs, hopanes, and elemental carbon (EC) generally showed a decrease with distance downwind of SH-71 (Table 7.1). Wang et al. compared the results of the model to field measurement data by varying the distance downwind of SH-71 and the chemical evolution of NO_x. Olson et al. found that, in Raleigh, North Carolina, the concentration of 55 VOCs (C_2–C_{12}) dropped exponentially with increasing distance (10–100 m) from a highway (Table 7.1). These studies showed much sharper gradients of air pollutants downwind of freeways, with concentrations above background levels only reaching up to 500 m at night and only 300 m during the day. Hu et al. observed that during the pre-sunrise hours of both the winter and summer seasons, air pollution was affecting a large area (~2600 m) downwind of Interstate 10 in West Los Angeles, California. Strong atmospheric stability, low wind speeds (0–1 m s^{-1}), cold temperatures (9–13°C), or high humidity (61%–79%) may be to blame for this phenomenon. These factors all work together to slow the movement of air pollutants before they dilute and disperse to background levels. Durant et al. concluded that variations in wind speed, surface boundary layer height, and highway traffic flow rate were responsible for the abrupt changes in pollutant gradients. The downwind concentrations of PM, CO_2, NO, and NO_2 were highest within 100–250 m of the highway before sunrise and peak traffic flow rates. Following sunrise, the surface boundary layer rose, wind speeds increased, and pollution levels drastically decreased, allowing for mixing with cleaner air above. The gradients also became less noticeable. These results imply that to protect public health from roadway emissions and make the best regulatory decisions, air quality policymakers must have adequate knowledge of local meteorology and traffic patterns.

Table 7.1. *Effects of Air Pollution on the Immune System*

Innate Immune System	Epithelial cells	↑ Pro-inflammatory cytokines: IL-1, IL-6, TNFα, GMCSF (O_3, NO_2, PM) ↑ ICAM-1 expression (NO_2) ↑ Leukotriene C4 (NO_2)
	Monocytes/Macrophages	↑ Carbon loading; ↓ Clearance capacity and phagocytosis; ↑ Pro-inflammatory cytokine (PM)
	Neutrophils	↑ Migration and activation (PM, O_3, NO_2)
	Eosinophils	↑ Migration and activation (SO_2, $PM_{2.5}$)
Adaptive Immune System	Immune tolerance	↓ T reg function; neo-antigens (ROS); ↑ antigens immunogenicity (PM); ↑ immunogenicity of antigens (PM)
	B cells	↓ IgA and ↑ IgE ($PM_{2.5}$)
	T cells	↓ Th1 and ↑ Th2 (NO_2, PM); ↑ Th17 response (PM); Dysregulated IFN-gamma and IL-17A production (PM)

O_3: Ozone; NO_2: Nitrogen Dioxide; PM: Particulate Matter; $PM_{2.5}$: Particulate Matter with an Aerodynamic Diameter of 2.5 μm; SO_2: Sulfur Dioxide; ROS: Reactive Oxygen Species; GM-CSF: Granulocyte-Macrophage Colony-Stimulating Factor; ICAM-1: Intercellular Adhesion Molecule-1; IFN: Interferon; IL: Interleukin; TNF-α: Tumor Necrosis Factor Alfa; ↑ Increase; ↓ Decrease.

Source: Dondi A, Carbone C, Manieri E, Zama D, Del Bono C, Betti L, Biagi C, Lanari M. Outdoor Air Pollution and Childhood Respiratory Disease: The Role of Oxidative Stress. International Journal of Molecular Sciences. *2023; 24(5):4345. https://doi.org/10.3390/ijms24054345.*

7.3.3. Key Air Pollutants Associated with Roadway Emissions

Road transport pollutants can be classified as either tailpipe or non-tailpipe emissions based on where they originate. The majority of emissions from tailpipes go toward UFPs (diameter <100 nm), VOCs and toxics, NO_x, CO, CO_2, O_3, and fine PM (<2.5 μm in aerodynamic diameter; $PM_{2.5}$). The majority of emissions from non-tailpipes go toward PM_{10} (<10 μm in aerodynamic diameter) and coarse PM (an aerodynamic diameter ranging from 2.5–10 μm; $PM_{2.5}$–PM_{10}), but they also include UFPs, $PM_{2.5}$, and VOCs.

Particles in the air with aerodynamic diameters smaller than 50 or 100 nm are referred to as ultrafine aerosol. UFPs make up the majority of ambient aerosol loading in terms of quantity, but they only make up a small portion of PM mass overall. It has been demonstrated that there is a strong correlation between harmful cardiopulmonary effects and elevated exposure to ultrafine aerosol from roadways. Furthermore, UFPs have the potential to expand quickly to larger sizes. The particles produced by these processes can make up as much as half of the global budget for cloud condensation nuclei, which has a significant impact on the ecosystem, the built and natural environments, and the formation of regional haze (Table 7.2).

Table 7.2. *Outdoor Air Pollutant Effects on Respiratory Disease and Related Issues: Prenatal Exposure, Short- and Long-Term Effects*

Prenatal Exposure	Short Term Exposure	Long Term Exposure
Reduced lung function and growth (TRAP, NO_2 PM_{10}, $PM_{2.5}$)	Bronchiolitis (PM_{10}, $PM_{2.5}$, NO, NO_2, UFPs) Severe Bronchiolitis ($PM_{2.5}$, PM_{10}, NO_2) RSV Bronchiolitis (NO, NO_2, $PM_{2.5}$, PM_{10}	Reduced lung function (NO_2, O_3)
Increased risk of Respiratory Tract Infection (NO_2)	Asthma exacerbation (UFPs, O_3, SO_2, NO_2, TRAP, BC, $PM_{2.5}$)	Reduced lung growth ($PM_{2.5}$, PM_{10}, NO_2)
Airway inflammation (NO_2) Higher ventilatory demand (PM_{10})	Pneumonia (UFPs, NO_2, PM_{10}, $PM_{2.5}$, TRAP), Severe pneumonia (O_3, $PM_{2.5}$)	Chronic Asthma ($PM_{2.5}$, TRAP)
Increased incidence of asthma (UFPs, $PM_{0.1}$, PM_1, $PM_{2.5}$)	Tuberculosis (PM_{10}, SO_2, NO_2)	COPD
Allergic rhinitis (TRAP, NO_2)	Upper Respiratory Tract Infection (UFPs, $PM_{2.5}$, PM_{10}, SO_2, NO_2, CO)	Allergic rhinitis (TRAP, PM_{10}, SO_2, NO_2)

TRAP: Traffic-Related Air Pollution; NO_2: Nitrogen Dioxide; PM: Particulate Matter; $PM_{2.5}$: Particulate Matter with an Aerodynamic Diameter of 2.5 μm; PM_{10}: Particulate Matter with an Aerodynamic Diameter of 10 μm; UFPs Ultrafine Particles; BC: Black Carbon; O_3: Ozone; SO_2: Sulfur Dioxide; CO: Carbon Monoxide; COPD: Chronic Obstructive Pulmonary Disease.

Source: Dondi A, Carbone C, Manieri E, Zama D, Del Bono C, Betti L, Biagi C, Lanari M. Outdoor Air Pollution and Childhood Respiratory Disease: The Role of Oxidative Stress. International Journal of Molecular Sciences. *2023; 24(5):4345. https://doi.org/10.3390/ijms24054345.*

Because they originate from a variety of sources, such as point, mobile, and area sources, VOCs are prevalent in urban environments. VOCs have varying effects on air quality in urban environments. They are significant O_3 precursors that support high O_3, peroxyacyl nitrate, and other oxidant concentrations in the troposphere. When combined, they cause photochemical smog and SOAs to form globally, which are the primary causes of haze formation in certain developing nations. Many VOCs linked to traffic emissions have been found and measured.

These include BTEX (benzene, toluene, ethylbenzene, and xylenes), carbonyl compounds (e.g., acetaldehyde, formaldehyde, propionaldehyde, and acrolein), as well as gaseous alkenes (e.g., ethane, propylene, acetylene, etc.). The US Environmental Protection Agency has also designated several compounds from traffic emissions as hazardous air pollutants, including benzene, toluene, ethylbenzene, m, p-xylene, o-xylene, formaldehyde, propionaldehyde, acetaldehyde, and acrolein (Table 7.1). These gaseous compounds have been shown in numerous toxicological studies to cause central nervous system injury, irritate human eyes and lungs, stimulate the respiratory system, and raise the risk of cancer. VOCs can also harm crops and lower agricultural productivity (Table 7.2). Among the many negative effects of NO_x emissions to the atmosphere are the creation of tropospheric O_3 and the particulate and gaseous nitrogen deposition products (i.e., NO_3^- particle and gaseous HNO_3). In contrast to CO_2, which is a significant greenhouse gas, CO is a spatially variable trace gas that, at ambient concentrations above approximately 35 ppm, is toxic to hemoglobin-containing animals. Numerous studies have shown that traffic-related NO_x and CO emissions are strongly linked to detrimental health outcomes, such as early death and morbidity from cardiovascular and respiratory diseases. To safeguard public health, the European Commission and the US Environmental Protection Agency have set limit values for CO and NO_2 in urban atmosphere (Table 7.2).

In both urban and rural settings, surface O_3 concentrations are common because O_3 is produced photochemically in the presence of NO_x, VOCs, and sunlight, especially in the summer when UV sunlight intensity is highest. Higher O_3 concentrations have also been reported in farther-flung locations that show upward mixing of O_3-rich air from above or long-range transport of O_3 from urban source regions. Surface O_3 is the third most significant greenhouse gas causing global warming and is essential to atmospheric chemistry. High concentrations of O_3 have negative impacts on ecosystem productivity and human health at the ground level (Table 7.2). Therefore, the US Environmental Protection Agency has set limit values for ground-level O_3 in National Ambient Air Quality Standards, which are reviewed every five years to assess and incorporate the best available scientific evidence, to protect human health and ecosystems. The O_3 level has decreased over the past few decades, from 0.080 ppm in 1997–0.070 ppm in 2015, in response to recent reviews. Directive 2008/50/EC, the European Union Framework Directive on Air Quality Assessment and Management, sets 120 $\mu g\ m^{-3}$ as the target value for daily 8-hour mean concentrations of O_3.

Road traffic tailpipe $PM_{2.5}$ is made up of a range of primary and secondary pollutants, including carbonaceous material, secondary ionic species like nitrate (NO_3^-) and sulfate (SO_4^{2-}), and trace metals (primary). The primary emissions from combustion, known as BC or EC, and organic matter (OM) make up the carbonaceous fraction. An increasing amount of data points to the possibility that $PM_{2.5}$ emissions from traffic could have complex interactions with climate change and be detrimental to ecosystems, construction, human health, atmospheric visibility, and traffic safety. Certain chemical components (e.g., PAHs and hopanes) from traffic emissions have been linked to markers of systemic inflammation in older adults, according to studies conducted in urban Los Angeles, California. A 2% reduction in China's total rice production and an 8% reduction in its total wheat production is anticipated as a result of haze formation, which is primarily caused by secondary aerosols (e.g., SO_4^{2-}, NO_3^-, and SOA) from traffic emissions. Furthermore, the primary cause of cultural heritage building damage in urban areas is black crusts resulting from vehicular $PM_{2.5}$ components. Developed nations have responded to these side effects by implementing effective migration policies and target regulations, which have led to significant decreases in road traffic exhaust emissions. Although tailpipe air pollutants in developed countries are beginning to pose a serious threat to public health, tailpipe air pollutants in developing countries are noticeably lower now than they were in previous periods. To successfully reduce vehicle emissions in their urban environments, strict regulations are being implemented in these developing nations (e.g., China and India). If the rules are successful, vehicle emissions shouldn't cause any more significant issues with air quality or health in the future.

The emissions from road wear, tire wear, brake wear, and road dust resuspension are typically classified as non-tailpipe emissions. As a result, trace metals (e.g., Cu, Zn, Ba, Sb, and Mn) in coarse PM are frequently used to identify non-tailpipe emissions. The emissions of trace metal markers, however, differ depending on the composition of the fleet; higher emission factors are recorded for certain components originating from heavy-duty vehicles. The primary determinants of non-tailpipe emissions are driving and traffic patterns, vehicle fleet characteristics, traffic volume and pattern, local geology and climate. Every part of the world has a different profile of trace metal concentrations in non-tailpipe PM because there are so many variables involved. With the current SA approaches, it is challenging to determine their contributions to ambient air quality with accuracy. Furthermore, there is currently no worldwide regulation of non-tailpipe emissions from road transportation, and most developed countries anticipate that by the end of this decade, these emissions will account for 80–90% of all traffic emissions. We can evaluate the relationship between emissions, concentrations, exposures, and health effects as well as determine the efficacy of potential remediation measures in the urban environment by having a thorough understanding of non-exhaust emissions, including the essential tracers of their source and their intrinsic toxicology.

7.4. EFFECTS OF AIR POLLUTION ON HUMAN HEALTH

Recent decades have seen a significant increase in air pollution, which poses a serious toxicological risk to both the environment and human health. The sources of pollution range in size from tiny amounts of cigarettes and natural sources like volcanic eruptions to massive amounts of emissions from industrial processes and car engines. Since it is generally acknowledged that air pollution has long-term effects on the development of diseases like cancer, cardiovascular dysfunction, respiratory infections and inflammations, air pollution is responsible for millions of deaths worldwide each year. According to a recent study, air pollution and male infertility are related.

In addition to worsening the symptoms of those who are already ill, air pollution raises the risk of lung cancer, heart disease, stroke, and respiratory infections. Air pollution poses varying health risks to different people based on factors such as age, location, underlying health conditions, and more.

7.4.1. Health Hazards

Any unusual suspended material in the air that interferes with an organ's ability to function normally is considered an air toxicant when it comes to health risks. Based on current information, the respiratory, cardiovascular, ophthalmologic, dermatologic, neuropsychiatric, hematologic, immunologic, and reproductive systems are the primary targets of air pollution exposure. On the other hand, over time, the toxicity to cells and molecules may also result in many cancers. Conversely, it has been demonstrated that even minute concentrations of air toxicants pose a risk to vulnerable populations, such as children, the elderly, and those with respiratory and cardiovascular conditions.

7.4.1.1. Respiratory Disorders

Since the majority of air pollutants enter the body through the airways, the respiratory system plays a primary role in the initiation and development of diseases linked to air pollution. The amount of harm that pollutants do to the respiratory system varies depending on the dose that is inhaled and how much of them is deposited in target cells. The first effect in the upper respiratory tract is irritation, particularly in the trachea, which causes dysphonia. One of the main environmental risk factors for several respiratory conditions, including lung cancer and asthma, is air pollution. The respiratory tract is seriously harmed by air pollutants, particularly PMs and other respirable chemicals like dust, O3, and benzene. Exposure to air toxicants can also lead to the development of asthma, a respiratory condition. Studies have confirmed links between elevated risk of COPD and air pollution from industry, traffic, and/or both. Treatment for respiratory conditions brought on by air pollution is comparable to that for conditions brought on by other harmful chemicals.

7.4.1.2. Cardiovascular Dysfunctions

Numerous epidemiological and experimental studies have demonstrated a link between air pollution exposure and illnesses related to the heart. Changes in white blood cell counts are linked to air pollution, and these changes may have an impact on cardiovascular health. Conversely, a study conducted on animal models indicated a strong correlation between exposure to air pollution and hypertension. Air pollution from traffic, particularly exposure to elevated NO_2 levels, has been linked to both left and right ventricular hypertrophy. Apart from the antidote therapy available for a limited number of cardiotoxic substances such as CO, the standard treatment for cardiovascular diseases ought to be administered.

7.4.1.3. Neuropsychiatric Complications

It has long been debated whether exposure to airborne suspended toxic materials affects the nervous system. On the other hand, it is currently thought that these poisonous compounds harm the nervous system. Air pollution has a toxic effect on the nervous system that can lead to mental health issues and neurological problems. The negative effects of neurological impairment can be disastrous, particularly in young children. Psychiatric disorders, on the other hand, can lead to antisocial and aggressive behaviors. Air pollution has been linked in recent research to neurobehavioral hyperactivity, criminal activity, and age-inappropriate behavior. Research has also demonstrated a link between air pollution and an increased risk of Parkinson's, Alzheimer's, and neuroinflammation. According to certain studies, there is a direct correlation between high levels of air pollution in megacities and aggression and anxiety.

7.4.1.4. Other Long-term Complications

The skin is the first organ to be contaminated by pollution and the body's first line of defense against foreign pathogens or infectious agents. The skin is a target organ for pollution, and the amount of pollutants absorbed from the environment through this organ is equal to what is inhaled through the respiratory system. Studies on the human skin have demonstrated that air pollutants linked to traffic, particularly PAHs, VOCs, oxides, and PM, impact the aging process of the skin and result in pigmented spots on the face.

In theory, breathing in harmful air pollutants or absorbing them through the skin can harm internal organs. Hepatocarcinogen chemicals make up a portion of these pollutants. Empirical evidence has demonstrated the impact of air pollution, particularly traffic-related air pollution, on the prevalence of autism and related disorders in children and fetuses. It has been suggested that one potential mechanistic pathway for autism or other neurological disorders is the disruption of endocrine function by chemical constituents of pollutants. According to certain research, there are connections between late-pregnancy fetal head size, fetal growth, low birth weight and air pollution exposure.

Numerous immune system dysfunction-related diseases can be influenced by a variety of environmental factors, including poor air quality. Serious immune system complications, including abnormally high serum levels of immunoglobulin (Ig), IgA, IgM, and the complement component C3 in humans, as well as chronic inflammatory respiratory diseases, can be brought on by poor air quality. In addition to causing immune dysfunction at various stages, exposure to these immunotoxicants may also raise the risk of several diseases, including neuroinflammation and altered brain innate immune response. Air pollution alters antigen presentation by upregulating costimulatory molecules on macrophages, such as CD80 and CD86.

The eye is an organ that is often overlooked but is extremely susceptible to the harmful effects of airborne pollutants, including those found in homes. Air pollution can have a range of clinical effects on the eyes, from dry eye syndrome to asymptomatic eye problems. Chronic air pollution exposure raises the risk of retinopathy and unfavorable eye outcomes. Furthermore, there is no evidence pointing to a connection between air pollution and major blinding, dry eye syndrome, and ocular irritation. Data show that there is a correlation between short-term increases in the number of visits to the ophthalmological emergency department and the level of air pollution.

7.4.2. Health Impacts of Near-roadway Exposures and Urban Air Pollution from Traffic Emissions

Acute toxicity and epidemiologic studies have extensively documented the role that traffic-related air pollutants play in the development of adverse health effects. In comparison to people who live farther away or near lower traffic flows, those who live near high traffic flows have a higher prevalence of most respiratory symptoms, a higher risk of developing leukemia and lung cancer, and a higher risk of developing asthma. Based on numerous studies conducted worldwide, epidemiological evidence regarding the impact of traffic sources on public health has been consistently positive (Table 7.3).

Table 7.3. *Summary of Major Epidemiological Studies on the Association of Ambient Air Pollution with Changes in Biomarkers Linked to Mental Health in Recent Years*

References	Air Pollutants	Study Design	Population	Major Exposure Metric	Outcome	Results	Study Area
Lane et al. (2023)	Traffic related air pollution	Cross-sectional	Community-based participant	1-year	Inflammatory biomarkers	Variance (p-value) for the association between traffic-related air pollution and inflammatory biomarkers: 0.76 (<0.01) for IL-6; 0.59 (0.02) for CRP; and 0.55 (0.02) for TNFRII.	Boston, Malden and Somerville, MA, USA
Du et al. (2023)	$PM_{2.5}$, UFP, BC, NO_2, and CO	Cross-over	College students	4 h	Exosomal lncRNAs	Differentially expressed lncRNAs (high air pollution vs. low air pollution): SNHG6 (FC = 6.31, FDR value = 2.42×10^{-15}).	Shanghai, China
Huang et al. (2022)	PM_1, $PM_{2.5}$, PM_{10}, NO_2	Panel study	Community-based participant	3-year	Methylation in *BDNF* promoter	Percent changes in methylation level at *BDNF* promoter for the 95th percentile of air pollution concentration against the threshold concentration: NO_2: 43.25% (95%CI: 13.10%, 73.40%) for average methylation; PM_1: 128.29% (95%CI: 43.27%, 213.31%) for average methylation; $PM_{2.5}$: 104.22% (95%CI: 34.34%, 174.10%) for average methylation; PM_{10}: 108.71% (95%CI: 36.60%, 180.81%) for CpG2 methylation.	Shijiazhuang, China
Sun et al. (2022)	$PM_{2.5}$	Cross-sectional	Preschoolers	30-day	Changes in the urinary acetic acid at different *OXTR* rs53576 genotypes	Absolute changes (μg/mg Cr) in urinary acetic acid per 1 μg/m^3 increase in $PM_{2.5}$ at different *OXTR* rs53576 genotypes: AA: −11.608 (95% CI: −21.685, −1.530); GG: −16.631 (95% CI: −26.949, −6.314); AG: −28.587 (95% CI: −36.078, −21.096).	Suzhou, China

Wei et al. (2022)	$PM_{2.5}$, PM_{10}	Panel study	Stable male schizophrenia patients	8-day	Oxidative stress biomarkers; schizophrenia relapse risk: evaluated by early signs scale (ESS), ESS-anxiety/agitation (ESS-A), ESS- excitability/disinhibition (ESS-D)	Absolute changes in cytokines per 10 μg/m³ increase in $PM_{2.5}$: CAT: −0.039 (95% CI: −0.060, −0.017) U/mL; SOD: −1.258 (95% CI: −1.975, −0.541) U/mL; T-AOC: −0.076 (95% CI: −0.126, −0.026) mmol/l. Regression coefficient for mediation of cytokines on ESS score changes associated with per 10 μg/m³ increase in $PM_{2.5}$: T-AOC: 0.019 (95% CI: 0.002, 0.040) for ESS-A; T-AOC: 0.013 (95% CI: 0.002, 0.030) for ESS-D.	Hefei, China
Gao et al. (2021)	$PM_{2.5}$, PM_{10}	Repeated measure	Stable schizophrenia patients	0–5-day	Immune cytokines; schizophrenia relapse risk: evaluated by ESS and ESS-A	Absolute changes (pg/mL) in cytokines per 10 μg/m³ increase in $PM_{2.5}$ at lag0: IL-6: 0.335 (95% CI: 0.045, 0.625); IL-17: 0.191 (95% CI: 0.144, 0.238); IL-2: 0.071 (95% CI: 0.030, 0.112); IL-12: 3.158 (95% CI: 1.719, 4.597); IFN-γ: 1.168 (95% CI: 0.645, 1.691). Regression coefficient for mediation of cytokines on ESS-A score changes associated with per 10 μg/m³ increase in $PM_{2.5}$ at lag0: IL-17: 0.202 (95% CI: 0.012, 0.392).	Hefei, China

Pergoli et al. (2017)	PM_{10}	Cross-sectional	Overweight or obese subjects	0–7-day	EV count and miRNA content of EVs	Percent changes in EV count per 10 μg/m^3 increase in PM_{10} at lag 1: EV count: 3.5%, $p = 0.0001$; miR-218-5p: −4.20% (95% CI: −1.87%, −6.47%); miR-143-3p: −2.75% (95% CI: −0.04%, −5.39%).	Lombardy, Italy
Zheng et al. (2017)	PM_{10}, $PM_{2.5}$, BC, elemental components	Repeated measure	Truck drivers and office workers	14-day	Histone H3 modification level	Percent changes in histone modifications per 1 μg/m^3 increase in 14-day average PM_{10} exposure: H3K27me3: −1.1% (95% CI: −1.6%, −0.6%); H3K36me3 levels: −0.8% (95% CI: −1.4%, −0.1%) in all participants Percent changes in histone modifications per 1 μg/m^3 increase in 14-day average BC exposure: H3K9ac: 4.6% (95% CI: 0.9%, 8.4%) in office workers; H3K36me3: 4.1% (95% CI: 1.3%, 7.0%) in truck drivers.	Beijing, China
Cantone et al. (2011)	Metal-rich PM	Cross-sectional	Steel workers	3-day	Histones H3K4me2 and H3K9ac	Standardized regression coefficient for changes in histone modifications associated with an increase equal to the difference between the 90th and 10th percentiles of nickel and arsenic exposures: H3K4me2 Nickel: 0.16 (95% CI: 0.01, 0.3); Arsenic: 0.16 (95% CI: 0.03, 0.29); H3K9ac Nickel: 0.24 (95% CI: −0.02, 0.51); Arsenic: 0.21 (95% CI: −0.06, 0.48).	Brescia, Italy

Abbreviations: BC: black carbon; BDNF: brain-derived neurotrophic factor; CAT: catalase; CO: carbon monoxide; CRP: C-reactive protein; EV: extracellular vesicles; FC: fold change; FDR: false discovery rate; H3K9ac: histone 3 lysine 9 acetylation; H3K4me2: histone 3 lysine 4 demethylation; IL: interleukin; IFN-γ: interferon-γ; NO_2: nitrogen dioxide; OXTR: oxytocin receptor; PM_1: particulate matter ≤ 1 μm in diameter; PM_{10}: particulate matter ≤ 10 μm in diameter; $PM_{2.5}$: particulate matter ≤ 2.5 μm in diameter; SOD: superoxide dismutase; TNFRII: tumor necrosis factor-α receptor II; UFP: ultrafine particles.

ıe Netherlands	Healthy, young, and non-smoking participants	Particle number concentrations, NO_2, and NO_x	Acute airway iı and impaire(function
, Sweden	Neonatal children	NO_2 and PM_{10}	Persistent whe
ın countries	Healthy volunteers	NO_2	Lung cancer

Source: Bai L, Wang K, Liu D, Wu S. Potential Early Effect Biomarkers for Ambient Air Pollution Related Mental Disorders. Toxics. *2024; 12(7):454. https://doi.org/10.3390/toxics12070454.*

Vineis et al. estimated exposures to environmental tobacco smoke and air pollution in non-smokers and ex-smokers in a large prospective study in 10 European countries (European Prospective Investigation into Cancer and Nutrition; n = 520,000) and found that 5%–7% of lung cancers in European non-smokers and ex-smokers were attributable to high levels of air pollution, as expressed by NO_2 or proximity to roads with heavy traffic.

In a prospective birth cohort (n = 4089) in Sweden, Nordling et al. assessed the health effects of children exposed to traffic-related NO_2. They concluded that early exposure to moderate levels of locally emitted air pollution from traffic appears to affect the development of airway disease and sensitization in preschool-aged children.

Healthy human volunteers were exposed to ambient PM at five different locations in the Netherlands (an underground train station, two traffic sites, a farm, and an urban background site) with varying PM characteristics. Strak et al. found associations between changes in particle number concentrations, NO_2, and NO_x, and lung function impairment as well as acute airway inflammation.

Beelen et al. looked at relationships between long-term exposure to exhaust air pollutants and natural-cause mortality in 367,251 participants from 22 European cohort studies. They found that there was a significantly higher hazard ratio for $PM_{2.5}$ of 1.07 (95% confidence interval: 1.02–1.13) per 5 μg m^{-3}.

Heo et al. presented data showing that in Seoul, Korea, $PM_{2.5}$, which originates from mobile sources, is linked to mortality from cardiovascular and respiratory diseases. According to Baumgartner et al., the health effects of BC from combustion emissions may be more pronounced in women who live close to a highway in Yunan, China, as they are exposed to high levels of motor vehicle emissions. They also suggested that BC is more strongly associated with blood pressure than PM mass.

Lelieveld et al. estimated the global burden of disease based on epidemiological cohort studies that link premature mortality to a variety of causes, such as the long-term health effects of O_3 and $PM_{2.5}$. They discovered that while emissions from traffic and power generation are more significant in most of the United States and a few other countries, emissions from residential energy use, such as cooking and heating, have the biggest effects on premature mortality.

The data, which came from 445,860 adults in 100 different US metropolitan areas, were used by Thurston et al. to assess historical relationships between long-term (from 1982–2004) PM2.5 concentrations and ischemic heart disease mortality. A significant correlation was found between diesel traffic-related EC and mortality from ischemic heart disease (hazard ratio = 1.03 per 0.26 μg m-3 EC increase; 95% confidence interval: 1.00–1.06). According to Kheirbek et al., the annual number of deaths and hospitalizations (440–1280) and ER visits (520–420) in New York City (NYC) caused by PM2.5 exposures is estimated to be 320 (95% confidence interval: 220–420) and 870 (95% confidence interval: 440–1280) respectively, resulting in a total of 5850 (95% confidence interval: 4020–7620) years of life lost. Ghosh et al. examined the burden of coronary heart disease caused by air pollution near roadways and contrasted it with the 2008 PM2.5 burden in the California South Coast Air Basin and a greenhouse gas reduction scenario for compact urban growth for 2035. Based on the findings, 1300 deaths in 2008 from coronary heart disease (6.8% of total deaths) were linked to traffic density, 430 deaths (2.4% of total deaths) to residential proximity to a major road, and 690 deaths (3.7% of total deaths) to EC. Because of population aging, the estimated numbers of deaths linked to each of these exposures are expected to rise to 2500, 900, and 2900 in 2035, respectively.

Residential locations are often characterized by their proximity to traffic in epidemiologic studies (e.g., within 20–500 m of a busy highway with traffic flow rates from 10,000–100,000 vehicles per day). More attention is being paid by researchers to the relationships between pollution levels in the environment and health consequences. The proximity exposure paradigm essentially holds that a person's residential location and their personal exposure to outdoor air pollution concentrations are correlated. However, since most people spend most of their time indoors, where certain air pollutants are less concentrated than outside, this is not true. Furthermore, the accuracy of this assumption could be affected by the length of the relevant exposure (weeks to decades), the type of health impact (acute vs. chronic), as well as the subjects' movability. Given that they don't spend the majority of the day in the residential area of interest, this assumption may present issues for working adults and school-age children. Moreover, numerous research projects have been created to highlight the connections between traffic-related air pollutants and the health consequences that follow. Differentiating between the health effects of indoor and outdoor exposures can be challenging when a traffic-related pollutant, such as CO, NO_2, and $PM_{2.5}$, also has significant indoor sources (e.g., cooking or biomass burning).

The answer is to distribute criteria air pollutants among various sources so that, even though the target pollutant's resulting spatial map is meant to represent that pollutant (alone), it actually includes all of the co-pollutants from the various source types involved. An accurate proxy for exposure to activities throughout the day is the personal exposure concentration, which takes into account concentrations indoors, outdoors, and near sources. Therefore, rather than focusing on the relationships between individual pollutants and ambient concentrations, air pollution control strategies may be more successful when they are based on the relationships between particular types of emission sources and personal exposure concentrations.

7.4.3. The Contributions of Mobile Sources to PM and O_3 in Cities around the World

The chemical makeup and health effects of gaseous pollutants and $PM_{2.5}$ related to traffic were covered in the section before this one. The contributions of road emissions to $PM_{2.5}$ and O_3 in urban settings worldwide will be examined in this section.

The SA technique is typically used to calculate the contributions of mobile sources to $PM_{2.5}$, which is a crucial task in the management, control, and formulation of policy options for air pollution. Receptor models are SA methods such as positive matrix factorization (PMF), principal component analysis, chemical mass balance (CMB), factor analysis with multiple regression, Unmix, and back-trajectory analysis that rely on statistical analysis of observations at monitor sites. A unique profile for every major source must be entered into the CMB model, but the PMF multivariate model can identify factors without requiring source composition data. Because PMF is practical and easy to use, it is the most widely used method for $PM_{2.5}$ SA.

A summary of the comparative description is provided in Table 7.4 to obtain a direct view of the recently reported contributions to $PM_{2.5}$ from mobile sources. Mobile source contributions ranged from 7.8%–44.6% of $PM_{2.5}$ in these 20 global cities, with a mass of 1.2–28.7 μg m^{-3}. Based on the findings, it was clear that various cities across all six continents shared traffic as a common source. The average mass of mobile source emissions in the megacities of the USA, Europe, and Australia contributed 1.2–4.9 μg m^{-3} of the $PM_{2.5}$ mass concentration, whereas in China, India, and Mexico, the mass of mobile source emissions ranged from 5.9–28.7 μg m^{-3}. This indicates that most megacities in developing nations still struggle with controlling mobile source emissions. A significant portion of $PM_{2.5}$ in certain cities in South America and Africa was attributed to mobile source air pollution, most likely as a result of antiquated car technology in those areas. Because human activities and economic development vary greatly among cities, it is evident that there is significant spatial variability in mobile source contributions within cities globally (Table 7.4).

Table 7.4. *Summary of Original Research on Cognitive Impact of Air Pollution Across Life Course*

SL. No.	Author (Year)	Study Site	Study Population	Exposure Studied	Outcome Variable Studied	Cognitive Impact
A	Studies from High Income Settings					
1	Lee et al. (2019)	South-eastern part of the United States	94 million follow-up records from fee-for-service Medicare records for 13 million Medicare beneficiaries of fee for service (FFS) residing in the southeastern United States (U.S.) from 2000–2013.	Spatially and temporally continuous $PM_{2.5}$ exposure data	Hospitalization rates for dementia	Long-term exposure to a high $PM_{2.5}$ levels associated with increased hospitalization with dementia per 1 μg/m^3 increase in annual $PM_{2.5}$, with higher risk for vascular dementia. Did not report analysis stratified by sex

2	Lo et al. (2019)	Taiwan	2241 community-dwelling, free-living elderly population with mean age at the time of recruitment 73.62 years; M:F = 57.5:42.5) followed from 1996–2007	PM_{10} Ozone	Short Portable Mental Status Questionnaire	Long-term exposure to PM_{10} and O_3 associated with cognitive impairment with greater impact of the joint effect of exposure to PM_{10} and O_3 Did not report analysis stratified by sex
3	Molina-Sotomayor et al. (2019)	Chile	181 older women, patients of the "La Estrella" Health Center of the Pudahuel commune, Metropolitan Region of Santiago de Chile, and patients of the Senior Centers of Viña del Mar City-Chile	Average PM_{10} $PM_{2.5}$ NO_2 SO_2 Ozone Concentration between 2012 and 2014 Cardio-respiratory Aerobic exercise indicators: Maximum oxygen uptake (VO_{2max}), estimated by the Six-Minute Walk Test (6mWT); heart rate (HR); and oxygen saturation (SpO_2).	Mini Mental State Examination (MMSE)	Significant differences ($p < 0.05$) between the Active Group residing in relatively less polluted areas versus sedentary group residing in more polluted areas on all the MMSE dimensions except "Registration," and in all the physiological variables (VO_{2max}, SpO_2, HR). Aerobic exercise may be a protective factor adverse cognitive effects of air pollution have on cognition probably due to improvement in the mechanisms of oxygen transport.

4	Andersson et al. (2018)	Umea, Sweden	Data of 1721 participants aged 55–85 years at baseline (Male: Female = 985:736 of Betula project, a longitudinal study of health and ageing aged 55–85 years at baseline	Estimates of annual mean levels of nitrogen oxides (NO_x) at the participants' residential address using a land-use regression model. Modelled data for road traffic noise levels at the participants' residential address	Dementia incidence	302 of 1721 participants at baseline, 302 developed dementia during the follow up period. Residing in the two highest quartiles of NO_x exposure was associated with an increased risk of dementia which was not modified by adjusting for noise. Did not report analysis stratified by sex
5	Berghuis et al. (2018)	Netherlands	101 children aged 13–15 years (M:F 55:46) participating in Development at Adolescence and Chemical Exposure (DACE)-study a follow-up of two Dutch birth cohorts.	Maternal pregnancy serum levels of PCB-153 and three OH-PCBs, 9 PCBs, 5 polybrominated diphenyl ethers (PBDEs), dichloroethene (DDE), pentachlorophenol (PCP) and hexabroomcyclododecane (HBCDD) in different parts of the cohort Maternal smoking and alcohol use	Wechsler Intelligence Scale for Children, third edition, Dutch version (WISC-III-NL) Dutch version of the Rey's Auditory Verbal Learning Test (AVLT) Movement Assessment Battery for Children (Movement-ABC)	Significant association between PCB-183 and lower total intelligence, HBCDD with lower performance intelligence and PBDEs with lower verbal memory Positive trends between OH-PCBs and verbal intelligence, and a negative trend between BDE-153 and fine motor skills was observed

						Several OH-PCBs associated with more optimal sustained attention and balance CONTD ON NEXT PAGE Boys had poorer outcomes on fine motor skills and better outcomes on ball skills with positive trend was seen between 4-OH-PCB-107 and fine motor skills In girls, Prenatal DDE levels were associated with (sub) clinical motor performance
6	Carey et al. (2018)	London, UK	Retrospective cohort of 130,978 adults aged 50–79 years	Average annual concentrations of NO_2, $PM_{2.5}$, ozone (O_3), traffic intensity, distance from major road and night-time noise levels	Clinical diagnosis of Dementia	Positive exposure-response relationship between dementia and all measures of air pollution except O_3. Did not report gender based findings

7	Cullen et al. (2018)	UK	86,759 middle- to older-aged adults from the UK Biobank	Cumulative impact of outdoor air pollution exposure (PM_{10}, $PM_{2.5}$, NO_2, NO_x)	Cognitive function including reasoning test, pairs matching test, reaction time, prospective memory, visuospatial memory, numeric memory	Weak association between air pollutant exposure and cognitive performance at baseline (dose-dependent lower reaction time, higher error rate on a visuospatial memory test and lower numeric memory scores) with no such association at 2.8 years follow-up. Did not report analysis stratified by sex
8	Guxens et al. (2018)	Netherlands	Data from population-based birth cohorts—GENERATION R (The Netherlands) (2002–2006) that recruited mother infant dyads including 8879 pregnant women and 1932 children born between April 2002 and January 2006 were taken of which 783 children between ages of 6–10 years participated in MRI sub-study	Prenatal exposure to air pollutants such as (NO_2, NO_x) in all regions and PM ($PM_{2.5}$, PM_{10} and PM_{coarse}) and $PM_{2.5}$ absorbance in a subgroup using land-use regression models for the period 2008–2011 and then modelled for air pollutant profile for the exact pregnancy periods using background monitoring sites.	MRI Brain Cognitive and psychomotor development	Air pollution exposure was not associated with global brain volumes Higher prenatal PM exposure per 5 $\mu g/m^3$ had a dose-dependent thinning of cortex increase in several brain regions of both hemispheres (e.g., cerebral cortex of the precuneus in Right hemisphere). Reduced cerebral cortex in precuneus and rostral middle frontal regions partially mediated

						the association between PM exposure and impaired inhibitory control. Did not report analysis stratified by sex
9	Kerin et al. (2018)	California, USA	327 children with Autism Spectrum Disorder (ASD) from the Childhood Autism Risks from Genetics and the Environment study	Exposure to NO_2, $PM_{2.5}$ and PM_{10}, ozone, and near-roadway air pollution in each trimester of pregnancy and first year of life.	Mullen Scales of Early Learning (MSEL), the Vineland Adaptive Behavior Scales (VABS), and the Autism Diagnostic Observation Schedule calibrated severity score.	ASD severity not associated with any air pollutant exposure. Prenatal and First year exposure to NO_2 associated with impaired cognitive performance, adaptive functioning and behavioral indices but not severity of ASD 3rd Trimester PM_{10} exposure was paradoxically associated with improved behavioral performance Did not report analysis stratified by sex
10	Oudin et al. (2018)	Umea, Norhern Sweden	1806 participants from Betula project from Umeå, Northern Sweden enrolled between (1993–1995) and followed upto 2010	Modelled levels of source-specific residential fine PM exposure to wood stoves or wood boilers and traffic	Validated data on dementia diagnosis	Increased dose-dependent risk for incident dementia (Vascular Dementia and Alzheimer's Disease) with local residential

						wood burning and traffic exhaust Did not report analysis stratified by sex
11	Ailshire et al. (2017)	USA	779 U.S. adults age ≥ 55 years from the 2001/2002 wave of the Americans' Changing Lives study	Annual average $PM_{2.5}$ concentration in 2001 in the area of residence by linking respondents with EPA air monitoring data using census tract identifiers. Exposure to neighborhood social stressors using perceptions of disorder and decay including subjective evaluations of neighborhood upkeep, presence of deteriorating/abandoned buildings, trash, and empty lots.	Error rate on Short Portable Mental Status Questionnaire (SPMSQ).	Association between higher rates of cognitive errors with high concentrations of $PM_{2.5}$ which was stronger in high stress neighborhoods indicating towards a possible role of social stressors and environmental hazards No statistically significant gender based differences found

12	Alvarez-Pedrerol (2017)	Barcelona, Spain	1234 children aged 7–10 years from 39 schools who commuted to school by foot	TRAP exposure (Average $PM_{2.5}$, Black Carbon (BC) and NO_2 concentrations) for the shortest walking route to school	Working Memory (the three-back numbers test) and inattentiveness (hit reaction time standard error of the Attention Network Test)	$PM_{2.5}$ and Black Carbon were associated with a reduction in the growth of working memory with no significant association of working memory with NO_2 Associations of TRAP concentrations (BC and $PM_{2.5}$) and working memory were stronger for males than females
13	Cacciottolo et al. (2017)	USA	3647 women aged 65–79 years from Women's Health Initiative Memory Study (WHIMS)	$PM_{2.5}$	Global cognitive decline and all-cause dementia APOE □4/4status	Residence in places with high $PM_{2.5}$ was associated with an increased risk for global cognitive decline and all-cause dementia by 81% and 92% respectively, with risk exacerbated by APOE □4/4 allele status

14	Chen et al. (2017)	Ontario, Canada	All Ontario residents who were 55–85 years old on 1 April 2001, Canadian-born, and free of physician-diagnosed dementia who were followed up to 2013	Long-term average residential exposure to Air Pollutants including $PM_{2.5}$, NO_s and ozone	Dementia incidence	Every interquartile-range increase in exposure to $M_{2.5}$ and NO_2 even at low levels of air pollution associated with higher incidence of dementia No statistically significant gender based differences
15	Colicino et al. (2017)	USA	428 older men in the Veterans Affairs (VA) Normative Aging Study	Black carbon	MMSE Score Telomere length C Reactive Protein	Black Carbon associated with lower cognition. Each doubling in BC level associated with 1.57 (95% CI: 1.20, 2.05) times higher odds of low MMSE scores in individuals with longer blood Telomere length (OR = 3.23; 95% CI: 1.37, 7.59; $p = 0.04$ for BC-by-TL-interaction).

16	Forns et al. (2017)	Barcelona, Spain	1439 of 2897 children recruited from 39 schools across Barcelona who participated in the BREATHE project 2012/2013	Composite exposure to indoor and outdoor levels of various TRAPs such as elemental carbon (EC), nitrogen dioxide (NO_2), $PM_{2.5}$ and ultrafine particles (UFP) at school	Working Memory	Slower development of working memory in children over 3.5 years period associated with higher schools based exposure to air pollution Did not report analysis stratified by sex
17	Lett et al. (2017)	USA	Sub population of Early Childhood Longitudinal Study, Birth Cohort (*n* = 4050)	Isophorone exposure using 2002 National Air Toxics Assessment levels Home learning environment assessed with a modified Home Observation for Measurement of the Environment (HOME) Inventory	Standardized math assessment scores as a measure of early cognitive skills.	High isophorone levels (>0.49 ng/m^3) and low HOME score were associated low math scale score. Decrement in math scale score was more than the additive effect of each exposure especially in male children.
18	Oudin et al. (2017)	Umeå, Northern Sweden	1469 participants aged 60–85 years from Betula project followed up every five years from 1988–2010	Cumulative annual residential mean of NO_x (marker of long-term exposure to TRAP)	Episodic memory	No association between long-term exposure to air pollution especially TRAP and episodic memory. Did not report analysis stratified by sex

19	Stingone et al. (2017)	USA	6900 children enrolled in the Early Childhood Longitudinal Study Birth Cohort	Residential concentrations of 104 ambient air toxins (including trichloroethylene, isophorone) from the National Air Toxics Assessment (2002) at age 9 months as per ZIP codes	Mathematics Tasks score	High isophorone levels (>0.49 ng/m^3) were associated with low mathematics task scores in urban and highly populated urban areas Did not report analysis stratified by sex
20	Sunyer et al. (2017)	Spain	2687 school children from 265 classrooms in 39 schools across Barcelona, Spain over one year from January 2012 to March 2013	Ambient TRAP exposure with daily levels of nitrogen dioxide (NO_2) and elemental carbon (EC) in $PM_{2.5}$ measured from $PM_{2.5}$ filters at a fixed air quality background monitoring station and in schools.	Computerized child Attention Network test (ANT)	TRAP was associated with attentional impairment Daily ambient levels of NO_2 and $PM_{2.5}$ contributed to an estimated equivalent to a 1.1 month delay in age appropriate improvement in response speed as part of natural developmental trajectory. Boys had worse cognitive performance on Hit Reaction Time with incremental increase of both Elemental Carbon and NO_2

21	Tallon et al. (2017)	USA	3377 participants aged 57–85 years (from Wave 2, August 2010 to May 2011 in National Social Life, Health, and Aging Project (NSHAP) cohort study	$PM_{2.5}$ exposure (estimated using GIS-based spatio-temporal models) and nitrogen dioxide (NO_2) exposures (obtained from EPA monitors).	Chicago Cognitive Function Measure (CCFM)	High $PM_{2.5}$ exposures associated with decrease in Chicago Cognitive Function Measure scores equivalent to aging by 1.6 years for $PM_{2.5}$ and 1.9 years for NO_2 exposure. Cognitive impact of $PM_{2.5}$ was modified by stroke, anxiety, stress and mediated by depression. Cognitive impact of NO_2 were mediated by stress with effect modification by impaired activities of daily living. Did not report analysis stratified by sex
22	Tzivian et al. (2017)	Bochum, Essen, and Mülhei, Germany	Heinz Nixdorf Recall population based cohort study with 4086 participants	Land use regression was used to assess long-term residential concentrations for size-fractioned PM and nitrogen oxides. Assessment of road traffic noise	Cognitive assessment using five neuropsychological subtests and an additively calculated global cognitive score	Association of air pollutants with cognitive dysfunction was amplified by higher noise exposure at high levels of exposure. Did not report analysis stratified by sex

23	Best et al. (2016)	USA	(*n* = 454; age ≥ 60 years from the 2001–2002 National Health and Nutrition Examination Survey).	Urinary 1-hydroxypyrene (indicator of PAH exposure)	Digit Symbol Substitution Test (DSST)	Dose-dependent 1% increase in urinary 1-hydroxypyrene associated with 1.8% poorer performance on DSST
24	Oudin et al. (2016)	Sweden	1469 persons aged 60–85 years at inclusion in the Betula project and followed up to 22 years, five years apart between 1988 and 2010	Exposure to traffic-related air pollution	Dementia incidence	Participants in the group with the highest exposure to TRAP more likely to be diagnosed with dementia. Did not report analysis stratified by sex
25	Porta et al. (2016)	Rome, Italy	474 children from a birth cohort of 719 newborns enrolled in 2003–2004 as part of the GASPII project evaluated at the age of 7 years	Air pollutants (NO_2, PM_{coarse}, $PM_{2.5}$, $PM_{2.5}$ absorbance) during pregnancy and at birth Maternal smoking	IQ assessed with Wechsler Intelligence Scale for Children-III	Traffic intensity in a 100 m buffer around home and an incremental 10 μg/m^3 higher exposure of NO_2 exposure in intra-uterine period was associated reduced verbal IQ and verbal comprehension IQ. Other pollutants also showed negative associations with much larger confidence intervals. Did not report analysis stratified by sex

26	Tzivian et al. (2016)	Bochum, Essen, and Mülhei, Germany	Heinz Nixdorf Recall population based cohort study with 4086 participants aged 50–80 years old	Land use regression was used to assess long-term residential concentrations for size-fractioned PM and nitrogen oxides. Assessment of road traffic noise	Cognitive assessment using five neuropsychological subtests and an additively calculated global cognitive score (GCS)	Long-term exposures to AP and traffic noise are negatively correlated with four subtests including memory and executive functions and GCS in dose-dependent relationship independent of noise exposure e.g., an interquartile range rise in $PM_{2.5}$ correlated with verbal fluency, labyrinth test, and immediate and delayed verbal recall. Did not report analysis stratified by sex
27	Tzivian et al. (2016)	Vide supra	Vide supra	Vide supra	Diagnosis of Mild Cognitive Impairment based on five neuropsychological tests (Vide supra) and subjective memory complaint	Positive dose-dependent association between long-term $PM_{2.5}$ exposure and mild cognitive impairment, mainly amnesic subtype (aMCI) Did not report analysis stratified by sex

28	Chen et al. (2015)	USA	1403 community-dwelling older women aged 65–80 years without enrolled in the Women's Health Initiative Memory Study (WHIMS), 1996–1998	Cumulative $PM_{2.5}$ exposure in 1999–2006 Based on given residential histories and air monitoring data	MRI Brain	Greater $PM_{2.5}$ exposures associated with significantly smaller white matter (WM) volumes in frontal and temporal lobes and corpus callosum (equivalent to 1–2 years of brain ageing), but not of gray matter or hippocampus
29	Harris et al. (2015)	Eastern Massachusetts, USA	1109 mother-child pairs in Project Viva, a prospective birth cohort study in eastern Massachusetts (USA)	Prenatal and childhood exposure to TRAPs including black carbon (BC) and $PM_{2.5}$ assessed by distance of residence from roadways and traffic density	Verbal and nonverbal intelligence, visual motor abilities, and visual memory assessed at mean age of 8 years	Children with a residence less than 50 m away from major highway had lower nonverbal IQ and lower verbal IQ and visual-motor abilities Equivocal findings for cross-sectional associations with major roadway proximity, prenatal and childhood exposure to traffic density and $PM_{2.5}$, third-trimester and childhood BC exposures Did not report analysis stratified by sex

30	Peterson et al. (2015)	New York, USA	40 children aged 7–9 years born to any of the 665 urban Latina (Dominican) or African American women 18–35 years old who were not cigarette smokers or users of other tobacco products or illicit drugs, with initial prenatal care by the 20th week of pregnancy, and who were free of diabetes mellitus, hypertension, and known human immunodeficiency virus recruited between 1998 and 2006 through the local prenatal care clinics who had completed survey and who had a full range of prenatal PAH exposure levels; no or very low prenatal exposure to environmental tobacco smoke (classified by maternal report validated by cotinine levels of less than 15 μg/L in umbilical cord blood) and low chlorpyrifos	Prenatal airborne PAH exposure by the sum of 8 Nonvolatile PAH: benzo[a] anthracene, chrysene/ iso-chrysene, benzo[b] fluoranthene, benzo[k] fluoranthene, benzo[a] pyrene, indeno [1,2,3c, d]pyrene, dibenzo [a, h] anthracene, and benzo [g, h, i]perylene. measured with personal air monitoring of the mothers over a 48-h period in the third trimester of pregnancy but assumed to be index exposure for the entire gestation Prenatal PAH exposures validated against concurrent 2-week monitoring of residential air samples in the final trimester of pregnancy Postnatal PAH exposure by spot	CBCL WISC-IV MRI Brain	A dose-response inverse relationship between prenatal PAH exposure and reductions of the white matter surface in most of the frontal, superior temporal, and parietal lobes, as well as the entire rostrocaudal extent of the mesial surface, in the left but not the right hemisphere of the brain and reduced white matter in later childhood associated with slower information processing speed during intelligence testing and more severe externalizing behavioral problems such as ADHD and Conduct disorder. Postnatal PAH exposure correlated with white matter surface measures in bilateral

			exposure (below 4.39 pg/g) 20 children were randomly selected from above the median PAH level group and 20 from below the median PAH level group	urine sample at age of 5 years for 9 metabolites (1-Hydroxynaphthalene, 2 Hydroxynaphthalene, 2-Hydroxyfluorene, 3-Hydroxyfluorene, 9-Hydroxyfluorene, 1 Hydroxyphenanthrene, 2-Hydroxyphenanthrene, 3-Hydroxyphenanthrene, and 4 Hydroxyphenanthren.)		dorsal prefrontal cortex bilaterally independent of prenatal PAH exposure Stronger inverse associations of prenatal PAH exposure with white matter surface measures in girls as compared to boys
31	Schikowski et al. (2015)	Ruhr Area and Borken, Germany	4874 women from the SALIA cohort (aged 55 years at baseline) enrolled between 1985 and 1994 and followed up in 2006 (*n* = 2116) and 2008 (*n* = 834) of which complete information available (*n* = 789)	Particulate matter (PM) size fractions and nitrogen oxides (NO_x) Traffic load	CERAD-Plus test Apolipoprotein E (ApoE) alleles	Air pollution was inversely related to visuospatial abilities on cognitive assessment with significant adverse association of traffic load in carriers of ApoE -4 risk alleles.

32	Sunyer et al. (2015)	Barcelona, Spain	2715 children aged 7–10 years from 39 schools in Barcelona	Chronic traffic air pollution [elemental carbon [(EC), nitrogen dioxide (NO_2), and ultrafine particle number (UFP; 10–700 nm)] measured twice during 1-wk campaigns both in the courtyard (outdoor) and inside the classroom (indoor)	*n*-back and the attentional network tests	Detrimental associations between Traffic related air pollution and cognitive performance were stronger in boys than in girls
33	Ailshire and Clarke (2014)	USA	Cross sectional data of 780 non-Hispanic black and white men and women aged ≥ 55 years from the 2001/2002 Americans' Changing Lives Study	$PM_{2.5}$ using PA air monitoring data linked to respondents using census tract identifiers.	Tests of working memory and orientation	Exposure to high $PM_{2.5}$ concentrations associated with 1.5 times greater error rate Male gender was associated with more cognitive errors
34	Ailshire and Crimmins (2014)	USA	13,996 men and women aged 50 years or older from the 2004 HRS survey	Residence in areas with higher $PM_{2.5}$ concentrations	Cognitive Function	Living in areas with higher $PM_{2.5}$ concentrations was associated with worse cognitive function especially episodic memory No statistically significant gender based differences

35	Gatto et al. (2014)	Los Angeles Basin, USA	1496 individuals (mean age of 60.5 years)	Air pollutants [O_3, $PM_{2.5}$ and nitrogen dioxide (NO_2)]	Six domains of cognitive function and global cognition	Increased exposure to $PM_{2.5}$ associated with lower verbal learning Exposure to $NO_2 > 20$ ppb associated with lower logical memory Exposure to O_3 above 49 ppb associated with lower executive function No air pollutant significantly associated with global cognition. No statistically significant gender based differences
36	Guxens et al. (2014)	Europe (Netherlands, Germany, France, Italy, Greece, Spain)	Mother-, infant pairs recruited between 1997 and 2008 yielding a total sample of 9482 children from 6 European population-based birth cohorts—GENERATION R (The Netherlands), DUISBURG (Germany), EDEN (France), GASPII (Italy), RHEA (Greece), and INMA (Spain)	Prenatal exposure to air pollutants such as (NO_2, NO_x) in all regions and PM ($PM_{2.5}$, PM_{10} and PM_{coarse}) and $PM_{2.5}$ absorbance in a subgroup using land-use regression models for the period 2008–2011 and then modelled for air pollutant profile for the exact pregnancy periods using background monitoring sites.	Assessment for cognitive and psychomotor development at 1 and 6 years of age.	Prenatal air pollution exposure during pregnancy, particularly NO_2, was associated with dose-dependent reduction in psychomotor development but not cognitive development. Did not report analysis stratified by sex

37	Tonne et al. (2014)	London, UK	2867 white men retired from work (mean age 66 years) from Whitehall II cohort	Particulate matter from traffic exhaust	Alice Heim 4-I test, 20-word free-recall test, semantic and phonemic verbal fluency	Higher $PM_{2.5}$ of 1.1 µg/m³ was negatively associated with reasoning and memory but not verbal fluency and significant 5-year decline in standardized memory score.
38	Loop et al. (2013)	USA	20,150 men and women enrolled in the REasons for Geographic And Racial Differences1 in Stroke (REGARDS) cohort	Satellite-derived estimate of $PM_{2.5}$ concentration map	Cognition	No consistent increase in odds of cognitive impairment with every 10 µg/m³ increase in $PM_{2.5}$ concentration. No statistically significant gender based differences
39	Power et al. (2013)	USA	628 men (mean age of 70 years) from the VA Normative Aging Study	TRAP exposure on cognitive function.	HFE C282Y variant (hemochromatosis gene polymorphisms) Cognitive function	Older adults lacking HFE C282Y variant had greater adverse cognitive impact of TRAP exposure

40	Weuve et al. (2012)	USA	19,409 US women aged 70–81 years from Nurses' Health Study Cognitive Cohort	Long-term exposure to higher levels of both $PM_{2.5-10}$ and $PM_{2.5}$	Cognition	Dose-dependent association between Long-term exposure to higher levels of both $PM_{2.5-10}$ and $PM_{2.5}$ and significantly faster cognitive decline at 2 years. Every 10 μg/m^3 increment in long-term (2 years) exposure to $PM_{2.5-10}$ and $PM_{2.5}$ cognitively equivalent to aging by approximately 2 years.
41	Power et al. (2011)	USA	680 men (mean ± SD, 71 ± 7 years of age) from the U.S. Department of Veterans Affairs Normative Aging Study	Traffic-related air pollution including Black carbon (BC) exposure	Mini Mental Status Examination (MMSE)	Significant association between Black carbon (BC) exposure and lower MMSE score

42	Edwards et al. (2010)	Krakow Poland	214 offspring of a cohort of pregnant, healthy, nonsmoking women of Krakow, Poland, between 2001 and 2006	Maternal 48-hr personal air monitoring to estimate foetal Polyaromatic Hydrocarbon air pollutant exposure and PAH estimation from maternal blood sample and/or a cord blood sample at the time of delivery	Raven's Colored Progressive Matrices (RCPM) at 5 years of age	Higher (above the median of 17.96 ng/m^3) prenatal exposure to airborne PAHs associated with decreased scores on RCPM corresponding to an estimated average decrease of 3.8 IQ points. Did not report gender-based differences
43	Freire et al. (2010)	Spain	210 boys from a population-based birth cohort from southern Spain	NO_2	Cognitive development	No significant association between NO_2 and cognitive development
44	Chen and Schwartz (2009)	USA	1764 adult participants (aged 37.5 ± 10.9 years) of the Third National Health and Nutrition Examination Survey in 1988–1991	Geocoded Residential Ambient annual PM_{10} and ozone	Neurobehavioral Evaluation System-2 (NES2) data (including a simple reaction time test [SRTT] measuring motor response speed to a visual stimulus;	Consistent associations between ozone and not PM_{10} and reduced performance on symbol-digit substitution test and a serial-digit learning test but not in simple reaction time test. Each 10-ppb increase in annual ozone

					a symbol-digit substitution test [SDST] for coding ability; and a serial-digit learning test [SDLT] for attention and short-term memory	was associated with equivalent to 3.5 and 5.3 years of aging-related decline in cognitive performance. Did not report analysis stratified by sex
45	Perera et al. (2009)	New York, USA	249 Children born to non-smoking black or Dominican-American women residing in New York City who had undergone prenatal ambient personal PAH monitoring	PAH exposure	Wechsler Preschool and Primary Scale of Intelligence-Revised (WPPSI-R)	High Prenatal Polyaromatic Hydrocarbons (PAH) levels (above the median of 2.26 ng/m^3) inversely associated with full-scale IQ and verbal IQ with a decrement of 4.31–4.67 points of IQ at age of 5 years. Did not report analysis stratified by sex
46	Ranft et al. (2009)	Germany	399 women aged 68–79 years	Traffic-related fine PM	Mild Cognitive Impairment (MCI)	Exposure to traffic-related fine PM consistent and significant risk factor for MCI

47	Lee et al. (2007)	USA	278 children aged 12–15 years included in the National Health and Nutrition Examination Survey 1999–2000	POPs such as 3, 3,'4, 4,' 5-pentachlorob-iphenyl, 1, 2, 3, 4, 6, 7,8-heptachlorodibenzo-p-dioxin (HPCDD), 1, 2,3, 4,6 ,7,8, 9-octachlorod-ibenzo-p-dioxin (OCDD), 1,2, 3, 4,6,7, 8-heptachlorod-ibenzofuran (HPCDF), beta-hexachlorocyclohexane, p, p'-dichlorodiphenyltrichloroethane, and trans-nonachlor.	Prevalence rates of learning disability (LD) and attention deficit hyperactivity disorder (ADHD), both of which are characterized by cognitive impairment	Direct association between POPs and LD/ ADHD. No statistically significant gender based differences found
B	Studies from Upper and Middle Income Settings					
1	Saenz et al. (2018)	Mexico	13,023 Mexican adults over age 50 participating in 2012 Wave of the Mexican Health and Aging Study	Indoor air pollution (Use of wood or coal as primary cooking fuel)	Verbal learning, verbal recall, attention, orientation and verbal fluency	Exposure to indoor air pollution associated with poorer cognitive performance Did not report analysis stratified by sex

2	Zhang et al. (2018)	China	25,486 individual respondents over age 10 in 2010 and 2014, from China Family Panel Studies (CFPS)	Cumulative and transitory exposures to air pollution	Verbal and Math tests	Adverse cognitive impact of air pollution on performance in verbal and math tests with greater deficits on verbal tasks as people aged. Significantly worse verbal tests performance in males on verbal tasks with short-term and cumulative air pollution exposure
3	Calderón-Garcidueñas et al. (2012)	Mexico	30 children (20 from Southwest Mexico City (SWMC) and 10 from Polotitlan)	High pollution versus low pollution areas	Frontal tau hyperphosphorylation with pre-tangle material amyloid-beta diffuse plaques	Nearly 40% of highly exposed children and young adults had frontal tau hyperphosphorylation with pre-tangle material and 51% had amyloid-beta diffuse plaques compared versus 0% of controls living in low pollution areas. Did not report analysis stratified by sex

5	Calderón-Garcidueñas et al. (2011)	Mexico	20 children from Mexico City (Mean age = 7.1 years, SD = 0.69) and 10 children from Polotitlán (Mean age = 6.8 years, SD = 0.66)	High pollution Areas versus Low pollution areas	MRI Brain Wechsler Intelligence Scale for Children-Revised (WISC-R) Inflammatory markers interleukin 12, cytokines and chemokines	Complex modulation of cytokines and chemokines influences children's central nervous system structural and volumetric responses and cognitive correlates resulting from environmental pollution exposures Did not report analysis stratified by sex Exposure high pollution associated with low white matter volumes especially in the right parietal and bilateral temporal areas irrespective of white matter abnormalities as well as progressive deficits on WISC-R Vocabulary and Digit Span subtests

5	Calderón-Garcidueñas et al. (2011)					White matter hyperintensities associated with evidence of inflammation, immunoregulation, and tissue remodeling on MRI. Children without white matter hyperintensities had raised levels of interleukin 12, pro-inflammatory cytokines, and low levels of neuroprotective cytokines and chemokines. Did not report analysis stratified by sex
6	Calderón-Garcidueñas et al. (2008)	Mexico	55 Children (mean age: 9.2 years) from Mexico City with high air pollution and 18 children (mean age: 10.5 years) from Polotitlán with low air pollution	Air Quality	Psychometric testing MRI Brain	Residence in high air pollution area associated with deficits in a combination of fluid and crystallized cognition tasks, high rates of prefrontal white matter hyperintense lesions Did not report analysis stratified by sex

Source: Chandra M, Rai CB, Kumari N, Sandhu VK, Chandra K, Krishna M, Kota SH, Anand KS, Oudin A. Air Pollution and Cognitive Impairment across the Life Course in Humans: A Systematic Review with Specific Focus on Income Level of Study Area. International Journal of Environmental Research and Public Health. 2022; 19(3):1405. https://doi.org/10.3390/ijerph19031405.

In urban environments, ground-level O_3 formation is caused by road traffic emissions, specifically NO_x, CO, and VOCs. In July of 1990, total road traffic emissions contributed more than 12% near the surface to the zonally averaged O_3 distribution in mid-latitudes and arctic latitudes of the Northern Hemisphere, according to a study that examined the global impact of road traffic emissions on ground-level O_3 formation using a general circulation model coupled with a chemistry module. In both the northern and southern extratropics, road traffic emissions contributed more than 8% of the zonally averaged O_3 distribution near the surface in January 1990. Summertime increases in the surface O_3 concentration at mid-latitudes in the Northern Hemisphere are usually between 5% and 15%, as a result of road traffic emissions in 1995. Estimating the effects of road traffic emissions in 1997, car emissions increase the summertime boundary layer O_3 by 20–40% in the Northern Hemisphere's urbanized regions, 25% in the Mediterranean, 40% in Southeast Asia and California, and 30% in Japan. However, during the winter months, O_3 falls by roughly 5%–20% in East Asia, California, and the Northeastern United States. The global average impact of road traffic emissions on changes in O_3 in 2000 indicated that approximately 36.7% of increases in O_3 were due to road traffic emissions. A 30% reduction in industrial and transportation sources had the greatest effects on O_3 concentrations, with a maximum decrease of 20 ppbv, in China's heavily industrialized and urbanized Beijing–Tianjin–Heibei Region. These modeling studies suggest that the current road vehicle fleet is having a moderate impact on changes in tropospheric O_3, especially in the Northern Hemisphere.

As the previous section demonstrated, mobile emissions have a well-documented impact on $PM_{2.5}$ mass and surface O_3 concentrations in all inhabited continents (Asia, Europe, North and South America, Africa, and Oceania), with significant impacts on both. With exhaust $PM_{2.5}$ and O_3 linked to heightened public health risks and the potential for climate change, cost-effective control measures and cutting-edge technology interventions that give priority to mobile source mitigation may also be more beneficial in preventing both public health issues and climate change.

7.5. IMPACTS OF AIR POLLUTANTS ON THE NATURAL AND BUILT ENVIRONMENTS

There is ample evidence of the detrimental effects that air pollutants from vehicle emissions have on ecosystems, built environments, and natural environments. Air pollution has significant negative effects on all types of ecosystems and poses threats to the conservation of biodiversity that are not unique to vehicle emissions. Even with frequent cleaning, the facades of both modern and culturally significant buildings in several globally significant cities still have dark, grimy appearances. The primary causes of this building's blackening are emissions from traffic and anthropogenic air pollutants from industrial processes. One of the main types of deterioration that leads to black soiling in these contemporary or cultural buildings is acknowledged to be black crusts. Black crusts are mostly formed on carbonate building materials, such as limestone and marble, which react with an atmosphere that is loaded with SO_2 to change calcium carbonate (calcite) into calcium sulfate dihydrate (gypsum).

Following its release from its primary sources (i.e., traffic and industrial emissions), SO_2 can go through many intricate processes, such as oxidation to sulfur trioxide and the aqueous formation of sulfuric acid while it is in the atmosphere. Either a wet or dry deposition process can be used to transfer sulfuric acid aerosols onto the building's surface, causing gypsum to form. Since soot particles embedded in black crusts contain transition metals (e.g., V, Ti, Fe, Mn, and Cu) necessary for the oxidation of SO_2, they can catalyze the destructive action during this process. These crusts are not only unsightly, but they also contain a variety of organic compounds and microorganisms. The latter may have a role in the calcareous substrata's bio-deterioration, which may yield large amounts of hydrated calcium oxalate. Nowadays, it is generally acknowledged that the black crusts that build up on buildings are made up of both noncarbonate and carbonate fractions. The organic carbon (OC) of biogenic and anthropogenic origin and the EC make up the non-carbonate fraction.

Del Monte et al. identified two varieties of these particles: (1) spherical in shape, originating from an oil combustion source, with a rough and irregular surface and high porosity; and (2) spherical in shape, originating from a coal combustion source, with a smooth surface. Using gas chromatography-mass spectrometry, Saiz-Jimenez examined the black crusts that covered the exteriors of buildings in Mechelen, Seville, and Dublin. He found that the petroleum derivatives included fatty acids, hopanes, and n-alkanes, among other organic molecular markers. Quantitative information about the composition of carbonate and non-carbonate fractions of black, sulfated crusts is generally agreed upon; however, only recently have data on the fractions of historical buildings and their environs been reported.

Carbon and sulfur were found to be the main elements in the black crusts of European cultural heritage buildings, as demonstrated by Bonazza et al. The Milan Cathedral in Milan, Italy; the S Maria del Fiore in Florence, Italy; the Corner Palace in Venice, Italy; the Vittoriano in Rome, Italy; St. Eustache Church in Paris, France; the Tuileries in Paris, France; the Seville Cathedral in Seville, Spain; the Tower of London in London, United Kingdom; and St. Rumbold's Cathedral in Mechelen, Belgium are some of these particular buildings. The range of the total carbon concentration was 1.3% (S. Maria del Fiore) to 7.7% (Tower of London), and from 0.14 (Milan Cathedral) to 8.37 (Tower of London) in terms of the OC/EC ratio. Sulfates were detected in every sample, ranging in concentration from 52 mg g^{-1} (London Tower) to 521 mg g^{-1} (Milan Cathedral). This indicates that sulfation resulting from SO_2-surface interactions remained a significant factor in the damage process. Belfiore et al. quantified the trace elements (e.g., Fe, Zn, Pb, and V) of the black crusts from the interior to the exterior layers of European cultural buildings (i.e., the Corner Palace, St. Rumbold's Cathedral, and St. Eustache Church) using laser ablation inductively coupled plasma mass spectrometry. They came to the conclusion that trace metals can be markers for assessing the variation of fuels used over time. According to Bergin et al., in areas with high aerosol loadings, the discoloration of the Taj Mahal in India and the natural and man-made surfaces of buildings may be caused by the deposition of light-absorbing PM (e.g., brown carbon).

The aforementioned contributing factors have made the cleaning of soiled historical buildings and monuments necessary. Lasers, micro-sandblasters, or chemical agents can be used for this type of cleaning, but the building's surface may be harmed by these techniques. To improve air quality and lower building soil rates, policies and regulations for traffic emissions abatement should be prioritized.

7.6. IMPACTS ON REMOTE SITES

Air pollution from mobile sources is more prevalent in urban areas than from area sources, and it has a detrimental impact on human health, decreases visibility, and alters the climate and ecosystems. Nonetheless, worries regarding how local mobile source pollution is affecting the state of the world's air quality are growing because these sources may have an impact on global meteorology, leading to the melting of glaciers, more frequent floods and droughts, and stronger hurricanes and other storms. To get a better understanding of the importance of controlling air quality in remote areas, it is crucial to quantitatively establish the temporal trends of mobile source emissions at remote (rural) sites.

The Interagency Monitoring of Protected Visual Environments (IMPROVE) monitoring network sites across the nation have been the site of extensive research by the US Environmental Protection Agency regarding the chemical composition of natural aerosols. The background concentrations of sea salt, biogenic OM, fugitive dust, and carbon from spontaneous wildfires are sufficiently measured at these locations to allow for the restoration of "natural conditions" in terms of visibility in 156 national parks and wilderness areas by 2064. This objective is crucial to the US Regional Haze Rule's implementation.

Furthermore, the IMPROVE database can be utilized to calculate the air pollution contributions from mobile sources to natural-condition $PM_{2.5}$ and O_3. These pertinent research findings have the potential to offer novel insights into strategy planning, which is essential for managing regional haze.

Using speciated $PM_{2.5}$ measurements from the Chemical Speciation Network (CSN) and IMPROVE monitoring networks spread throughout the state of Minnesota, Chen et al. applied the CMB method. The findings showed that while gasoline and diesel contributed 44% of the $PM_{2.5}$ mass concentration at urban sites, their contributions to $PM_{2.5}$ at rural Minnesota IMPROVE sites were negligible (<4%). Using IMPROVE samplers, Kim and Hopke gathered $PM_{2.5}$ samples at a rural monitoring site in Brigantine, New Jersey. They then used the PMF method to quantify source contributions to $PM_{2.5}$. Sulfate-rich secondary aerosols contributed 48% of $PM_{2.5}$ during the monitoring period, with traffic emissions contributing 16%. Sulfate-rich secondary aerosols had a regional impact, according to potential source contribution function analysis, whereas traffic emissions were a local source. At this monitoring site, the most significant impacts from mobile sources were probably brought about by a nearby highway with heavy traffic (Table 7.5 *gives an example*).

Table 7.5. *Indoor Pollutants, Sources, and Health Effects in Various Rural and Urban Areas of Pakistan*

Region in Pakistan	Indoor Setting	Source	Pollutant	Concentration	Health Hazards
Lahore	Living area	Traffic/seasonal changes	$PM_{2.5}$	>25 μg/m³	Not reported
Karachi	Restaurants/cafés/clubs	Indoor smoking; heating; cooking; hospitality	$PM_{2.5}$	25–390 μg/m³	Not reported
Rural areas of Pakistan	Living area/Kitchen	Use of biomass fuel; indoor cooking	PM_{10}, $PM_{2.5}$, PM_1	4000–8555 μg/m³	Not reported
Rural and urban areas of Pakistan	Living area/Kitchen	Indoor biomass burning	$PM_{2.5}$, PM_{10}	PM_{10} (>50 μg/m³); $PM_{2.5}$ (>25 μg/m³)	Not reported

Haripur city	Living area/ Kitchen	Outside traffic	$PM_{2.5}$, PM_{10}	$PM_{2.5}$ 23.7–126.0 μg/m³; PM_{10} 39.0–166.3μg/m³;	Not reported
Lahore	Living area/ Kitchen	Outside air due to industrial and traffic pollution	$PM_{2.5}$, PM_{10}, CO, NO_2, SO_2, O_3	Not reported	64% of asthma patients
Northern Pakistan	Living area/ Kitchen	Brick kilns	PM_1, $PM_{2.5}$, and PM_{10}	PM_1, $PM_{2.5}$, and PM_{10} concentrations 3377, 2305, and 3567.67 μg/m³, respectively.	Not reported
Rural areas	Living area/ Kitchen	Indoor dust exposure/ingestion	Polybrominated diphenyl ether, polychlorinated biphenyl, tri-(2-butoxyethyl) phosphate, triphenyl phosphate	~15.2 ng/kg bw/day	Not reported

Lahore, Faisalabad, Bahawalnagar	Living area/Kitchen	Indoor dust; polychlorinated biphenyl concentrations	Polychlorinated biphenyls	~34.39 ng/g, 9.94 ng/g, and 8.79 ng/g in Lahore, Faisalabad, Bahawalnagar, respectively	Not reported
Rural regions	Living area/Kitchen	Unhealthy fuel (wood, crop residues, charcoal, coal, kerosene, and animal dung)	Exposure to CO_2, CO	Not reported	Respiratory infections in children
Rural regions	Living area/Kitchen	Indoor biomass; fuel burning	PM, CO_2, CO	PM level 200–5000 μg/m^3 carbon monoxide (CO) level ~29.4 ppm	Not reported
Bungalows and apartments	Living area/Kitchen	Gas-fired kitchens	CO	CO concentrations in the range 2.13–5.29 ppm	Not reported
Lahore	Living area/Kitchen	Tobacco smoke	PM; CO_2 emissions	Not reported	Coughing/sneezing; eye irritation
Pakistan cities	Indoor restaurants, bars, and cafes	Indoor smoking	PM, CO_2 emissions	High $PM_{2.5}$ level ~1745 μg/m^3	Not reported

Urban/ Rural areas	Living area/ Kitchen	Indoor bacteria	Bioaerosols	Concentration high i.e., 14,650 cfu/m^3	Not reported
Lahore	Kitchens and living rooms	Microflora concentrations	Microflora	~9829–14,469 cfu m^{-3}	Not reported
Rawalpindi and Islamabad	Living area/ Kitchen	Indoor air	Polyaromatic hydrocarbon	~2132 pgm^{-3}; in dust ~90.0 $ng{\cdot}g^{-1}$	Respiratory system/ cancer risks
Urban areas	Offices	Indoor air	Brominated or phosphorated flame retardants	Flame retardants concentration 128,000 $ng{\cdot}g^{-1}$	Loss of indoor working productivity
Lahore	Hospitals; operating theatres	Indoor air	PM	PM concentrations 757–970 µg/m^3	Repeated infections in patients

Source: Kausar A, Ahmad I, Zhu T, Shahzad H. Impact of Indoor Air Pollution in Pakistan—Causes and Management. Pollutants. 2023; 3(2):293-319. https://doi.org/10.3390/pollutants3020021.

Understanding the relationships between air pollution and climate change in European nations was the goal of the European Integrated Project on Aerosol Cloud Climate and Air Quality Interactions (EUCAARI). Within this framework, various sites across Europe, including urban, rural, and remote areas as well as high-altitude sites, were used to perform PM_1 (an aerodynamic diameter less than 1 µm) SA using PMF with the multilinear engine on Aerodyne aerosol mass spectrometer data. Between 6% and 12% of PM_1 mass at urban sites was attributed to traffic emissions, whereas at remote and rural sampling sites, the contributions to total mass ranged from 0.8% to 5%. Additionally, at some of the remote and rural sampling locations, the contributions of traffic emissions were not detected.

Lefohn et al. examined rural O_3 trends in the USA from 1994–2008, a period of 15 years. They discovered that while NO_x and VOC emissions decreased by

40% and 47%, respectively, most rural sites did not experience a statistically significant drop in O_3 concentration.

Wilson et al. evaluated, using the European scale 3D chemistry-transport model (CTM) CHIMERE, the effects of varying anthropogenic emission tracers on surface European rural background O_3 concentrations from 1996–2005. With the exception of Austria and Hungary, they found that anthropogenic NO_x or VOC reductions had no discernible impact on the observed annual mean rural background O_3 trends in most of Europe. The findings from European research support those from US studies, indicating that the effects of mobile sources on the quality of the air at remote locations are not as great as those at urban locations **(Table 7.6)**.

Table 7.6. *Summary of Particulate Matter Health Effects on Young Children's Respiratory Symptoms and Diseases Extracted from Selected Articles Based on Eligibility Criteria and Best Judgment of Pediatric Experts*

Health Effects	Author	Country	Location of Study	Number of Children
Respiratory symptoms and inflammatory response (shortness of breath, wet and dry cough, wheezing, chest tightness, throat congestion)	Kumar et al. (2007)	India	Indoor & outdoor	394
	Siddique et al. (2010)	India	Outdoor	16,164
	Barakat-Haddad et al. (2012)	Canada	Outdoor	395
	Oluwole et al. (2013)	Nigeria	Indoor	59
	Nunes et al. (2015)	Portugal	Indoor & outdoor	N/A
	Watanabe et al. (2016)	Japan	Indoor & outdoor	360
Wheezing and poor respiratory health status	Rodriguez et al. (2010)	Colombia	Indoor & outdoor	768
	Nandasena et al. (2012)	Sri Lanka	Indoor &outdoor	N/A
	Pedersen et al. (2013)	Europe	Outdoor	N/A

Asthma, Asthma and bronchiolitis, atopy and airways inflammation, Hospital admissions and emergency department visits for asthma	Gehring et al. (2010)	Germany	Outdoor	3863
	Butz et al. (2011a)	USA	Indoor	126
	Butz et al. (2011b)	USA	Indoor	41
	Robinson et al. (2011)	Peru	Indoor	1441
	Lewin et al. (2013)	Canada	Outdoor	N/A
	Madureira et al. (2015)	Portugal	Indoor	1600
	Lim et al. (2016)	Global	Indoor & outdoor	N/A
Deterioration and changes of lung function	He et al. (2010)	China	Outdoor	983
	Siddique et al. (2011)	India	Indoor & outdoor	5671
	Zora et al. (2013)	USA	Outdoor	N/A
	Buonanno et al. (2013)	Italy	Indoor & outdoor	103
	Ferreira & Cardoso (2014)	Portugal	Indoor & outdoor	1019
	Paulin & Hansel (2016)	Global	Indoor & outdoor	N/A
	Rice et al. (2016)	USA	Indoor & outdoor	614
Acute lower respiratory tract infections	Ramesh et al. (2012)	India	Indoor	202
	Sram et al. (2013)	Czech Republic	Outdoor	1492
Increasing risk of hospitalization for all diagnoses	Marchettia et al. (2014)	Italy	Outdoor	3854
Onset of skin, mucosal and general symptoms	Zhang et al. (2014a)	China	Indoor	2134

Source: Liu H-Y, Dunea D, Iordache S, Pohoata A. A Review of Airborne Particulate Matter Effects on Young Children's Respiratory Symptoms and Diseases. Atmosphere. 2018; 9(4):150. https://doi.org/10.3390/atmos9040150.

Asian nations are recognized for producing significant volumes of anthropogenic pollutants because of their high industrial activity densities and rising energy consumption rates. Many international projects, such as the Asia-Pacific Regional Aerosol Characterization Experiment (ACE-Asia), the East Asia/North Pacific Regional Experiment (APARE), the Pacific Exploratory Mission-West (PEM-WEST), and the Atmospheric Brown Cloud (ABC) program, were launched to formulate effective control strategies for ambient PM to estimate aerosol compositions and sources in this region. SA studies at Asian background sites were conducted with funding from these grants to comprehend $PM_{2.5}$ sources and their roles in local air quality. Stone et al. measured the contributions of traffic emissions to $PM_{2.5}$ in the Himalayas, which could be one of South Asia's background sites, using the CMB receptor model. Diesel engines contributed an average of 5% of 26 μg m^{-3} to $PM_{2.5}$, which is much less than the average vehicle contribution of 17% of 168 μg m^{-3} in India's megacities. A size-resolved PM SA study was carried out by Han et al. in Gosan, South Korea, a background site that is representative of East Asia. Traffic contribution accounted for a small proportion of $PM_{0.56–2.5}$ mass (7% of 32 μg m^{-3}), which was much lower than the results (23% of 42.6 μg m^{-3}) from one SA study conducted in Seoul, South Korea. Huaniao Island, China, which is located 66 km east of Shanghai and serves as a downwind backdrop site for East Asian continental outflow, was the subject of an investigation by Wang et al. regarding the effects of seasonal atmospheric transport of land-based PAHs on air quality. With an estimated 27% contribution, the annual average concentration of PAHs in $PM_{2.5}$ is 5.24 ± 5.81 ng m^{-3}, suggesting that exhaust emissions from China's Yangtze River Delta may have an impact on this background site. With the exception of China, the effects of mobile sources on air pollution at isolated locations in Asia are similar to those in the USA and Europe. With so little data, it is impossible to conclude scientific studies on $PM_{2.5}$ SA at the Chinese background sites. These studies have rarely been reported.

It should be acknowledged that this review of the literature is comparatively narrow, upon which these conclusions are based. Biased measurement techniques, non-representative measurements, and shifting scientific hypotheses in response to new data are all possible. As a result, there is a good deal of uncertainty surrounding this assessment. To ensure that the effects of mobile sources on regional air quality in both developed and developing nations are fully understood, ongoing efforts are essential.

REFERENCES

1. Abdel-Shafy, H. I., & Mansour, M. S. M. (2016). A review on polycyclic aromatic hydrocarbons: Source, environmental impact, effect on human health, and remediation. *Egyptian Journal of Petroleum, 25*(1), 107–123.
2. Alam, M., Ferreira, J., & Fonseca, J. (2016). Introduction to intelligent transportation systems. In *Intelligent Transportation Systems* (pp. 1–17). Springer International Publishing.
3. Bagloee, S. A., Tavana, M., Asadi, M., & Oliver, T. (2016). Autonomous vehicles: Challenges, opportunities, and future implications for transportation policies. *Journal of Modern Transportation, 24*(4), 284–303.
4. Beelen, R., Hoek, G., van den Brandt, P. A., Goldbohm, R. A., Fischer, P., Schouten, L. J., et al. (2008). Long-term effects of traffic-related air pollution on mortality in a Dutch cohort (NLCS-AIR study). *Environmental Health Perspectives, 116*(2), 196–202.
5. Bishop, J. D. K., Molden, N., & Boies, A. M. (2017). Real-world environmental impacts from modern passenger vehicles operating in urban settings. *International Journal of Transport Development and Integration, 1*(2), 203–211.
6. Brunekreef, B., Beelen, R., Hoek, G., Schouten, L., Bausch-Goldbohm, S., Fischer, P., et al. (2009). Effects of long-term exposure to traffic-related air pollution on respiratory and cardiovascular mortality in the Netherlands: The NLCS-AIR study. *Research Report Health Effects Institute, 139*, 5–71.
7. Chen, T. M., Gokhale, J., Shofer, S., & Kuschner, W. G. (2007). Outdoor air pollution: Nitrogen dioxide, sulfur dioxide, and carbon monoxide health effects. *American Journal of the Medical Sciences, 333*(4), 249–256.
8. Christiansen, L. L., Frederick, H., Knechel, K., & Mussman, E. L. (2016). Meeting the needs of modern transportation researchers by transforming the Iowa Department of Transportation Library—Early efforts & results. *Transportation Research Board 95th Annual Meeting*, No. 16-3827.
9. Cofala, J., Amann, M., Klimont, Z., Kupiainen, K., & Hoglund-Isaksson, L. (2007). *Atmospheric Environment, 41*(37), 8486–8499.
10. Cohen, A. J., Anderson, H. R., Ostro, B., Pandey, K. D., Krzyzanowski, M., Kunzli, N., Gutschmidt, K., Pope, A., Romieu, I., Samet, J. M., & Smith, K. (2005). *Journal of Toxicology and Environmental Health, Part A, 68*(13, 14), 1301–1307.
11. Dargay, J., Gately, D., & Sommer, M. (2007). *Energy, 28*, 143–170.

12. Doniger, D., Friedman, D., Hwang, R., Lashof, D., & Mark, J. (2002). *Dangerous Addiction: Ending America's Oil Dependence.* National Resources Defense Council and Union of Concerned Scientists.
13. dos Santos, L. H. M., Kerr, A. A. F. S., Verı´ssimo, T. G., Andrade, M. d. F., de Miranda, R. M., Fornaro, A., & Saldiva, P. (2014). *Journal of the Air & Waste Management Association, 64*(5), 519–528.
14. Ehsani, M., Hoelscher, D., Shidore, N., & Asadi, P. (2003). Impact of hybrid electric vehicles on the world's petroleum consumption and supply. *SAE Future Transportation Technology Conference*, Paper No. 2003-01-2310.
15. Ellingsen, L. A. W., Singh, B., & Strømman, A. H. (2016). The size and range effect: Lifecycle greenhouse gas emissions of electric vehicles. *Environmental Research Letters, 11*(5), 054010.
16. Ergas, C., Clement, M., & McGee, J. (2016). Urban density and the metabolic reach of metropolitan areas: A panel analysis of per capita transportation emissions at the county-level. *Social Science Research, 58*, 243–253.
17. Faiz, A., Weaver, C. S., & Walsh, M. P. (1996). *Air Pollution from Motor Vehicles: Standards and Technologies for Controlling Emissions* (1st ed.). World Bank Publications.
18. Ferguson, C. R., & Kirkpatrick, A. T. (2001). *Internal Combustion Engines—Applied Thermo-Sciences* (2nd ed.). John Wiley & Sons.
19. Fontaras, G., Franco, V., Dilara, P., Martini, G., & Manfredi, U. (2014). *Science of the Total Environment, 468, 469*, 1034–1042.
20. Gao, Y., & Ehsani, M. (2002). An investigation of battery technologies for the Army's hybrid vehicle application. In *Proceedings of the IEEE 56th Vehicular Technology Conference*, Vancouver, British Columbia, Canada.
21. Guido, C., Beatrice, C., & Napolitano, P. (2013). *Applied Energy, 102*, 13–23.
22. Hagler, G. S. W., Baldauf, R. W., Thoma, E. D., Long, T. R., Snow, R. F., Kinsey, J. S., Oudejans, L., & Gullett, B. K. (2009). *Atmospheric Environment, 43*(6), 1229–1234.
23. Hakami, A., Henze, D. K., Seinfeld, J. H., Chai, T., Tang, Y., Carmichael, G. R., & Sandu, A. (2005). *Journal of Geophysical Research: Atmospheres, 110*, D14031.
24. Hogerwaard, J., Dincer, I., & Zamfirescu, C. (2017). Thermodynamic and environmental impact assessment of NH3 diesel–fueled locomotive configurations for clean rail transportation. *Journal of Energy Engineering, 143*(5), 04017018.

25. Hu, X., Murgovski, N., Johannesson, L. M., & Egardt, B. (2014). Comparison of three electrochemical energy buffers applied to a hybrid bus powertrain with simultaneous optimal sizing and energy management. *IEEE Transactions on Intelligent Transportation Systems, 15*, 1193–1205. http://dx.doi.org/10.1109/TITS.2013.2294675.

26. Jedwab, R., & Moradi, A. (2016). The permanent effects of transportation revolutions in poor countries: Evidence from Africa. *Review of Economics and Statistics, 98*(2), 268–284.

27. Jerrett, M., Finkelstein, M. M., Brook, J. R., Arain, M. A., Kanaroglou, P., Stieb, D. M., et al. (2009). A cohort study of traffic-related air pollution and mortality in Toronto, Ontario, Canada. *Environmental Health Perspectives, 117*(5), 772–777.

28. Johnson, D., Ambrose, S., Bassett, T., Bowen, M., Crummey, D., Isaacson, J., et al. (1997). Meanings of environmental terms. *Journal of Environmental Quality, 26*(3), 581–589.

29. Kakouei, A., Vatani, A., & Idris, A. K. (2012). An estimation of traffic-related CO2 emissions from motor vehicles in the capital city of Iran. *Iranian Journal of Environmental Health Science & Engineering, 9*(1), 13.

30. Karner, D., & Francfort, J. (2007). Hybrid and plug-in hybrid electric vehicle performance testing by the US Department of Energy Advanced Vehicle Testing Activity. *Journal of Power Sources, 174*(1), 69–75. https://doi.org/10.1016/j.jpowsour.2007.06.069.

31. Ke, W., Zhang, S., He, X., Wu, Y., & Hao, J. (2017). Well-to-wheels energy consumption and emissions of electric vehicles: Mid-term implications from real-world features and air pollution control progress. *Applied Energy, 188*, 367–377.

32. Kim, E., & Hopke, P. K. (2005). *Journal of the Air & Waste Management Association, 55*(10), 1456–1463.

33. Lawry, M., Mirza, A., Wang, Y. W., & Sundaram, D. (2017). Efficient transportation—Does the future lie in vehicle technologies or in transportation systems? *International Conference on Future Network Systems and Security* (pp. 126–138). Springer, Cham.

34. Leary, P. J., Kaufman, J. D., Barr, R. G., Bluemke, D. A., Curl, C. L., Hough, C. L., et al. (2014). Traffic-related air pollution and the right ventricle: The multi-ethnic study of atherosclerosis. *American Journal of Respiratory and Critical Care Medicine, 189*(9), 1093–1100.

35. Manuela, C., Francesco, T., Tiziana, C., Assunta, C., Lara, S., Nadia, N., et al. (2012). Environmental and biological monitoring of benzene in traffic

policemen, police drivers, and rural outdoor male workers. *Journal of Environmental Monitoring, 14*(5), 1542–1550.

36. Newman, N. C., Ryan, P., Lemasters, G., Levin, L., Bernstein, D., Hershey, G. K., et al. (2013). Traffic-related air pollution exposure in the first year of life and behavioral scores at 7 years of age. *Environmental Health Perspectives, 121*(6), 731–736.
37. Ren, W., Xue, B., Geng, Y., Lu, C., Zhang, C. Y., Zhang, L., Fujita, T., & Hao, H. (2016). Inter-city passenger transport in larger urban agglomeration area: Emissions and health impacts. *Journal of Cleaner Production, 114*, 412–419.
38. Roy, M. M., Calder, J., Wang, W., Mangad, A., & Diniz, F. C. M. (2016). *Applied Energy, 180*, 52–65.
39. Saeb, K., Maryam, M., & Saeed, K. (2012). Air pollution estimation from traffic flows in Tehran highways. *Current World Environment, 7*, 1–6.
40. Schauer, J. J. (1998). Source contributions to atmospheric organic compound concentrations: Emission measurements and model predictions (Doctoral dissertation, California Institute of Technology).
41. Schneider, S. H. (1989). The greenhouse effect: Science and policy. *Science, 243*, 771–781.
42. Shafie-Pour, M., & Tavakoli, A. (2013). On-road vehicle emissions forecast using IVE simulation model. *International Journal of Environmental Research, 7*, 367–376.
43. Sosale, S. S. (2017). Performance analysis of various 4-wheelers with IC engines for hybridization.
44. Tam, W. W., Wong, T. W., Wong, A. H., & Hui, D. S. (2012). Effect of dust storm events on daily emergency admissions for respiratory diseases. *Respirology, 17*, 143–148.
45. Timmers, V. R. J. H., & Achten, P. A. J. (2016). *Atmospheric Environment, 134*, 10–17.
46. Vallero, D. (2007). *Fundamentals of Air Pollution* (4th ed.). Academic Press.
47. Van Hee, V. C., Adar, S. D., Szpiro, A. A., Barr, R. G., Bluemke, D. A., Diez Roux, A. V., et al. (2009). Exposure to traffic and left ventricular mass and function: The multi-ethnic study of atherosclerosis. *American Journal of Respiratory and Critical Care Medicine, 179*, 827–834.
48. Volk, H. E., Lurmann, F., Penfold, B., Hertz-Picciotto, I., & McConnell, R. (2013). Traffic-related air pollution, particulate matter, and autism. *JAMA Psychiatry, 70*, 71–77.

49. Wakefield, E. H. (1994). *History of the Electric Automobile: Battery-Only Powered Cars.* Society of Automotive Engineers (SAE).
50. Wakefield, E. H. (1998). *History of the Electric Automobile: Hybrid Electric Vehicles.* Society of Automotive Engineers (SAE).
51. Weiner, E. (2016). *Urban Transportation Planning in the United States: History, Policy, and Practice.* Springer.
52. Zielinska, B., Sagebiel, J., McDonald, J. D., Whitney, K., & Lawson, D. R. (2004). *Journal of the Air & Waste Management Association, 54*(9), 1138–1150.

Chapter 8

Water and Resource Use in EV Production

Contents

INTRODUCTION

Natural resources are needed in significant quantities for the production of electric vehicles (EVs), from the extraction of raw materials to the manufacturing procedures. These resources are primarily used in the manufacturing of lithium-ion batteries, which power electric vehicles. Critical components for these batteries include lithium, cobalt, nickel, and graphite, which are frequently mined in environmentally delicate regions. Energy-intensive procedures used in the extraction of these materials have the potential to cause deforestation, water pollution, and significant carbon emissions. The supply chains for these materials are frequently intricate and require lengthy transit routes, which adds to the carbon footprint of EV manufacturing.

Furthermore, rare earth metals are needed in the production of electric motors and magnets, among other parts of the EV. These metals are not only rare, but their extraction also has an adverse effect on the environment. Large amounts of energy-intensive metals like steel and aluminum are also needed in the production of EVs, which adds to the vehicles' overall environmental impact. Positively, the switch to EVs has sparked advancements in environmentally friendly production techniques. In an effort to lessen the impact of production on the environment, businesses are putting more and more emphasis on recycling and material reuse. One way to lessen the need for new mining operations is through the development of battery recycling technologies, which aim to recover valuable materials from used batteries. Manufacturers are also looking into less resource-intensive and environmentally friendly substitute materials and designs.

The amount of resources used in the production of EVs is still a major concern despite these advancements. As EV adoption rises, there will likely be a greater need for raw materials, which, if not managed sustainably, could worsen environmental degradation. In order to produce EVs in a more sustainable manner and to guarantee that the whole lifecycle of these vehicles, from production to disposal, is environmentally responsible, the industry must thus continue to invest in research and development. The move to EVs presents a chance to reconsider how the automotive sector uses resources, but to make this change truly sustainable, manufacturers, governments, and consumers must work together.

8.1. EV PRODUCTION

The manufacturing of EVs, which are distinguished by their reliance on cutting-edge technology and specialty materials, marks a dramatic change in the automotive industry. In contrast to conventional cars with internal combustion engines, EVs need lithium-ion batteries, which are the most important component. Lithium, cobalt, nickel, and graphite are just a few of the many raw materials needed to produce these batteries, and their extraction involves mining techniques that can have a negative influence on the environment. Since these materials are frequently taken from areas with lax regulatory oversight, questions regarding ethical labor practices and environmental degradation are raised.

In order to reduce battery weight and improve vehicle efficiency, lightweight materials like carbon fiber and aluminum are also used in the assembly of electric vehicles. The energy-intensive process of producing these materials adds to the total carbon footprint associated with EV production. Nonetheless, the industry is making progress in lessening these effects by developing innovative sustainable production methods, like employing recycled materials and enhancing manufacturing processes' energy efficiency.

The production of electric vehicles still requires a lot of resources, despite these efforts. The complicated global supply chain for EV components frequently necessitates long-distance travel, which worsens the environmental impact. Furthermore, the supply of essential raw materials has been strained by the sharp rise in demand for EVs, raising questions about the sustainability of the present production techniques.

Enhancing the sustainability of EV production is a top priority as the industry expands. This include researching alternate manufacturing methods that lower energy consumption and waste, creating new battery technologies that use fewer essential materials, and improving recycling procedures to recover valuable resources from used batteries. In an effort to create more resilient supply chains that put social and environmental responsibility first, manufacturers and policymakers are collaborating.

Making the switch to EVsis essential to cutting greenhouse gas emissions worldwide and implementing a more environmentally friendly transportation system. Nevertheless, addressing the resource and energy demands of EV production is crucial to achieving the full environmental benefits of EVs. Through persistent innovation and enhanced production methodologies, the industry can guarantee that the expansion of EVs makes a positive impact on environmental sustainability and the global economy.

8.1.1. EV Conversions

A number of conversion vehicles must be available to complement the original equipment manufacturers (OEM) offerings due to the limited availability of

EV models. The most cost-effective method for creating a limited run of EVs based on a specific conventional vehicle is conversion from a complete vehicle. The cost of conversions can be decreased by producing a special glider once the number of vehicle conversions reaches a significant level (a few hundred per year). A conventional car can be converted to an EV by removing the internal combustion engine and all of its supporting components, swapping it out for an electric motor and controller, and adding enough battery capacity to allow for a respectable range. Typically, there is space for the batteries in the passenger compartment, the luggage or cargo area, the engine compartment, or a combination of these. Having to take up space from other designated areas—the luggage or cargo area being the least important—is the primary drawback of converting conventional vehicles to electric vehicles. This is because there is no dedicated space for the energy storage system. The cost of manufacturing an EV is far higher than that of an equivalent vehicle built "from the ground up" due to a number of inefficiencies introduced by the conversion process. On the other hand, an original design for a relatively small number of vehicles cannot be justified. Therefore, when the quantity of units to be produced is very low, conversions have their proper place, despite the inherent inefficiencies. The primary inefficiencies in this process are related to redundant tasks and resources. The conventional powertrain that came with the original vehicle must be removed and swapped out for an electric powertrain. Due to its brief usage and occasional damage, the removed material loses value during the process and cannot be reclaimed in full. Furthermore, it is frequently necessary to make changes to the car and/or its parts, which can be expensive due to the labor-intensive nature of the work and the low volume involved. Finally, because the components were not made for the purpose, extra parts—which are typically not needed—often need to be used in order to make the components fit correctly (adaptations). Gliders are a useful tool for reducing or eliminating many of these inefficiencies.

Heavy-duty truck manufacturers are where the word "glider" first appeared. With the exception of the engine, transmission, and rear axle, most OEMs sell gliders as complete trucks. In this market, gliders are meant to offer an alternative to wrecked truck repair. When a heavy truck is involved in a significant accident (like rolling over), the cab is typically destroyed and the chassis frame is frequently irreparably damaged. Frames that have been bent cannot be straightened because they are heat-treated. The rear axle, engine, and transmission do, however, typically survive. Given that these parts are extremely valuable and account for at least a third of the truck's total cost, it makes sense to salvage them and install them in a new glider, effectively creating a brand-new vehicle. The similarities between EV gliders and trucks are readily apparent. A glider should ideally have every standard component that an EV can use and leave out the ones that can't. Actually, a glider might have a lot of unnecessary parts, either because it is hard to omit them from the assembly process or because they are integrated with other parts that are required. The instrument panel and controls area, for example, is one system that may be particularly difficult to sort out. Still, we can assume that a glider for EVs will come without the following systems:

- Engine (including electronic controls);
- Engine accessories (alternator, power steering pump, air-conditioning compressor);
- Transmission;
- Exhaust system (including the catalytic converter);
- Fuel storage and evaporative emissions control system;
- Engine cooling system; and
- Starter, battery, and charging control system

However, some of the systems required by the EV might only be partially present because some of them might be integrated into major parts that aren't included (like the engine). The following are examples of such systems:

- Power steering (the pump is normally mounted on and driven by the engine);
- Heating, ventilation, and air-conditioning system (the heat source is the engine, and the air-conditioning compressor is also mounted on and driven by the engine); and
- Braking system (the booster is powered by vacuum from the engine).

Lastly, it's possible that some of the systems or specific parts of the conventional (IC-engine) car aren't the best fit for the EV and that they need to be adjusted or modified. Depending on how the logistics are worked out, special components may be installed directly at the OEM assembly line and may later need to be exchanged after assembly. Examples of this type of component include the following:

- Suspension springs and shocks;
- Wire harnesses;
- Battery trays or attachment brackets; and
- Motor mounting frame.

The amount of demand determines whether or not the glider is really optimized for EV conversion. A "deleteonly" variant of the glider would be an ordinary IC-engine car with a few key systems removed (engine, transmission, exhaust, fuel tank, etc.) if there is only a small annual demand for the vehicle type. But, if the annual demand is between 3,000 and 5,000 units, it would be prudent to design and create a unique version of this platform with a separate "bill of materials" that includes the appropriate kind of suspension, wire harnesses, instrument panels, HVAC parts, brackets for electric vehicle components, etc.

The latter type of glider is the most economical to manufacture because it would only require the assembly of its unique electric traction and energy storage components. A vehicle OEM (like Chrysler, with its electric minivan) would more likely produce something similar in-house for their own production than something like this glider. The glider that would be sent to an independent converter would most likely be in the middle, primarily a "delete-only" model with a few easy-to-replace special parts like wheels and tires, suspension shocks and springs, etc. Unless the volumes are particularly compelling, one generally does not expect the automakers to create new platform iterations, retool their assembly lines, or extensively retrain their workforce to handle these special models. Instead, the automakers aim to reduce any adverse effects on the productivity of their assembly lines. Even the removal of components during the assembly process results in inefficiencies because it is difficult to replace or reallocate the workers who would have performed the removed tasks. Thus, there won't be much of a reduction in the labor cost of assembly; instead, the savings will primarily come from the cost of the parts that were eliminated. Even so, a subpar glider would still eliminate a great deal of the conversion process's inefficiencies, such as the majority of material duplication and the extra work needed to remove the components connected to the IC engine. It would be necessary to carry out all or most of the necessary modifications at the OEM level to eliminate most or all of the conversion's inefficiencies; essentially, this entails creating a special glider bill of materials with all the necessary components for an EV.

8.1.2. Direct OEM Production

When the OEMs decide to build their own electric vehicles, they will have to decide between designing a vehicle from the "ground up" and altering an already-existing one. If the production volume were high enough to warrant it, an original design would be the better option in terms of both cost and performance. However, it is anticipated that EV production volumes will be, at most, modest in the early years of their introduction. It is therefore anticipated that the majority of OEMs will choose an EV based on an existing vehicle; the only uncertainty is the extent to which their chosen designs will differ. Except for the powertrain, Chrysler, for example, offers an electric vehicle (EV) based on its minivan. Conversely, Honda's first EV model in California, the CUV-4, was based on the Civic hatchback model but had a heavily altered body floor to make room for a battery compartment beneath the passenger cabin. While the Honda vehicle needs a completely new floor and extensive modifications to many of the external main panels, the Chrysler vehicle only has minor body modifications. There are substantial investment differences between these two strategies because body tooling is very costly, even when it is only used for small-scale prototype dies.

Since initial demand is predicted to be very low, the OEMs need to figure out how to introduce EV technology without having to invest heavily in high-volume tooling. There is a very small market for reasonably EVs made from fleet vehicles, such as minivans and compact pickup trucks, which don't require much modification. The OEMs will eventually have to release an EV that the average person can purchase; otherwise, the demand will never materialize to its full potential. A completely new design, like the GM EV-1, comes with hundreds of millions of dollars in tooling and development costs.

Such an investment in tooling would purchase approximately 50,000 units of production annually, far exceeding the initial requirement. The Honda example approach appears to call for a significantly smaller investment. The investment required for an additional model, like a convertible or station wagon based on an existing platform, would be similar to that of an EV produced as a variant of an existing vehicle. The majority of the extra money would go toward the necessary body modifications, with a smaller amount going toward new, specialized parts. Instead of hundreds of millions, the required investment may then be in the tens of millions. The majority of the chassis parts, almost all of the interior parts, and many of the body-in-white parts of the conventional vehicle would be used in the EVs. Though perhaps not on the same assembly line, the electric cars could be put together in the same facility as their conventional counterparts.

Depending on the volume needed, up to three distinct production systems may be used to manufacture EVs. It's possible for these three strategies to coexist for a while: conversions, glider conversions, and full OEM production from the ground up. In the long run, EVs should be produced in the same manner as any other vehicle when demand reaches a critical mass. The process of converting from conventional vehicles is only appropriate for very small quantities—roughly 100–400 units annually. It might be better to convert from a glider for small volumes (between 500 and 4,000 units annually). Based on an existing vehicle, full OEM production would need to be done in fairly large quantities—5,000–40,000 units annually. Full OEM production from a completely unique design would need to produce more than 50,000 units annually.

These three processes would not always result in vehicles that are equal. It is obvious that there will be some major restrictions on the conversions from CVs, especially in terms of luggage and passenger capacity. EVs built on current car platforms have the potential to be very competitive, at least initially. It will take an innovative design to produce an EV that is genuinely optimized and has a lightweight body. Should the specific electric vehicle parts (traction motors, controllers, batteries, etc.) from private vendors become affordable for OEMs and converters, the real disparities in EV costs between the three manufacturing systems probably won't be as large as the price difference between CV and EV. OEM production will eventually overtake other modes as demand volume increases dramatically due to its lower cost (and potentially higher quality).

8.1.3. Electric Car Production: Key Stages and Environmental Implications

The manufacture of electric cars goes through several crucial phases, each with unique environmental effects. As raw materials are needed to produce batteries, electric motors, and other components, the process starts with their extraction and processing. For the batteries, essential components include lithium, cobalt, nickel, and graphite; for the electric motors, rare earth metals are needed. These materials require a lot of energy to mine and process, which frequently leads to serious environmental problems like deforestation, water pollution, and greenhouse gas emissions. Moreover, the extraction of these resources can lead to social and ethical concerns, particularly in regions where mining practices are poorly regulated.

The production of lithium-ion batteries, the vital component of EVs, begins with the acquisition of raw materials. One of the phases of the EV production process that uses the greatest resources and energy is the battery production process. Large energy requirements can result in a significant increase in carbon emissions, depending on the energy mix of the manufacturing region. Batteries also produce chemical waste, which needs to be properly managed to avoid contaminating the environment.

There are multiple important phases in the production of electric vehicles, each with unique environmental consequences. The assembly of the vehicle, which entails building the body, mounting the electric motor, and integrating the battery system, is the following phase in the production of electric cars. Materials like steel, plastic, and aluminum are also used in this step. The energy-intensive process of producing these materials increases the vehicle's overall carbon footprint. However, the environmental impact of vehicle assembly is being lessened thanks to improvements in manufacturing techniques, such as the use of lightweight materials and more effective production processes.

8.1.3.1. Raw Material Extraction and Processing

Lithium-ion batteries and rare earth metals are two materials that are vital to electric vehicles. However, the extraction and processing of these materials come with environmental concerns.

- **Lithium-ion Batteries:** Lithium, cobalt, and nickel must be extracted in order to produce lithium-ion batteries, an essential part of electric vehicles. Water pollution, habitat destruction, and possible social problems in mining communities are all possible outcomes of mining these materials.
- **Rare Earth Metals (REEs):** They are used in the magnets and batteries of BEVs. The earth's crust contains a moderate amount of REEs, some of which are more abundant than copper, lead, gold, and platinum. The majority of REEs, however, are not concentrated enough to be readily exploited economically. As a byproduct of their production, rare earth elements are frequently found in combination with other elements like copper, gold, uranium, phosphates, and iron. Neodymium and dysprosium, two rare earth metals, are also used in the motors and magnets of electric vehicles. Rare earth metal extraction frequently entails energy-intensive procedures that can contaminate water and soil.

8.1.3.2. Component Manufacturing and Assembly

Electric vehicle production and assembly involve several procedures that add to their environmental impact.

- **Steel and Aluminum Production:** Large amounts of energy are needed and greenhouse gases are released during the production of steel and aluminum, which are widely used in the construction of electric vehicles. Mitigating the environmental impact of electric car manufacturing requires reducing the emissions linked to the production of steel and aluminum.
- **Battery Manufacturing:** Lithium-ion battery production entails energy-intensive procedures, including the use of chemicals. The battery manufacturing process involves energy consumption and chemical use, which adds to the overall environmental impact of electric vehicles.

8.1.4. Carbon Footprint of Electric Car Manufacturing

The greenhouse gas emissions produced during the production process are included in the carbon footprint of the production of electric cars.

8.1.4.1. Greenhouse Gas Emissions in the Production Phase

Greenhouse gases are released during the energy consumption, material extraction, and manufacturing processes involved in the production of electric vehicles. The production of components, the extraction and processing of raw materials, and the electricity used in manufacturing facilities are the primary sources of emissions.

Producing these batteries takes a significant amount of energy, especially in areas where the primary energy sources are coal or other carbon-intensive sources. For instance, the energy required to refine raw materials and assemble battery cells can contribute significantly to the CO_2 emissions associated with the production of a single lithium-ion battery. Because of this, the battery production stage of the entire EV lifecycle is a significant hotspot for GHG emissions.

Emissions of greenhouse gases are also caused by the manufacturing of the electric motor and other components, in addition to the battery. Rare earth metals are used in the production of parts like electric motors and their extraction and processing need a lot of energy. Moreover, since aluminum and steel require a lot of energy to produce, the body of the car produces a lot of emissions during the manufacturing process.

The actual process of assembly, which involves assembling the various parts and finishing the car, increases greenhouse gas emissions, albeit less so than the manufacturing of the batteries. The manufacturing facilities' energy efficiency and the electricity source they utilize have an impact on the emissions at this point. Compared to plants that use fossil fuels, those that use renewable energy sources emit fewer greenhouse gases.

Even though EVs produce more greenhouse gas emissions than conventional internal combustion engine vehicles, throughout their lifetime, EVs typically have a lower total carbon footprint. This is because, when in use, EVs emit no tailpipe emissions and are frequently powered by cleaner electricity, resulting in reduced emissions. Even so, there is still much to be done to improve the production stage. Reducing the carbon intensity of battery manufacturing, improving the energy efficiency of production processes, and increasing the use of renewable energy in manufacturing facilities are all being pursued.

8.1.4.2. Comparison with Internal Combustion Engine (ICE) Vehicles

Compared to conventional vehicles, electric vehicles have a higher carbon footprint during the manufacturing process, but overall emissions are greatly reduced during operation. According to studies, compared to ICE vehicles, electric vehicles can offset their higher manufacturing emissions within a few years of operation, leading to lower lifetime emissions.

8.1.4.2.1. Production Phase

Compared to ICE vehicles, the production of EVs, especially their lithium-ion batteries, typically results in higher greenhouse gas emissions. This is mostly because producing batteries requires a large amount of electricity, as well as energy-intensive processes for the mining and processing of raw materials like nickel, cobalt, and lithium. The carbon footprint of creating an EV can be relatively high in areas where fossil fuels are used to power the electrical grid. On the other hand, less energy-intensive processes are typically used in the production of ICEs. This is because ICEs' engines are built from conventional materials like steel and aluminum, which require less energy-intensive processing than EV batteries.

8.1.4.2.2. Use Phase

When comparing EVs to ICE vehicles, the former typically have a far smaller carbon footprint during use, despite producing higher emissions. When an EV is operating, it doesn't release any carbon dioxide (CO_2), nitrogen oxides (NOx), or particulate matter (PM). Instead, it produces zero tailpipe emissions. When EVs are charged with renewable energy, their environmental benefits escalate and their overall greenhouse gas emissions are further reduced. However, because ICE vehicles run on fossil fuels to produce power, they continuously release large volumes of CO_2 and other pollutants into the atmosphere during their operation.

8.1.4.2.3. Lifecycle Emissions

EVs typically have a lower overall carbon footprint than ICE vehicles when taking into account the whole lifecycle of both types of vehicles, from raw material extraction, manufacturing, and operation to end-of-life disposal. When EVs are used, their lower emissions during the use phase eventually balance out their higher emissions during the production phase. According to studies, EVs can reduce their overall lifecycle emissions by 30–60% when compared to ICE vehicles, depending on the energy mix of the electricity used to charge them.

8.1.4.2.4. End-of-Life Considerations

There are also notable differences between EVs and ICE cars at the end of their useful lives. The comparatively well-established process of recycling ICE vehicles entails removing metal from the body and engine. Recycling lithium-ion batteries is a challenge as well as an opportunity for EVs. Recovering valuable materials like lithium, cobalt, and nickel can lessen the need for new mining operations and lessen the environmental impact of future electric vehicle production. Battery recycling is still in its infancy.

8.1.4.2.5. Environmental and Social Impacts

The effects of both car types on the environment and society extend beyond GHG emissions. The extraction of raw materials for electric vehicle batteries, especially in areas with lax environmental regulations, can result in habitat destruction, water pollution, and violations of human rights. Analogously, the extraction and refinement of oil for internal combustion engine vehicles leads to environmental deterioration and can have noteworthy social consequences, such as disputes over resource allocation and health issues linked to pollution.

8.1.5. Resource Consumption in Electric Car Production

The production of electric cars requires the use of several resources in addition to carbon emissions.

- **Water Usage and Pollution:** Water is needed for chemical reactions, cleaning, and cooling during the production of electric vehicles. To reduce water use and avoid pollution, proper water management is essential.
- **Land Use and Ecosystem Disruption:** Elimination of habitats and disturbance of ecosystems can result from the extraction of raw materials like lithium and rare earth metals. Initiatives for land reclamation and ethical sourcing can help lessen these effects.
- **Metal and Mineral Extraction:** The industries that extract minerals and metals are under pressure from the need for materials in the production of electric cars. Recycle programs, sustainable mining techniques, and the creation of substitute materials can all lessen the negative effects of resource extraction on the environment.

8.1.6. Mitigating Environmental Impact in Electric Car Manufacturing

There are various strategies that can be put into practice to address the environmental impact of manufacturing electric cars. These include:

- **Sustainable Material Sourcing and Design:** Manufacturers can give recycling programs, the creation of alternative materials, and sustainable material sourcing top priority. Recyclable and disassemble electric vehicle designs can aid in the recovery of valuable materials.
- **Renewable Energy Integration in Manufacturing Facilities:** The carbon footprint of producing electric cars can be greatly decreased by switching to renewable energy sources like solar and wind for the electricity used in manufacturing facilities. Investing in on-site renewable energy infrastructure and collaborating with renewable energy providers are crucial first steps.
- **Reducing Carbon Emissions in Production:** Manufacturing facilities can drastically lower their carbon intensity by switching to renewable energy sources. To power their manufacturing facilities, numerous automakers are making investments in wind, solar, and other renewable energy projects. For example, certain factories are built to run entirely on renewable energy, significantly reducing greenhouse gas emissions related to the manufacturing process. Emissions can also be further decreased by increasing energy efficiency in factories through the use of energy-efficient machinery and process optimization.
- **Advancements in Battery Technology:** The development of new battery technologies is essential to reducing the negative environmental effects of EV production. Scientists are working on new battery chemistries that use less rare and harmful materials to the environment. For instance, solid-state batteries promise longer lifespans and a reduction in the requirement for materials like cobalt, which can lower the frequency of battery replacement and, in turn, the demand for raw materials. In addition, research is being done on the use of substitute materials like silicon or sulfur to make batteries that are more effective and sustainable.
- **Lightweighting and Material Efficiency:** Lightweight materials like carbon fiber and aluminum can be used to reduce the weight of EVs, which will increase vehicle efficiency and lessen production's negative environmental effects. But because it takes a lot of energy to produce these materials, research is also being done to create new materials that have the same advantages but don't harm the environment as much. There is continuous research being done on bio-based composites and advanced polymers with the goal of producing materials that are lighter and more eco-friendly.

8.2. ELECTRIC VEHICLES AND WATER CONSUMPTION

For a future where greenhouse gas emissions are reduced, electric vehicles are essential, but what about the amount of water used in the energy production process? As freshwater becomes increasingly scarce, we need to take this into account in every aspect of our lives. The water-energy network is briefly touched upon in the following.

8.2.1. Electricity

Water resources are heavily utilized in the production of electricity, even though this isn't always obvious. Water requirements vary depending on the method used to produce the electricity. For instance, the production of nuclear power requires large amounts of water, whereas wind energy uses very little. Other electricity-producing methods that use a lot of water include hydropower and biomass-based systems. Hydropower needs enormous amounts of water right away, while the production of bioenergy uses enormous amounts of water to grow crops. The need for water in the electricity generation sector is anticipated to increase over the next few decades due to the anticipated expansion of the electrified economy, particularly in developing nations with expanding populations. Two concepts that need to be understood and distinguished are water consumption and water withdrawals before we talk about the water demand intensities by various electricity generation methods. Water withdrawal is the process of taking water from a concentrated source and returning it to it, whereas consumption is the taking of water from a concentrated source and not returning it directly. It's critical to distinguish between these two concepts since they will affect water quantity and quality in different ways (King, Holman, and Webber, 2008).

The majority of energy produced in the US comes from thermoelectric power generation, which has a wide range of water demand intensities. The plant's various cooling systems affect the thermoelectric water demand. Comparatively speaking, once-through cooling systems use a lot more water than evaporative or dry cooling systems. Closed-loop cooling is an alternative technique for cooling thermoelectric power plants that use less water than the once-through system. However, since the cooling is accomplished through evaporation, the water consumption is higher overall because the water is not returned directly to the source. When access to a large body of water is restricted, closed-loop cooling is employed, which removes half as much water as the once-through method (King and Webber, 2008). The last possible technique is air-cooling, which involves using a fan to produce cooling. Although this method doesn't use any water, it is less cost-effective and energy-efficient, which makes it a less desirable choice for energy producers (King, Holman and Webber, 2008). The type of inputs also affects the water intensity of thermoelectric power generation. Rio Carillo and Frei (2009) draw attention to the fact that certain thermal plants rely on fossil fuels, which, depending on the kind of plant, require different amounts of water. According to King, Holman, and Webber (2008), the production of thermoelectric power accounts for up to 40% of all US freshwater withdrawals, despite producing 91% of the nation's electricity. According to King, Holman, and Webber (2008), the way these plants use water results in thermal pollution and water withdrawals, which harm aquatic ecosystems by releasing water that is warmer than the surrounding water.

Since it has been demonstrated that producing electricity uses a lot of water, it is crucial to think about how this will affect the electrification of economies around the world (Payne, 2010; Sharma, 2010; Kyle et al., 2013). This also considers the possibility of electric cars replacing gasoline-powered cars. Due to the water cooling techniques used in power plants, electric miles are observed to consume three times more water than conventional gasoline vehicles and to require 17 times more water withdrawals (King and Webber, 2008). While electric vehicles are effective in lowering greenhouse gas emissions, vehicles that run on electricity and hydrogen have significant effects on water (King, Holman, and Webber, 2008). On the other hand, electric vehicle water intensity would be minimal if electricity were produced using solar and wind power (King, Holman, and Webber, 2008). With these previously mentioned problems, the interplay between water and energy is beginning to become apparent, and this interplay will be explored further in this chapter. The water footprint of producing fuel for energy-related vehicles, including gasoline and electric ones.

8.2.2. Fuels

Water resources are heavily utilized in the production of electricity because electricity is necessary for human domestic and industrial needs. Nonetheless, humans also rely significantly on fuels like diesel, ethanol, and gasoline, all of which need a lot of water for production, processing, extraction, and refining (King, Holman, and Webber, 2008). Since producing ethanol requires watering crops, it is one of the biggest users of water; however, the water requirements of ethanol are similar to those of producing electricity (Scown, Horvath, and McKone, 2011). According to projections, the energy sector may use 3–4% more water by 2035 than it does now (Hightower, 2009), moving up to the second-highest sector in the US (Elcock, 2010).

Turning our attention to the distinctions between gasoline and electric vehicles, King, Holman, and Webber (2008) show that the US would see a

0% increase in water demand if most cars were converted to electric power and produced using current technology. As a result, it has been shown that the production of fuel energy is less dependent on freshwater at this time. In the United States, 95% of energy comes from the mining and petroleum industries, combined for which 5% of freshwater withdrawals are made (Scown, Horvath, and McKone, 2011). Freshwater availability could be seriously threatened if transportation were to depend more on water-intensive industries like electricity and biofuels (Scown, Horvath, and McKone, 2011). But it's important to remember that conventional gasoline-powered cars release greenhouse gases into the atmosphere, which could counteract any benefits from reduced water use in terms of environmental effects. Nevertheless, the use of freshwater will probably rise for every alternative fuel that is being considered for transportation. Alternative transportation fuels need to transition to less water-intensive energy production methods, like wind power, and account for more effective freshwater use (Hightower, 2009).

8.3. CHASSIS AND BODY STRUCTURE TECHNOLOGY

In the market for electric vehicles, two different technological approaches are probably going to coexist, albeit probably in different market niches. The first involves producing models that can be configured as either ICEVs or EVs or converting current vehicles to electric propulsion. The second route entails using cutting-edge materials and technology to produce EVs that are specifically designed and optimized. The introduction of new materials and their impact on various technical aspects, including weight reduction, aerodynamic performance, crashworthiness, endurance, and recyclability, will have an impact on the competitive potential of EVs in both scenarios, especially in the latter.

8.3.1. Purpose-built EVs: Materials and Manufacturing Techniques

When compared to steel, the weight of automobiles is significantly reduced through the use of plastics and light metal alloys. For instance, replacing steel with aluminum that has an equivalent structural strength results in a weight reduction of about 55%; magnesium and composite plastic would yield reductions of 75% and 65%, respectively. Realistically speaking, a production car that heavily uses structural composites would weigh about 30% less than a steel one. The trend of less steel being used in car bodies and chassis is probably going to pick up steam as electric vehicles become more and more common. In the 1950s, glass fiber-reinforced plastic parts were first used in mass-produced automobiles. Since then, the proportion of plastic parts in cars has progressively increased to help meet increasingly stricter fuel economy regulations and to keep costs low. For instance, in 1973, plastics made up only 2.9% of the material used in Japanese cars; by 1989, however, that percentage had risen to 7.5%. In a similar vein, BMW raised the proportion of plastics in the overall weight of its vehicles from roughly 5% in 1970 to roughly 10% in 1990. An average car contains about 300 pounds of plastic, divided into 20 different kinds.

Automobile manufacturers are starting to use more plastic in non-structural applications as well as body parts like engine parts, fenders, door panels, and bumpers. While some of these parts, like plastic bumpers, can withstand more impact than steel, they still adhere to safety regulations. However, more crash testing and the new materials' economic viability are still necessary before using composites for crucial structural components. For instance, materials such as aluminum, magnesium, titanium, ceramics, polymers, carbon fiber, Aramid, and Kevlar are frequently utilized in the construction of prototype and limited-series automobiles; however, due to technical and financial constraints, these materials are typically inappropriate for mass-produced automobiles, which typically employ simpler parts and less costly materials.

Beyond their relatively light weight and resistance, the new composite materials have additional benefits. Compared to steel components that are prone to corrosion, they last longer. Because parts that are typically bolted or welded together can be integrated into a single part, they speed up and simplify assembly. Consequently, the conventional 300–400 car body parts can be lowered to as few as 3 or 4. Certain materials possess the ability to be tailored to various uses and can also integrate pigmentation into their formula, obviating the necessity for traditional exterior paint. Furthermore, compared to steel-stamping tools, composite materials are molded in less expensive dies.

8.3.2. Impacts of Plastics and Composites on Automobile Manufacturing and Recycling

The design, manufacture, and assembly of cars are altered by the use of composites. It takes time and multiple steps for molding and curing to produce strong, intricate composite parts for structural uses. As a result, specialty automakers are the main users of them in small production runs. For instance, the Lotus Elan and Esprit, BMW Z1, Alfa Romeo SZ, and Dodge Viper are all made using a process called resin transfer molding, or RTM. The majority of composites used on a large scale are produced via compression molding, which is quick and affordable but restricts the parts' strength and complexity. However, an increase in the study, creation, and testing of novel materials points to their potential as practical solutions for a wide range of additional automotive applications. Strong and lightweight body materials are even more important for electric vehicles because of their limited range. Due to initial market size limitations, it is also likely that small production runs will be used in the early years of EV production. This implies that early in the production process, plastics, composite materials, and light metal alloys may play a significant role in the bodies of electric vehicles.

Environmental concerns are also raised by the expanding range of automotive materials. Energy-intensive processes are used in the creation of light alloys made of magnesium and aluminum. Petroleum is the source of many composites, glues, and plastics, some of which contain hazardous ingredients. The recycling process is also made more difficult by the new materials. Even though disassembling cars is one of the earliest commercial recycling practices, plastics are essentially excluded from the current system. **Hence the increased use of plastics creates a major disposal problem, since they are almost undegradable and more than 11 million cars are scrapped annually in the U.S. alone.**

Reduced use of various plastics (mixed plastics cannot be recycled with metal, increased use of materials like thermoplastic (a polypropylene that can be remelted and formed back into the same part), adoption of universal recycling codes on all plastic components, and incentives for automakers to buy back recycled products and for end users to return their vehicles for recycling are some potential ways to lessen the issue. Numerous automakers have made progress in this area. For instance, Fiat, BMW, Nissan, Saab, and other automakers as well as American automakers are implementing plastic part coding; Nissan is stepping up its research into the use of thermoplastics; and Chrysler, Ford, and GM have established a consortium to create vehicle recycling procedures.

Given that conversions rather than specially designed EVs may serve a sizeable portion of the initial electric vehicle market, the technical issues covered above are probably of secondary importance. Most on-road electric cars are converted from gasoline to run on electricity. Because conversion leverages pre-existing, well-proven platforms, it may be selected because it entails less risk in terms of technology, finances, and marketing. There is also a claim that using "gliders," or whole cars without an engine, transmission, fuel, or exhaust system, as a platform for electric vehicles would be more cost-effective due to a "deletion cost refund." Nevertheless, automakers frequently refuse to provide "gliders," so the unnecessary parts are either taken out or remain unutilized. The experience of France and England demonstrates that these economies are rarely achieved because the amount refunded is typically far less than the cost of the non-installed items. Advocates of gasoline van conversions for the fleet market believe that this market niche—large fleets in particular—will be crucial to the introduction of electric vehicles. The reasons for this strategy include:

- A higher degree of alignment between the performance capabilities of current EVs and fleet vehicle requirements compared to private vehicle requirements;
- Fleet cars are typically kept in central garages, which makes battery replacement, EV servicing, and training easier;
- When making purchase decisions, fleet buyers are more likely to consider the vehicles' entire life-cycle costs than their initial cost;
- Large fleets have more flexibility to operate with a mix of electric and conventional vans and thus have a higher propensity to buy EVs than smaller fleets;
- Large fleets are better targets for marketing efforts because of the higher sales potential per customer;
- Large fleets are further incentivized to invest in environmentally friendly technology by their increased public profile. Because of their great visibility, they also offer a chance to demonstrate the technology; and
- Governmental organizations and utility companies that are interested in promoting electric vehicles make up a large portion of the fleets.

Proponents of this approach assert that it would create the initial demand for EVs by encouraging commercial fleets to use them, with some government support. Due to production economies of scale brought about by this demand, EV costs would eventually decline to the point where they would be completely competitive with traditional vans. Electric vans could eventually be able to compete in the market for personal cars as well as passenger fleet vehicles thanks to further production increases and technological advancements.

8.3.3. Arguments for Purpose-built Electric Vehicles

Purpose-built EV proponents contend that the conversion process results in technological, production, and financial compromises that, at the very least, preserve the shortcomings of both systems and produce an automobile that is unduly heavy, complex, and expensive. It is illogical to expect an efficient product from a simple drivetrain and energy source substitution; product optimization necessitates control over every aspect. Purpose-built electric vehicles, according to them, inspire far more confidence in the new technology than do existing demonstration programs and inadequate ICE platform adaptations. Essentially, proponents of purpose-built vehicles aim to optimize their products, contending that electric propulsion can only reach its maximum potential if every aspect of vehicles is designed to best utilize this technology.

The risks and expenses of competing in an EV market that was created specifically for them would be the biggest issue for small makers. To create a product that would be commercially viable, significant research and development work would need to be done, and this expenditure would need to be spread over a sizable number of production runs. Both the cost of tools and the amount of engineering work would be decreased by conversion. Because of this, even after the introduction of purpose-built EVs, the conversion of new and used ICEVs for market segments like small city cars for private transportation and minivans for the fleet market is likely to continue.

8.4. VEHICLE AND BATTERY PRODUCTION

The manufacturing and assembly of vehicles and batteries is the stage of the equipment life cycle. While many parts of an automobile are similar in BEVs and ICEVs, there are a few key differences between the two technologies, such as propulsion, braking, and energy storage components. Though some models use lightweight materials to offset the effects of heavier batteries, most vehicle body and auxiliary system components are the same, and manufacturers can take advantage of existing production lines to benefit from economies of scale. The environmental impact of producing batteries, and other BEV-specific parts, and using alternative materials is different from that of manufacturing traditional ICEVs. The varying environmental impact profiles of BEV technologies during production can be largely attributed to the increased need for electricity and other energy sources needed for battery production.

8.4.1. Factors Affecting the BEV Production Stage

LCA results for vehicle production can be impacted by various factors in addition to the mix of electricity generation employed during production. These consist of driving distance over the course of a lifetime, fuel consumption, component material composition, vehicle mass, and powertrain. Generally speaking, the larger the vehicle, the more materials and energy are needed for it throughout its life cycle. Modifications to the composition of the materials may result in higher energy consumption during the manufacturing process, but they may also lighten the vehicle and use less fuel while it is in use. The longer the vehicle and battery lifetime, the less of an impact production-related emissions have and the more of an impact emissions during the use stage have on the overall life cycle effects. The size of the battery, the chemistry and configuration of the battery, and the manufacturing efficiencies are some other factors influencing LCAs of BEV production. The performance characteristics of different battery chemistries vary, and some batteries need more energy-intensive materials or production processes. Energy consumption per vehicle or battery produced may be decreased by manufacturers who can utilize production plants' full capacity and take advantage of economies of scale.

8.4.2. Environmental Assessment of Battery Manufacturing

The production of batteries is found to have the potential to account for the majority of energy consumption and the ensuing environmental effects during the manufacturing stage in many life cycle assessments (LCAs) of BEV technologies. Depending on the methodology used and the method of generating electricity (e.g., coal-fired, natural gas-fired, or renewable), estimates vary from 10% to 75% of manufacturing energy and 10%–70% of manufacturing GHG emissions. Regarding the other BEV components, LCAs estimate that the production of the electric motor contributes between 7% and 8% of all production-related emissions (which include raw material extraction and processing) because of its high copper and aluminum content, and the power train contributes between 6% and 8% because of its high aluminum content.

The manufacture of cells, the assembly of several cells into a battery pack, and the preparation of anode and cathode materials are all steps in the production of lithium-ion batteries, the technology that is the subject of this report. Asia is where most cells are made (South Korea and Japan, for example, are net exporters of battery packs). Pack assembly requires less energy and is less complicated than cell manufacture. Packs are either put together by automakers directly, or are assembled by cell manufacturers and shipped to automakers.

LCA practitioners typically find that the impact of BEV production is greater than that of ICEV production when comparing the GHG emissions from the production of BEVs and ICEVs (including those emissions from raw material extraction and processing). The GHG emissions of BEV production are typically estimated to be between 1.3 and 2 times those of ICEV production when GHG emissions of similarly sized ICEVs and BEVs are compared during the production phase.

Furthermore, a lot of LCAs state that the production of BEVs produces NO_x, SO_2, and PM emissions that are roughly 1.5–2.5 times higher than those of ICEVs. This is mostly because the production of batteries is an energy-intensive process, and the manufacturing sector currently uses a variety of sources to generate electricity. The consequences of this increased energy consumption on ecosystem health and human health are more widespread (depending upon fuel source and pollution controls).

8.5. COSTS OF EV BATTERY PRODUCTION

The performance characteristics of batteries for electric vehicles have been extensively documented, but information regarding the materials and production costs is harder to come by. A large portion of the information is confidential, and the designs and procedures are still being refined. Nonetheless, research on the effects on health and the environment offers some helpful information on battery composition, and material costs can be approximated. The costs of materials and total costs for four types of advanced electric vehicle batteries—advanced lead-acid (Pb-acid), sodium-sulfur (Na-S), nickel-cadmium (Ni-Cd), and nickel-metal hydride (Ni-MH)—are compiled in this section.

The batteries will account for a sizeable portion of the vehicle's cost and mass (between 20 and 40%). The fact that some of the batteries will probably need to be replaced at least once due to their predicted shorter lifespans than the vehicles will increase the impact. Except advanced lead-acid, all types of electrode and electrolyte materials are nonstandard automotive materials (though some cadmium has been used in coatings and pigments), about which not much is known. Nonetheless, casings, separators, and connectors—which are typically composed of more widely used materials like steel and polypropylene—make up a sizeable portion of the battery mass. The degree of uncertainty in battery material costs is decreased because these are well characterized.

8.5.1. Battery Materials and Their Costs

The majority of batteries' primary cost component is materials. To calculate the energy needs and emissions from manufacturing and recycling, Argonne previously estimated the material composition of various battery types (Gaines and Singh, 1995). The costs of battery materials are estimated in this analysis using the previously established material compositions. While the prices of some of the materials are known and standard commodities, others are new products whose prices are likely to decrease as the infrastructure and processes necessary for their production are developed. Another significant factor influencing future battery material prices is supply. More thorough material pricing data for Pbacid and Ni-MH batteries has been made available since this information was gathered; interested readers are directed to Lipman (1999).

8.5.2. Total Costs

Materials plus manufacturing, overhead, etc. are included in the total costs. Many researchers have offered cost estimates, but there is a wide range due to uncertainties and variations in the designs, materials, and procedures taken into consideration. Furthermore, a variety of bases—which aren't always clear—are employed, such as the duration of the project, its developmental stage, and—above all—its production scale. Each of these elements significantly affects the estimated expenses. The majority of the estimates are given in their entirety, without any breakdown. It is not always clear what the assumed production scale or level of technological development is. One source did provide a breakdown of costs by contribution for three different battery types based on high volume production and technological advancements (*see* Figure 8.1), while another provided a detailed estimate of materials for a single battery type. The cost breakdown provides several insights. First, materials are the best area to cut costs because they account for the largest portion of overall costs. The costs of materials vary greatly in absolute terms, but they are relatively constant in percentage terms, making up between 50% and 70% of the total cost of batteries. The cost of manufacturing makes up a comparatively small portion of the overall cost and is not anticipated to vary significantly depending on the type of battery.

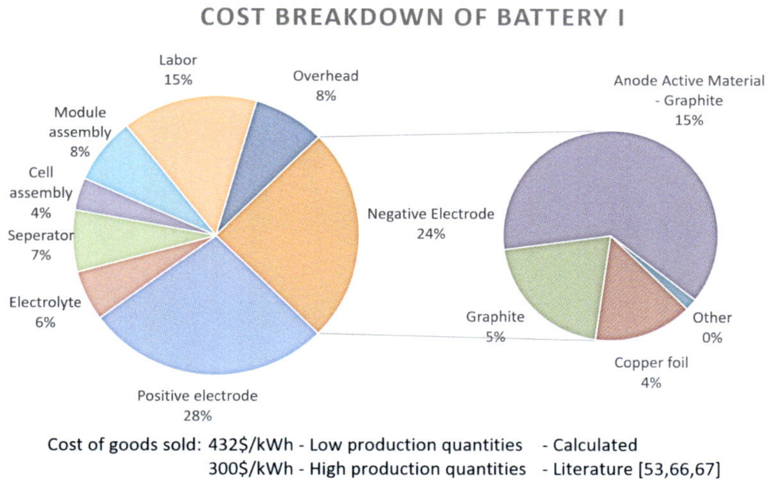

Figure 8.1. *Elements of battery cost.*

Source: Berckmans G, Messagie M, Smekens J, Omar N, Vanhaverbeke L, Van Mierlo J. Cost Projection of State of the Art Lithium-Ion Batteries for Electric Vehicles Up to 2030. Energies. *2017; 10(9):1314. https://doi.org/10.3390/en10091314.*

To get the total costs of the materials, we estimated using two different methods. In the first, researchers assumed that 60% of the total cost would come from raw materials. Klein and Salkind (1993) employed a comparable technique for Ni-MH batteries. According to their estimates, the cost of fabricated materials was 1.5 times higher than that of raw materials, and the cost of manufacturing batteries from the fabricated materials was multiplied by 1.5 times more. At that point, 44% of the total cost was made up of the raw materials. While actual manufacturing costs were probably going to be roughly the same for all battery types, this method increased the costs of battery types with high raw material costs much more (on an absolute basis) than those with low material costs. The materials cost was increased by a set amount in the alternative estimation technique.

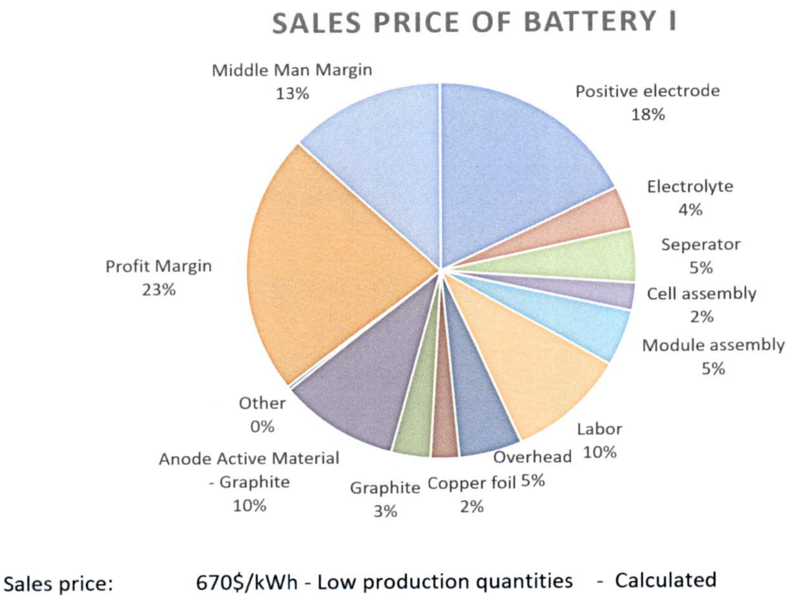

Figure 8.2. *Sales price of a battery.*

Source: Berckmans G, Messagie M, Smekens J, Omar N, Vanhaverbeke L, Van Mierlo J. Cost Projection of State of the Art Lithium-Ion Batteries for Electric Vehicles Up to 2030. Energies. *2017; 10(9):1314. https://doi.org/10.3390/en10091314.*

It's evident that the lead-acid battery is the cheapest and always will be. Because of the high cost of the ceramic electrolyte, the Na-S battery would seem to be very expensive; however, there is no reason why the cost cannot be significantly reduced. Although the Ni-Cd and Ni-MH batteries are less costly, there is no chance that material costs will drop significantly with them. It is evident that only the advanced Pb-acid battery (and possibly the Na-S battery) would be likely to meet the mid-term cost target of $150/kWh set by the Advanced Battery Consortium (ABC), despite the wide range of cost estimates for each battery type. When comparing batteries, one must factor in the likelihood that the lead-acid battery will need to be replaced one or more times during the vehicle's lifetime; that is, the actual cost of the battery must be multiplied by at least two. Figure 8.2 illustrates the profit margins of the producer and middle man.

One may compare the estimated total expenses of these batteries with the total expenses of the cars that will utilize them. The total cost of a battery pack could range from $3,750 at the very least, assuming some technology could meet the ABC goal, to $12,500 (at $500/kWh) or more, if a standard battery size of 25 kWh (typical of a small car battery) is selected. To minimize the number of battery replacements that will be necessary, it is crucial to concentrate R&D on both lowering battery costs and extending battery lifespan, as these expenses can be comparable to those of the rest of the vehicle (Gao and Wnfield, 2012).

REFERENCES

1. Athanasopoulou, L., Bikas, H., & Stavropoulos, P. (2018). Comparative well-to-wheel emissions assessment of internal combustion engine and battery electric vehicles. *Procedia CIRP, 78*, 25–30.
2. Auxer, W. (n.d.). Sodium sulfur battery–Status of development. *Silent Power, Inc.*, Press pack, Wayne, PA.
3. Bohnes, F., Gregg, J., & Laurent, A. (2017). Environmental impacts of future urban deployment of electric vehicles: Assessment framework and case study of Copenhagen for 2016–2030. *Environmental Science and Technology, 51*, 13995–14005.
4. Cook, M., & Morrow, H. (1992). Market prospects for nickel-cadmium batteries in electric vehicles. *Metal Bulletin Monthly*, April.
5. Cooper, A., & Moseley, P. T. (1998). Lead acid electric vehicle batteries–Improved performance of the affordable option. In *Proceedings of the Fifteenth International Electric Vehicle Symposium and Exhibition*, Brussels, Belgium.
6. Crossin, E., & Doherty, P. J. B. (2016). The effect of charging time on the comparative environmental performance of different vehicle types. *Applied Energy, 179*, 716–726.
7. Dunn, J. B., Gaines, L., Kelly, J. C., James, C., & Gallagher, K. G. (2015). The significance of Li-ion batteries in electric vehicle life-cycle energy and emissions and recycling's role in its reduction. *Energy and Environmental Science, 8*, 158. (Updated in 2019 using the U.S. Department of Energy, Argonne National Laboratory (ANL), GREET® 2018 dataset).
8. Faria, R., Moura, P., Delgado, J., & de Almeida, A. T. (2012). A sustainability assessment of electric vehicles as a personal mobility system. *Energy Conversion and Management, 61*, 19–30.
9. Freire, F., & Marques, P. (2012). Electric vehicles in Portugal: An integrated energy, greenhouse gas and cost life-cycle analysis. In *Proceedings of the 2012 IEEE International Symposium on Sustainable Systems and Technology (ISSST)* (Boston, MA, USA, 16–18 May 2012; IEEE: Piscataway, NJ, USA, 2012).
10. Gaines, L., & Singh, M. (1995). Energy and environmental impacts of electric vehicle battery production and recycling. *SAE 1995 Total Life Cycle Conference and Exposition*, Vienna, Austria.
11. Gao, L., & Winfield, Z. C. (2012). Life cycle assessment of environmental and economic impacts of advanced vehicles. *Energies, 5*, 605–620.
12. Helmers, E., Dietz, J., & Hartard, S. (2017). Electric car life cycle

assessment based on real-world mileage and the electric conversion scenario. *International Journal of Life Cycle Assessment, 22*, 15–30.

13. Hirabayashi, T., Furuta, S., & Satou, H. (1992). Cost estimation on advanced batteries for electric vehicles. *1992 ISATA Proceedings*, SAE Paper 920238.
14. Kalhammer, F. R., Kozawa, A., Moyer, C. B., & Owens, B. B. (1995). *Performance and availability of batteries for electric vehicles: Report of the Battery Technology Advisory Panel.* California Air Resources Board, El Monte, CA.
15. Klein, M., & Salkind, A. J. (1993). Design and cost analysis of a nickel-metal hydride EV battery. In *Extended Abstracts, 183rd Electrochemical Society Meeting*, Honolulu, Hawaii (May).
16. Lipman, T. E. (1999). The cost of manufacturing electric vehicle batteries (Report No. UCD-ITS-RR-99-5). *University of California at Davis*, Davis, CA.
17. Morrow, H. (1994). Nickel-cadmium batteries: The performance choice for electric vehicles. *11th International Seminar on Primary and Secondary Battery Technology and Applications*, Deerfield Beach, FL (March 1).
18. Varta. (1995). Varta R&D on a wide range of battery types for EVs. *Electric Vehicle Progress, 17*(5), 1 (March 1).
19. Vyas, A. D., Ng, H. K., Santini, D. J., & Anderson, J. L. (1997). *Electric and Hybrid Electric Vehicles: A Technology Assessment Based on a Two-Stage Delphi Study* (Report No. ANL/ESD-36). Argonne National Laboratory, Argonne, IL.

Chapter 9

Environmental Impacts of EV Charging Infrastructure

Contents

INTRODUCTION

The use of electric vehicles (EVs) has emerged as a viable strategy to combat climate change and lower greenhouse gas emissions. However, it is crucial to assess the environmental effects of EV charging because the source of electricity and the infrastructure for charging them both have a significant impact on their overall sustainability.

While EVs do not emit any emissions while they are in operation, it is important to recognize the environmental effects associated with the processes involved in charging them. The electricity used to charge electric vehicles usually comes from conventional power plants, which rely primarily on fossil fuels such as coal and natural gas. Consequently, the emission of greenhouse gases from these power plants offsets the reduction in emissions achieved by electric vehicles.

Environmental effects are also a result of EV battery production and disposal. Deforestation, water pollution, and habitat destruction are possible outcomes of the mining and processing of raw materials used in the production of batteries, such as lithium, nickel, and cobalt. Furthermore, incorrect battery disposal can result in hazardous waste and contaminate the environment.

Installing environmentally friendly infrastructure for EV charging has several advantages for both EV owners and the environment. First off, by lowering reliance on fossil fuels for the production of electricity, it lowers greenhouse gas emissions. This decrease is in line with international initiatives to tackle climate change and move toward a low-carbon future. Additionally, environmentally friendly charging infrastructure helps to improve air quality because cleaner transportation results from the combination of renewable energy sources and EVs' zero tailpipe emissions. A healthier living environment and improved public health are the results of this cleaner air.

Although home charging currently satisfies the majority of the demand for charging, publicly available chargers are becoming more and more necessary to offer the same degree of accessibility and convenience as refueling conventional vehicles. Infrastructure for public charging is essential for encouraging the use of electric vehicles, especially in densely populated areas where access to home charging is more constrained. By the end of 2022, there were 2.7 million public charging stations globally, with over 900,000 of those being installed in 2022—roughly a 55% increase over the stock in 2021 and similar to the 50% pre-pandemic growth rate between 2015 and 2019.

9.1. ELECTRIC VEHICLE CHARGING INFRA-STRUCTURE OVERVIEW: TYPE, LOCATION, AND SPEED

In general, it is evident that having access to charging stations is a crucial component that makes EV deployment possible. Additionally, there are various locations, prices, and speeds for EV charging; these must be adjusted to meet the needs of drivers. To maximize the deployment of charging stations, policymakers and market players should take into account the strategic roles played by different charging modes, specifically home charging, charging at work, and charging in public spaces.

9.1.1. Charging Modes

The definition of charging levels varies greatly between countries and regions as well as within academic, policy, and market literature. Here, we refer to charging ratings as Direct Current Fast Charging (DCFC), Level 1 (slow charging), or Level 2 (slow to fast charging using AC). Alternately, charging modes 1 (slow), 2 (slow to semi-fast using AC), 3 (semi-fast to fast using AC), and 4 (DC fast charging) are referred to by a different rating system that is more widely used in Europe (Figure 9.1) (Meyer and Wang, 2018).

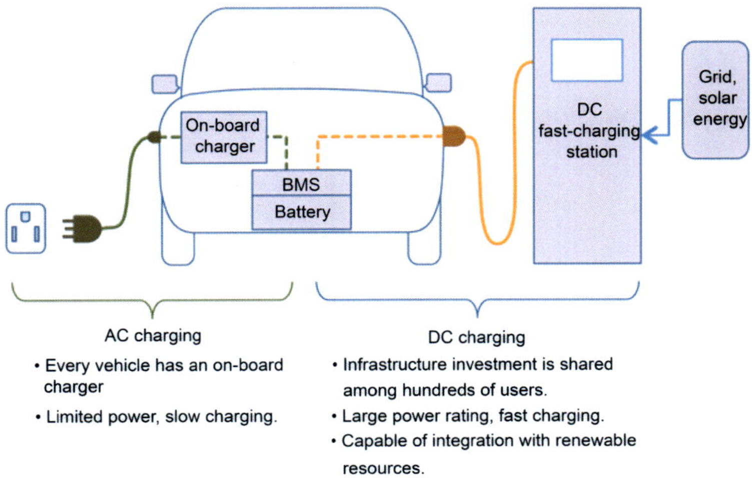

Figure 9.1. *Schematic representation of differences in on-board chargers and off-board chargers (or DC-fast charging station).*

Source: Dini P, Saponara S, Colicelli A. Overview on Battery Charging Systems for Electric Vehicles. Electronics. *2023; 12(20):4295. https://doi.org/10.3390/electronics12204295.*

9.1.2. Home Charging

The most popular method for recharging an EV is usually at home, where level 1 or level 2 chargers are used. To encourage EV buyers to install a Level 2 charger in their homes, several nations frequently offer subsidies. Drivers must have access to a designated off-street parking space, usually a driveway or garage, to install home charging. For instance, only 50% of the general population lives in single-family homes, whereas about 80% of Swedish electric car owners do.

In the current market, EV drivers rely largely on private charging. There is one public charger for every 19 EVs on the road in Norway, the country with the most developed electric vehicle market in terms of market share. Home charging is preferred over other locations, according to behavioral research and overall data on charging locations.

Nonetheless, the majority of EV owners are found in urban or densely populated areas. This is the case in Norway, where 90% of EV owners belong to this demographic, and in the US, where 82% of EV sales in 2015 took place in the 50 most populous metropolitan areas. According to surveys conducted in the UK, while roughly 70% of homes have garages or off-street parking, this suggests that residential infrastructure could provide a significant portion of the country's future charging demand. In urban areas, however, this percentage drops to just 30%. A high percentage of city dwellers in apartment buildings may also result in limited availability of EV charging stations and off-street parking, which could impede the adoption of EVs.

9.1.3. Public Charging

When public charging stations are few and far between, home charging has proven to be a practical and effective option, especially for early adopters of electric vehicles. But policymakers shouldn't assume that public charging infrastructure is unnecessary for uptake just because home charging is so common in the current market. In fact, focusing solely on the percentage of charging events that take place in each kind of location may undervalue the importance of workplace and public charging in the early stages of the market. Instead, the frequency of charging events (as opposed to volume) can demonstrate that even drivers who primarily charge at home occasionally use public charging networks. Level 2 and DC fast chargers are the two types of public charging stations. Fast chargers are more commonly found along highway corridors, while Level 2 charging stations can be found in places where car owners leave their vehicles parked for extended periods of time, such as shopping malls, airports, hotels, government offices, and other businesses. With an indicative ratio of one charger for every 10 electric cars, EU members were required to set deployment targets for publicly accessible chargers for 2020, 2025, and 2030 under the Alternative Fuels Infrastructure Directive (AFID) (EU/2014/94). In the European Union, there were about 225,000 publicly accessible charging stations as of the beginning of 2020, or nine cars per station. However, a 2021 review of the directive discovered that charging infrastructure was not implemented uniformly throughout the EU. Consequently, a new regulation is proposed that mandates Member states to make sure that a total power output of at least 1 kW is provided through publicly accessible recharging stations for each battery electric light-duty vehicle registered in their territory, and a total power output of at least 0.66 kW is provided through publicly accessible recharging stations for each plug-in hybrid light-duty vehicle registered in their territory.

Qualitative interviews conducted with experts on EVs in the Nordic region revealed that public charging infrastructure played a significant role in driving uptake of EVs in all four study countries, for both genuine need and peace of mind reasons. According to Figenbaum and Kolbenstvedt, the majority of EV owners in Norway used slow chargers to charge their vehicles at home or at work. However, the surveys also revealed that 10% of battery electric vehicle owners used public Level II charging at least once a week, 28% used fast charging at least once a month, and 60% of battery EV owners used Level II charging at least monthly or annually. Public charging will become more crucial for this user group as more drivers who live in multi-unit apartment buildings or other locations without access to dedicated private parking buy EVs.(J. Kester, L. Noel, G. Zarazua de Rubens, B.K. Sovacool, 2018)

A paradoxical problem arises from the low use of public charging in comparison to other charging locations, such as homes and workplaces. Because charging at home is affordable and convenient, drivers prefer to do so. However, drivers still rely on public charging networks to give them the assurance that they will have access to charging should they need it, even though their actual charging needs may be largely met by home charging. Neubauer and Wood discovered that while access to Level 2 public charging infrastructure (at 6.5 kW DC power) significantly increased total distance traveled and allowed many drivers to achieve nearly 100% vehicle utility, increasing home charging power beyond Level 1 charging (at roughly .15 kW DC) had little effect on vehicle utility. Simultaneously, low utilization rates pose a challenge to the profitability of charging infrastructure operations.

Because of the current difficulty in attaining profitability, governments that wish to support the development of public charging networks must take into account ways to facilitate effective, strategic investments that provide the necessary infrastructure. As public charging infrastructure becomes more widely available and as driver needs and characteristics change with growing EV utilization, the high rate of home charging may change. Empirical research indicates that when EV adoption rises, there are measurable changes in the use of various charging infrastructure types. Therefore, ongoing research on driver characteristics and habits ought to assist in informing charging expenditures.

9.1.4. Workplace Charging

For EV drivers who might not have easy access to home charging, charging at work can offer a vital opportunity. The most prevalent kind of chargers installed in offices are Level 2 stations. According to a comprehensive US survey, 30% of EV drivers charged their vehicles exclusively at work most days. Employer-located charging stations could be especially helpful in encouraging adoption among drivers without access to off-street parking for home charging when primary or sole reliance on workplace charging is feasible. A useful summary of workplace charging initiatives is given by Hall and Lutsey. These initiatives include up to 50% funding from the province of Quebec for employers who offer free charging; 40% of the costs per charge point for workplace charging or public charging on-site for French employers; and up to £300 in rebates for UK employers who install workplace or fleet charging. Workplace charging can also be helpful in promoting electric vehicle uptake: Garas et al. note the value of peer-to-peer communication as a driver of technology diffusion in general, and note that charging infrastructure located in social settings such as the workplace could help to facilitate EV adoption.

Olexsak presents survey results from the US Department of Energy's Workplace Charging Challenge, showing that workers who work for organizations that offer infrastructure for charging are 20 times more likely to buy an electric car.

Workplace charging should also take into account the possibility that it will lead to a rise in the use of renewable energy, especially in areas where there is an excess of clean power generation during the day. California is one state where widespread distributed solar generation has resulted in significant drops in net power demand during the day, which is then sharply ramped up to meet peak load in the evening. According to a high-EV deployment scenario for California, controllable EV charging might prevent the curtailment of about 2 GW of renewable energy. Likewise, Fang et al. Utilize both average marginal pricing and marginal rates of carbon and pollutant emissions to demonstrate how the frequency of charging events—that is, the times when fossil fuels are used to generate electricity—affects the social cost of charging. It is important to remember that charging at the workplace during the day can encourage people to use less energy. This can happen if the price of electricity is set to reflect the social cost of carbon. An alternative perspective is provided by Chakraborty et al. who find that free workplace charging may shift home charging to the workplace, which in turn can cause congestion in the network (Chakraborty, Lee,. Tal, 2019).

9.1.5. Role of Fast-Charging

The location of chargers—at home, at the office, or in public—is just one factor to take into account when evaluating the utility of charging. Optimizing utilization and appropriately coordinating charging networks with driver requirements also depend on the charging time and its correlation with current driver preferences and behaviors. In the network of charging options available to electric vehicle drivers, level 2 public charging stations are crucial. This is especially true when drivers can use these stations for up to an hour or longer without compromising their ability to travel. For example, charging while working or shopping could fall under this category. Fast charging is more useful, though, for drivers who plan to travel longer distances—up to 100 miles or more—during which they might need to refuel. In order to satisfy regular driving needs, it can also be helpfully aligned with current behavioral profiles. For example, fast-charging stations can be placed where drivers would normally run a quick errand.

Additionally, fast charging has the ability to more closely mimic the process of refueling a conventional car, which could make the switch to electric vehicles easier for drivers. This will be particularly true as technology advances to the point where fast charging equipment, which can currently charge between 50 and 120 kW, can charge 400 kW or more, allowing for meaningful refueling in 10–25 minutes. Here, too, the economics of providing fast-charging can be difficult because, while users value fast-charging stations greatly, low usage rates can result in low profitability. Hundt and associates cite a number of obstacles to the widespread adoption of public charging, including ambiguity surrounding its use, hardware costs, installation complexity and expense, and challenges evaluating revenue models. These issues are all made worse by the comparatively high initial cost of fast-charging stations.

Fast charging, regardless of its initial cost, is undoubtedly going to be essential to the ongoing adoption of EVs.

One worry about fast charging is the possibility of overcycling leading to battery degradation. Depending on variables like temperature and charge level, lithium-ion batteries typically lose capacity when they are frequently charged and discharged over time. Fast charging is now available, which has raised concerns that it could hasten the rate at which batteries lose their driving range. Tesla, the maker of electric vehicles, has admitted that to maintain vehicle range, its battery management system may limit the rate of EV charging under a variety of circumstances, including those that indicate high use of fast charging. However, behavioral research indicates that EV owners do not use fast charging in a way that compromises battery performance, so it does not seem that this effect is typical for the majority of EV drivers. According to a 2015 NREL study, drivers could receive up to 41.5% of their energy from fast charging, or an average of 7.6%. Fast chargers are not frequently used, so their effect on battery capacity loss was minimal.

A small but significant field test was also carried out by Idaho National Laboratory to investigate the impact of fast-charging on battery life issues. Two 2012 Nissan Leafs were charged twice a day using Level 2 charging, and the other two were charged twice a day using a 50 kW fast-charger. There was only a 2% difference in battery capacity between the Level 2 and DC fast charged Leafs after 50,000 miles driven, with no discernible difference observed in capacity after 10,000 miles driven.

9.2. THE ENVIRONMENTAL IMPACT OF EV CHARGING: ANALYZING THE CARBON FOOTPRINT OF CHARGING STATIONS

Globally, the popularity of electric vehicles (EVs) is rising as more people become aware of the advantages they have over conventional gasoline-powered cars. In addition to being more eco-friendly than conventional cars, EVs also offer a quieter, more comfortable ride and require less maintenance. It's critical to evaluate the environmental effects of EVs and charging stations, particularly the carbon footprint of the latter given the rise in EV adoption.

EVs' reduced carbon footprint is a major factor in their growing popularity. When comparing gasoline-powered vehicles to EVs, people who want to lessen their carbon footprint will find EVs to be a more appealing option. On the other hand, the emissions from the electricity needed to charge an electric car also contribute to the carbon footprint of an electric vehicle.

An essential factor in EVs' environmental impact is charging stations. Electricity for EVs is provided by charging stations, and the carbon footprint of these facilities can be ascertained by the source of that electricity. The carbon footprint will be minimal if the power is generated by solar or wind power. However, there will be a significant carbon footprint if the electricity is generated using fossil fuels like coal or natural gas.

To make sure that EVs continue to be an environmentally friendly mode of transportation, it is crucial to evaluate the environmental impact of charging stations. We can pinpoint areas where changes can be made to lessen the carbon footprint by examining the carbon footprint of charging stations. By examining EV chargers' carbon footprints, we will investigate how EV charging stations affect the environment in this blog. We'll also talk about ways to lessen the carbon footprint of the charging stations and comprehend how the government can support environmentally friendly transportation.

The quantity of greenhouse gas emissions connected to the generation and distribution of electricity used to charge electric vehicles at the charging stations is known as the "carbon footprint" of EV charging stations. An essential component to take into account when assessing the overall environmental impact of electric vehicles is the carbon footprint of charging stations.

9.2.1. Factors Contributing to the Carbon Footprint of EV Charging Stations

The carbon footprint of charging stations is influenced by various factors include:

- The source of the electricity that powers the charging station comes first. The carbon footprint of electricity produced using renewable energy sources, like solar or wind power, is smaller than that of electricity produced using fossil fuels, like coal or natural gas.
- The charging station's efficiency comes in second. An electric vehicle's carbon footprint is decreased by using more energy-efficient charging stations, which use less electricity to charge vehicles.
- Third, the charging station's location may have an impact on its carbon footprint. The charging station's carbon footprint will be reduced if it is situated in a region with a high concentration of renewable energy sources.

When comparing the carbon footprint of different types of charging stations, there are several key factors to consider.

- Level 1 charging stations (AC chargers), which are typically used in residential areas have the lowest carbon footprint because they rely on the existing electrical infrastructure and can be powered by renewable sources such as solar panels.
- Level 2 charging stations (Statiq Pillar and Statiq AdWall), which are commonly found in public places such as parking lots and shopping centers, have a slightly higher carbon footprint due to their higher power requirements.
- DC fast charging stations, which can charge an EV in as little as 30–45 minutes, have the highest carbon footprint of all charging

stations due to their high-power requirements and reliance on the electricity grid.

Overall, a charging station's carbon footprint can be influenced by its location, efficiency, and electricity source. The carbon footprints of Level 1 and DC fast charging stations are typically the lowest and higher, respectively.

We can lessen the carbon footprint of EV charging stations and encourage the use of electric vehicles as a more environmentally friendly mode of transportation by emphasizing the use of renewable energy sources and increasing the efficiency of charging stations.

9.2.2. Environmental Impact of EV Charging

Compared to nations that rely on renewable energy sources like solar and wind power, those that primarily rely on coal, natural gas, and oil for electricity generation will have a greater carbon footprint.

Because EV charging stations emit fewer greenhouse gases and other air pollutants than conventional gasoline-powered vehicles, they may help to improve the quality of the air. It is contingent upon the source of the electricity utilized for vehicle charging, though. Positive effects on air quality will result from using electricity generated by renewable energy sources like solar or wind power. However, there will be a negative environmental impact on air quality if the electricity is generated using fossil fuels like coal or natural gas. Water quality is usually not directly impacted by charging stations. However, if the electricity is produced by thermal power plants or other sources that need water for cooling, it may lead to pollution of the water supply.

The local ecosystems may be impacted by the placement of EV charging stations. Natural habitats and ecosystems may be harmed by the construction of new charging stations, particularly if those locations are close to delicate ecosystems. Energy grids may be impacted by EV charging. There is a chance that the energy grid will become overloaded as the number of EVs on the road rises, particularly during peak charging hours. Infrastructure for charging needs to be built to handle the rising demand for electricity to lessen this. This might necessitate building new infrastructure as well as making improvements to the current infrastructure.

9.2.3. How to Reduce the Carbon Footprint of Charging Stations?

Promoting the widespread use of electric vehicles requires minimizing the environmental impact of EV charging. The following tactics can lessen the carbon footprint that charging stations leave behind.

9.2.3.1. Prioritize the Use of Renewable Energy Sources

Using renewable energy sources, like solar and wind power, to power charging stations is one of the best ways to lessen their carbon footprint.

- **Pros:** This strategy helps to reduce the carbon footprint of charging stations and promotes the use of renewable energy sources.
- **Cons:** The cost of installing on-site solar panels can be high, and not all locations are suitable for solar power.

9.2.3.2. Improve Charging Station Efficiency

An electric vehicle's carbon footprint can be decreased by enhancing the efficiency of charging stations, which require less electricity to charge an electric vehicle. This can be accomplished by using smart charging technology, which optimizes charging times to lower electricity demand during peak hours, and more efficient charging equipment.

- **Pros:** This strategy helps reduce the carbon footprint of charging stations and can save money on electricity costs.
- **Cons:** Upgrading charging infrastructure can be expensive, and not all charging stations may have the necessary infrastructure to support smart charging technology.

9.2.3.3. Encourage the Use of Off-Peak Charging

Utilizing renewable energy sources and lessening the burden on the electricity grid are two benefits of charging electric cars during off-peak hours. Time-of-use pricing that encourages charging during off-peak hours can help achieve this.

- **Pros:** This strategy helps to reduce the strain on the energy grid and promotes the use of renewable energy sources.
- **Cons:** The effectiveness of this strategy depends on the availability of time-of-use pricing and may not be feasible for all electric vehicle owners.

9.2.4. Future of EV Charging and Sustainability

Convenience and sustainability are anticipated to be the main priorities for EV charging in the future. There will be a greater focus on sustainable charging infrastructure, which is fueled by renewable energy sources like solar, wind, and hydroelectric power, as more nations pledge to reduce their carbon footprint and meet climate goals.

It is also anticipated that the market for electric vehicles will continue to grow, propelling the advancement of quicker and more effective charging technologies. Long-distance drivers will find EVs more convenient and useful as battery technology and charging infrastructure continue to advance. It is anticipated that charging times will continue to drop and ranges will continue to rise.

All things considered, convenience and sustainability will take precedence over emissions reduction and increased efficiency when it comes to EV charging in the future. We may anticipate more developments and breakthroughs in EV charging technology as the market for electric cars expands, which will make EVs an even more alluring and environmentally friendly mode of transportation.

In summary, EV popularity is growing, which is a significant step in the direction of lowering transportation-related greenhouse gas emissions. To make sure that EVs and charging stations continue to be an environmentally friendly choice, it is crucial to evaluate their effects on the environment.

We can pinpoint areas where improvements can be made to lower the carbon footprint by examining the charging station's carbon footprint. Governments and companies can encourage the use of renewable energy and energy-efficient equipment, which will significantly reduce the carbon footprint of charging stations.

9.3. IMPACT OF RENEWABLE ENERGY ON EV CHARGING STATIONS

Renewable energy sources include biomass, solar, wind, hydropower, geothermal, and other energy sources that can be naturally and sustainably replenished. Because they don't use up natural resources, emit harmful emissions or waste, and have smaller carbon footprints than non-renewable energy sources like fossil fuels, they are crucial to the switch to sustainable energy systems. One exciting advancement in sustainable energy systems is the combination of renewable energy sources and EV charging stations. They can supply the electricity required to refuel electric cars, lowering transportation's carbon footprint.

EV batteries can also function as energy storage devices, storing excess renewable energy for use when needed and assisting in grid balancing. But the infrastructure supporting renewable energy sources and the grid face additional difficulties as a result of the increased demand for electricity to run these cars (Figure 9.2). (Schäfer et al., 2017).

Figure 9.2. *Solar-powered EVs.*

Source: By Hideki Kimura, Kouhei Sagawa – Own work, CC BY 3.0, https://commons.wikimedia.org/w/index.php?curid=8341063.

9.3.1. Opportunities for Integrating Renewable Energy Sources in EV Charging Stations

The transition to a cleaner, more sustainable energy system can be greatly aided by the many opportunities that integrate renewable energy sources with electric vehicle charging. These include:

- **Enabling a Cleaner and More Sustainable Transportation System**: It is feasible to drastically lessen the transportation sector's dependency on fossil fuels, which are a primary source of greenhouse gas emissions, by providing renewable energy to power electric vehicles. Because renewable energy sources are clean and sustainable—like solar, wind, and hydropower—EVs can be powered by them without producing more emissions or negatively affecting the environment.
- **Reducing Greenhouse Gas Emissions and Air Pollution**: Combining EV charging with renewable energy sources can aid in lowering air pollution and greenhouse gas emissions. This is so because the operation of renewable energy sources produces no greenhouse gases or air pollutants. The transportation sector's

overall carbon footprint can be decreased by powering EVs with renewable energy.

- **Enhancing Energy Security and Resilience**: Energy security and resilience can be improved by lowering reliance on fossil fuels and diversifying the energy mix. This is so that they are less susceptible to changes in price and disruptions in supply due to the distributed and frequently decentralized nature of renewable energy sources.
- **Leveraging the Flexibility and Controllability of EV Charging for Grid Balancing**: Flexible and controllable EV charging allows it to be tailored to the availability of renewable energy sources. This makes it possible to use it for grid balancing, which contributes to grid stabilization and lessens the requirement for peaking power plants that run on fossil fuels.
- **Unlocking New Business Models and Revenue Streams**: Combining EV charging with renewable energy sources has the potential to generate new revenue streams and business models. For instance, energy providers can charge EV owners for the excess energy stored in their batteries by selling it back to the grid, and EV owners can profit from this service.

All things considered, there are a lot of opportunities to lower emissions, improve energy security and resilience, and develop new business models and revenue streams by combining renewable energy sources with EV charging stations.

9.3.2. Challenges of Integrating Renewable Energy Sources

Incorporating renewable energy sources with electric vehicle charging poses various obstacles that need to be overcome to guarantee the effective acceptance and application of this technology. Some of the key challenges are:

- **Variability and Intermittency of Renewable Energy Sources**: The availability of renewable energy sources, like wind and solar power, can be unpredictable due to their variability and intermittent nature. This makes integrating them with EV charging difficult because it can be challenging to balance the supply and demand of renewable energy with EV charging requirements.
- **Matching the Demand and Supply of EV Charging and Renewable Energy**: It may be difficult to balance the supply of renewable energy with the demand for EV charging due to its high degree of variability. To manage and coordinate the charging of EVs with the availability of renewable energy sources, complex systems

and technology are needed.

- **Grid Integration and Stability Issues**: Because of the increased load of EVs and the variability of renewable energy, the integration of renewable energy sources may cause problems with grid integration and stability. To make sure that the grid can manage the increased load and variability, careful planning and grid infrastructure investments are necessary.
- **Cost and Complexity of Infrastructure and Equipment**: It costs a lot of money to integrate renewable energy sources with EV charging infrastructure and equipment. This can be costly and complicated, particularly in places where grid infrastructure needs to be updated and renewable energy sources are not easily accessible.
- **Regulatory and Policy Barriers**: integration of renewable energy sources may be hampered by the absence of laws and regulations that support the installation of EV charging infrastructure and renewable energy sources. To get beyond these obstacles, stakeholders—including legislators, regulators, and business leaders—must work together in a concerted effort.

9.3.3. Overcome the Challenges

Here are the possible solutions to overcome the challenges of integrating renewable energy sources with EV charging:

- **Advancing Renewable Energy and Energy Storage Technologies**: More dependable and reasonably priced renewable energy to power EV charging may be made available through the development of renewable energy sources like solar and wind power, as well as the advancement of energy storage technologies like batteries and pumped hydro storage.
- **Implementing Smart Charging and Vehicle-To-Grid (V2G) Systems**: When renewable energy sources become available, smart charging systems can automatically modify the rate at which EVs are charged. To further enhance grid stability, V2G systems can also allow EVs to return energy to the grid during times of high demand.
- **Coordinating and Optimizing EV Charging and Renewable Energy Management**: By coordinating EV charging with the availability of renewable energy sources, fossil fuel-based power plants may become less necessary. Ensuring the efficient and effective utilization of renewable energy sources can also be achieved by optimizing their management.

- **Developing Supportive Policies and Regulations**: Governments have the power to create laws and policies that promote the use of EVs and the development of renewable energy sources. For instance, they could mandate the use of renewable energy in particular industries or offer incentives for the production of renewable energy and the purchase of electric vehicles.
- **Encouraging Collaboration and Partnerships Among Stakeholders**: The development of integrated renewable energy and EV charging solutions can be aided by cooperation and partnerships between providers of charging infrastructure, producers of renewable energy, and other stakeholders. These kinds of partnerships can also guarantee that the requirements and viewpoints of all parties involved are taken into consideration.

9.3.4. Future Prospects

With increased renewable energy deployment, advancements in energy storage technology, and advancements in EV charging infrastructure, the future looks bright for integrating renewable energy sources with EV charging stations. Decarbonizing transportation is a goal that many companies and governments have set, which will encourage the use of EVs and renewable energy sources.

The integration of renewable energy and electric vehicle charging will be made possible by the continued development of smart charging and vehicle-to-grid systems. All things considered, integrating renewable energy sources with EV charging stations can be extremely important to building a more resilient and sustainable energy system (Meyer and Wang, 2018; Anwar et al., 2022).

9.4. THE CHARGING INFRASTRUCTURE MARKET

Historically, a variety of distinct market actors, including public or semi-state enterprises as well as private businesses, have developed, owned, and operated charging infrastructure. Here, we look at the value chain involved in creating and managing EV charging infrastructure, including the services that the market offers and the roles that various businesses play in it. Generally speaking, we only take into account strategies used by parties who have created or provided publicly accessible networks for charging, as opposed to those who merely supply or install hardware as needed.

9.4.1. Key Functions

While the business model for providing charging infrastructure is discussed in general in several studies, no clear model has yet been developed for providing this service, and only a small number of studies thoroughly assess the models that are currently in use. To provide clarity when addressing charging companies and the networks that are connected to them:

- **Manufacturing:** The original equipment manufacturer that produces charging equipment such as mounted home chargers and standalone public charging stations.
- **Installation/Development:** Arranges and executes installation and sources financing
- **Network Operation:** Manages and provides maintenance and customer-facing services for the physical network and payment platform.
- **Sales & Marketing:** An entity that hosts infrastructure, sets payment structure and prices, and/or collects or shares revenues from the resale of electricity

Businesses can be structured around one, two, or all four of the aforementioned essential roles. Regarding market participants with a single function, certain electronics manufacturers are exclusively involved in the manufacturing function of EV charging by manufacturing and providing hardware, such as charging stations. Others, such as Charge Your Car in the UK, just offer a network operating platform; they do not develop, acquire, or sell charging products or services to consumers directly (Madina, Zamora, and Zabala, 2016).

9.4.1.1. Charging Company Business Models

When it comes to assessing business models and potential policy interactions, asset ownership is a crucial factor for companies operating in various market functions. Some businesses concentrate on setting up and managing a network of charging stations, but they do not keep ownership of the system. Garas et al. label this approach the network-operator model, in which the company develops and maintains a network, but sells the actual hardware to host sites, which then manage billing and access to charging stations. With this model, host sites are free to determine their own fees and payment schedules. They can also keep the revenue they earn after deducting operator fees. Under such circumstances, the network operator transfers a significant amount of business risk to host entities, who bear the risk of future revenue fluctuations in addition to the initial cost of the EVSE.

On the other hand, some businesses—like EVGo and Blink Network—create, run, and possess their own charging networks. Garas et al. call this approach the "owner-operator model," representing a vertically integrated set of functions in which the charging company supplies and owns hardware, manages billing and access, and collects revenues from the infrastructure. The owner-operator model bears resemblance to a conventional mobile telecommunications enterprise, which incurs costs for infrastructure and bills customers for the services rendered by it. In these situations, the network operator might collaborate with a private host to locate charging stations, but they would still be in charge of setting the price and bear the risk of losing money.

There are still more integrated approaches beyond the owner-operator model. Examples of these include Tesla and Aerovironment, which operate in all four market functions—manufacturing equipment, setting up and maintaining networks, and collecting revenue—though not always in an integrated manner. Companies are required by these agreements to recoup their infrastructure costs through user fees. As an alternative, Tesla provides free charging by factoring the expense of creating a network of charging stations into the initial price of its cars. Big equipment producers that create, own, and run charging stations might also discover that running a fully integrated EV value chain in tandem with conventional operations is not a good fit. For instance, in June 2018, Aerovironment sold its electric vehicle charging business to German auto supplier Webasto in order to continue concentrating on its primary business of unmanned aircraft and missile systems.

Apart from these methods, Tesla and Better Places conducted experiments in the initial years of the previous decade utilizing the separation model. In this business model, the EV and battery are kept separate; the vehicle is bought, but the battery is leased. There is no waiting involved because this battery can be replaced with a fully charged one as soon as it gets to a service station when it needs to be recharged. The battery is now merely a standardized container that is rented out as needed, and the consumer essentially just purchases the charge itself. Chinese automakers NIO and BIAC are reexamining this strategy in an effort to implement the model domestically. As China reduces its state subsidies for EV makers and EVSE suppliers, market watchers look for new, more lucrative strategies to control the market. Yang along with others argues that the separation model might be a workable approach, and as a result, it suggests more collaboration between the state and the participating companies as well as standardization in battery design. Although this model is still in its early stages, this radical approach may at least be worth further investigation given that China is the world's largest EV market, still faces profitability problems, and is seeing a reduction in state financial aid. There are several distinct business models in use in China. The owner-operator model is used by the second-largest provider there, the state-owned China Grid Corporation. On the other hand, TGOOD, the biggest supplier of EV charging stations, is similar to Tesla in that it produces, installs, and runs stations and charging apparatus. Since TGOOD controls a large portion of the Chinese market, it is probably among the biggest suppliers of this kind of service worldwide. While they control a much smaller portion of the market, major automakers like BYD and SAIC also adhere to the Tesla model. Although there are some regional monopolies, the Chinese market is still rather crowded. The main distinction between business models in China and the rest of the world is the massive backing provided by the national, provincial, or local governments, which can take the form of subsidies, laws, or state-owned businesses. Additionally, State Grid and Potevio, the second and fourth-largest EVSE providers, are owned by the state.

9.4.1.2. Ownership of Charging Point Host Sites

Many different commercial or public actors who want to offer charging access as a perk for their staff, as an amenity on their property, or both can host charging points. Retail and city center locations accounted for a combined 24% of publicly-accessible charging stations in New York state, according to an early survey of charging locations. Commercial offices, universities, hospitals, and transport centers had less than 7% of charging stations each. According to the survey, the majority of charging stations (41%) were found at car dealerships or gas stations.

According to Bloomberg New Energy Finance (BNEF), there has been a significant amount of consolidation among EV charging companies as major energy companies buy out or heavily invest in infrastructure players. Examples of this include the acquisition of 20,000 charging points by Dutch charging company NewMotion by Royal Dutch Shell, investments made in Chargepoint by Statoil, EVBox by Engie, and eMotorWerks by Enel. Utilities are especially well-suited for some parts of the EV charging market because of their institutional experience in creating and managing electricity distribution networks. In Germany, power companies are in charge of 35% of all public charging stations. Meanwhile, in California, the electricity regulator recently approved three major utilities' $738 million investment in transport electrification projects. Similar utility proposals have been approved or are being considered by other US state regulators.

Through significant joint partnerships, big automakers are contributing significantly to the development of the charging infrastructure alongside electricity providers. For instance, the Ionity consortium, which provides an integrated European fast-charging network, is made up of BMW, Daimler, Ford, and Volkswagen. Since the joint initiative's official announcement in November 2017, about half of the 400 EV fast-charging stations that were intended to be installed throughout Europe have been installed, with an average of six chargers per station. Other consortia of a similar nature are the NEXT-E charging network, which includes participation from manufacturers Nissan and BMW, as well as the Hungarian oil and gas company MOL and the German utility E.ON, and Ultra-E, of which manufacturers Audi, BMW, and Renault are among the core partners. Customers who drive their cars can access free charging stations provided by automakers Nissan and Tesla.

In addition to offering US Leaf buyers and lessees the opportunity to participate in an on-site charging network hosted by numerous Nissan dealerships, Nissan also offers a two-year free charging program that entitles drivers to a membership card that can be used at a variety of public charging network providers. Tesla has instead established its own proprietary fast-charging network for the exclusive use of Tesla drivers, with locations in public areas close to major highways in relevant markets, while Nissan has taken a proactive approach to supporting infrastructure buildout. Apart from its Supercharger network, Tesla also runs a program called Destination Charging, which allows businesses to host Tesla charging stations. Tesla installs these facilities for free, subject to certain requirements. This strategy adheres to the vertical integration model for every function in the EVSE value chain. The extent to which Tesla infrastructure may support a thriving EV charging market is constrained by the proprietary features of Tesla's network, which also present issues with market fragmentation and interoperability. Given the role that inadequate charging can play in discouraging customers from adopting EVs, investments in building out the charging infrastructure would seem to be strategically important as automakers continue to bolster their offerings of electric vehicles.(H. Bonges and A. Lusk, 2016; Meyer and Wang, 2018; Anwar et al., 2022)

9.4.1.3. Industry Lessons Learned

There are some unsuccessful approaches in the EV charging market that can serve as a useful lesson when deciding which future models should receive support. According to Garas et al. the Blink network once used a leasing model in which host sites signed contracts to rent charging equipment from a supplier. Because it required them to perform the installation and development function with regard to preparing sites for the construction of charging facilities—a task for which they were typically ill-equipped—the arrangement proved unpopular with potential host entities. For example, zoning, grid connection, and planning can pose major risks and obstacles to the successful installation of EVSEs.

Another failed attempt at creating EV charging infrastructure was made by the Israeli company Better Place, which shut down in 2013. As part of the Better Place concept, a network for battery swapping and recharging was to be developed concurrently with the company maintaining ownership of the battery vehicle and charging a mileage-based fee from customers. Noel and Sovacool investigate the main causes of Better Place's eventual bankruptcy. Although they note that a variety of social, technological, political, and managerial factors played a role in Better Place's demise, they finally come to the conclusion that EV adoption in general was too low in Israel and Denmark, the company's two main Better Place markets, to justify the venture.

9.4.2. Economics of Charging Infrastructure

The fundamental cost drivers for charging electric vehicles are a mix of one-time installation and recurring operating expenses. The sub-segments of the EVSE value chain with the biggest potential to reduce overall costs can be inferred from the evaluation of these components.

9.4.2.1. Upfront Costs

Hardware (the actual charging station itself), labor, permits, and installation—which may require significant public works efforts—are among the upfront costs associated with EVSEs. The connection and upgrades to the local distribution network may also be included in the upfront costs, depending on the location and the planned infrastructure's charging capacity. Securing charging station access locations, such as parking spots on public roads or in private parking lots, may also incur non-hardware installation costs.

Trying to categorize the various kinds of EV charging technology becomes more difficult due to its rapid development. Generally speaking, during the previous 10 years, DC Fast (50 kW) chargers were introduced, and AC chargers with Level One (<11 kW), Level Two (11 kW–22 kW), and Level Three (22 kW–44 kW) charging rates were the norm. The more recent arrival of DC Ultra-Fast (350 kW) and DC Rapid (150 kW) chargers, however, drastically changes the balance. These chargers are much faster and more expensive than the older AC chargers, but as the cost of DC Fast chargers has decreased, it is anticipated that the cost of these chargers will also decrease. These days, chargers come in a variety of forms and sizes. From the most basic 3 kW Level One AC chargers, which are frequently included with an EV purchase, to 350 kW monsters that can charge an EV battery in minutes and can cost several times the EV itself. Here, we have chosen just three distinct types in order to keep the table more manageable and to address limitations in the current research that frequently group different but similar speeds as one type. Three charging speeds: slow, medium, and fast. Slow chargers, also known as Level One chargers, are unique in that they are primarily used at night in homes and are also significantly less expensive. Level Two, Level Three, and DC Fast chargers currently occupy the middle ground in terms of pricing and charging speeds. Due to their high prices and incredibly quick charging rates, DC Rapid and DC Ultra-Fast stand alone as distinct categories.

As previously mentioned, installing Level 1 charging for public access can cost anywhere from €150 to €10,000, depending on the location of the new outlet in a commercial setting. Level 2 and DCFC installations can necessitate significant capital expenditures due to the need for electrical upgrades and public works, including fortifying local distribution networks. Naturally, costs can differ significantly depending on the site's characteristics, location, and grid conditions. However, when installing multiple chargers at a single station, economies of scale may lower the installation costs per unit. Nevertheless, it may be challenging for a charging station to turn a profit due to high installation costs. Ownership of EV charging assets is not profitable under the current model, according to calculations made by Nigro and Frades based on cost assumptions at the time. For example, a single DC fast charger can cost a developer $44,000 over a ten-year period, while a Level 2 charger with five connections can cost more than $26,000.

9.4.2.2. Operating Costs

After they are installed, EV charging stations need to pay for ongoing services like operations and maintenance as well as the cost of buying electricity from a utility or supplier. One of the main factors influencing the cost of EV charging business models is the cost of electricity and the makeup of regional electricity tariffs. This is accurate in terms of how electricity pricing structures affect the ability to provide EV refueling services for both residential and commercial installations.

Significant variation in home charging costs can be caused by the various residential retail rate structures. According to a 2015 study commissioned by the New York State Energy Research and Development Authority (NYSERDA), for instance, depending on overall usage patterns, households that charge their electric vehicles at home could save as much as $400 by switching to a time-of-use rate instead of the standard rate.

The economic feasibility of charging may be restricted for commercial charging station hosts, such as shopping centers, supermarkets, and gas stations, by a tariff structure based on demand charges. These charges can account for as much as 50% or more of a typical commercial customer's monthly electricity bills. Demand charges for EV charging businesses and individual hosts offering 50 kW+ of charging capacity can be especially severe due to the high power rating of fast chargers. An energy-only tariff from a California utility was compared to a demand charge plus energy-only bill. The energy-only tariff realized an average cost of less than $5 per charge with only 25 charges, while the energy-plus-demand charge rate required 150 charges per month. Customers could save between $1125 and $1220 a month, or 58%–73%, on their overall bills with critical peak pricing proposals, according to research.

Infrastructure for charging is still expensive today, both in terms of the initial installation cost and possibly ongoing costs based on retail tariff structures. Although not covered in detail here, utilization rates are frequently fairly low. According to a report by Helmus et al. on the public charging infrastructure in the Netherlands, in recent years, 4–5 unique users have been charging for 40–60 hours on average per charger per week. Low utilization rates consequently result in low potential for cost recovery for many individual charging stations (mixing high- and low-utilization chargers in station portfolios can help guarantee total costs are covered).

9.4.2.3. Payment and Fee Structures

The relatively high cost of hardware installation must be covered by charging service providers, especially for fast-charging facilities. Other expenses include continuing maintenance and the cost of purchasing electricity for resale. Current market players frequently use payment plans that include one or more different kinds of fees. A fixed fee, such as one that is paid annually or monthly, a volumetric fee that charges customers for each kWh used, a time-based fee that charges customers for the minutes or hours spent charging, and/or a per-session fee that charges customers a set amount to connect to a charger are examples of these. Pricing is determined by the charging station owner in these situations rather than by a corporate entity like US-based Charge point, which oversees networked charging stations owned by individual charging station hosts. Other businesses, such as the Irish utility ESB, first offered charging at no cost because the infrastructure had been paid for by the government, but they later switched to membership and pay-as-you-go payment plans.

Roughly one-third (7 companies) of the 21 charging companies that were reported included a fixed fee. About half of the plans had some kind of volumetric charge, about a fifth had a time-based charge, and about 10% (2 companies, mostly DCFC) had a flat connection fee. The remaining businesses employed different fee structures, such as free charging or rates determined by the proprietors of individual stations. The variety of payment options suggests that there is currently no clear or universally applicable model for charging for services; instead, flat fees are being replaced by subscription, volumetric, or time-based fees. Additionally, structures occasionally might mirror terms unique to the legal environment in which businesses operate. For instance, almost all of the payment plans with a per-kWh cost for the businesses in this sample were based in Europe. This might be because US states and European nations have different laws governing the resale of electricity.

More generally, the underdeveloped state of the EV market as a whole, which means that low or irregular utilization rates force businesses to use subscription fees to ensure steady revenues, is probably the driving force behind the prevalence of the subscription model—present in a variety of payment plans. Kley and associates suggest that home-charging as a core business would be a more appropriate source of funding for these assets given the high importance but low utilization of public charging infrastructure during early market conditions. Payment schemes, however, will probably keep changing as EV ownership grows in popularity and the amount of public charging power rises. A standardized approach to refueling costs is essential from the standpoint of protecting consumers, since a variety of rates and billing formats result in a lack of clarity and comparability for users.

As in the case of EV drivers who own home charging infrastructure to make their equipment available for public use, either for free or for a usage fee, charging services are also provided as part of emerging sharing-economy business models. In certain instances, this has been implemented as a specialized platform, such as the Swedish ElBnB, introduced by Renault in 2016, that enables drivers to find and get in touch with the owner of a charging station, and Chargie, which operates on a similar principle in the UK. Another strategy makes use of already-existing sharing-economy platforms to create alliances between players in the market, like Airbnb and Tesla.

9.5. THE IMPORTANCE OF CHARGING INFRASTRUCTURE TO THE ELECTRIC VEHICLE REVOLUTION

Infrastructure for EV charging that satisfies consumer demands must be established in the UK. As charging trends, EV consumption, and demand for EV charging become more apparent, such a network is currently being implemented. But to meet targets aligned with the UK government's strategy, the electrification of transportation must happen at the rate that charging points must be installed—ahead of demand and throughout the UK. To provide this infrastructure and guarantee that drivers have access to dependable charging regardless of the UK region, the public and private sectors must collaborate.

The infrastructure needs are largely determined by how well EV batteries perform. For example, larger batteries with higher capacity and improved performance will extend the electric vehicle's range, requiring a less dense network of charging infrastructure than would otherwise be necessary. Similarly, faster charging batteries would make it much more convenient to refuel EVs during lengthy trips, which would raise demand for public rapid chargers. The Faraday Institution's research will therefore have a direct impact on the features of the UK charging infrastructure network needed for the 2030s and beyond. This research aims to improve battery performance, understand battery degradation processes, and develop next-generation battery chemistries (Meyer and Wang, 2018; Anwar et al., 2022).

9.5.1. Different Types of Public Charging Points

Since EV owners use their vehicles in a variety of ways, the UK has already seen the emergence of a wide range of infrastructure and charging capabilities. Charging point technology is standardizing because investors, businesses, and consumers may find the variety in type and performance confusing. For instance, the Office for Zero Emission Vehicles (OZEV) has categorized charging stations into eight distinct categories and established a minimal technical standard for workplace charging. These definitions will probably change as the market grows and technology advances.

Although descriptions of public charging stations that aim to provide EV users with options for charging differ greatly between operators, they usually employ a larger number of aggregate categories than the eight OZEV workplace definitions. Table 9.1 compares between off-board and on-board charging systems for electric vehicles.

Table 9.1. *Comparison Between Off-Board and On-Board Charging Systems for Electric Vehicles*

Features	Off-Board Charging Systems	On-Board Charging Systems
Availability of charging stations	Wide presence of public stations in urban areas and along highways.	Accessible wherever there is an electrical outlet, but requires the installation of a dedicated home charging station.

Types of charging stations	Public Charging Stations, Rapid Charging Stations, Induction Charging, Home Charging Stations.	Mainly Home Charging Stations.
Charging power	High-power DC charging available at many public stations. Rapid Charging Stations offer high charging speeds.	Slower charging compared to high-power public stations.
Connector compatibility	Various connector types used but often compatible with a wide range of vehicles.	Standard connection to home outlets or dedicated charging stations with a specific vehicle connector.
Charging time	Rapid charging times (usually less than an hour) at high-power public stations.	Longer charging times (varying from several hours to overnight) depending on the charging power.
Charging costs	Some public stations may charge usage fees.	No usage fees at home charging stations.
Initial installation	Requires no additional installation.	Requires the installation of a dedicated home charging station, incurring associated costs.
Advantages	Fast charging, wide availability, ideal for long trips.	Maximum convenience, lower long-term costs, the opportunity to take advantage of cost-effective energy rates, always ready for use.
Disadvantages	Potentially higher costs due to usage fees, congestion in some areas.	Slower charging, requires the installation of a charging station, limited to home charging station availability.

Source: Dini P, Saponara S, Colicelli A. Overview on Battery Charging Systems for Electric Vehicles. Electronics. 2023; 12(20):4295. https://doi.org/10.3390/electronics12204295.

Slow chargers are typically used at home or at work, where users don't mind waiting for extended periods of time to finish their charging, either overnight or during off-peak electricity demand. Fast chargers are frequently found in places like supermarkets, retail malls, entertainment venues, hotels, airports, train stations, general parking lots, and park and ride programs, where cars are likely to park for a few hours. Additionally, charging points are categorized as alternating current (AC) or direct current (DC). Batteries can only be charged with DC power; however, since the power supplied by the national grid is AC, a rectifier is required to convert the AC power.

EVs must be charged while traveling long distances, necessitating the use of even faster chargers, also referred to as rapid chargers and ultra-rapid chargers. These chargers are usually found at motorway service stations and deliver power faster but at much higher tariffs. Although opportunities to "open up and increase charging competition within motorway services" have recently been identified, customers are willing to pay a premium to save time, and there are frequently few alternatives. In addition, compared to fast charging, rapid charging allows delivery vans to travel longer distances or spend more time on the road.

The voltage and current of the supplied electricity directly correlate with a charger's specified power. The maximum practical charging rate is limited because thicker cables are needed to increase power by increasing current. However, raising the voltage allows for quicker charging without adding thickness to the cable. The majority of cars today use 400V battery packs, but more recent models—like the Hyundai IONIQ 5 and Porsche Taycan—are coming out with 800V battery packs that allow for faster charging times with extremely quick chargers.

9.5.2. Key Characteristics of the UK's Public Charging Infrastructure

Transportation in the United Kingdom is rapidly electrifying; by 2022, there will be 492,000 battery-electric vehicles and 361,000 plug-in hybrid vehicles on UK roads. Zero emissions vehicles (ZEVs) are predicted to make up about 24% of all cars on the road in the UK by 2030, and they are predicted to account for at least 80% of all new car sales in the country.

The network of charging infrastructure in the UK is expanding quickly as well. In the UK, there were about 400,000 installed charging points at homes and places of business as of the end of August 2022. An additional 35,000 charging points were spread across 21,000 public charging locations (Figure 9.3).

The coverage of charging points varies greatly throughout the United Kingdom. Currently, areas like London and the South East, which have high rates of EV adoption along with high population densities, have a disproportionate amount of charging infrastructure installed. Scotland stands out as an exception, possessing substantial infrastructure per individual despite having a lower population density. To raise the number of chargers per head of population, regions like Northern Ireland, the North West, Yorkshire, and the East of England require attention.

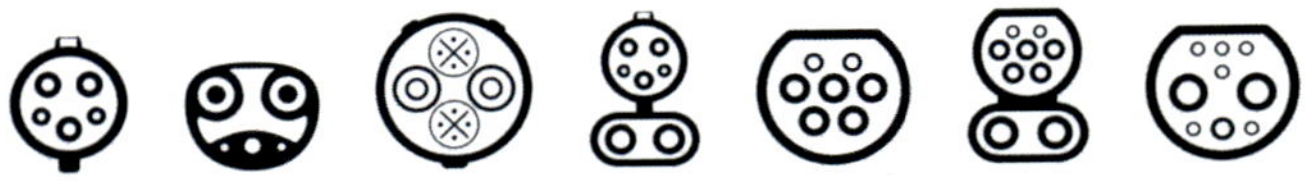

	J1772	Tesla	CHAdeMO	CCS	Type 2	CCS (2)	GB/T
N. America	AC/DC	AC/DC	DC	DC			
Europe	AC	AC/DC	DC		AC	DC	
Japan	AC		DC				
China							AC/DC

Figure 9.3. *Type of connector and their usage in related areas.*

Source: Dini P, Saponara S, Colicelli A. Overview on Battery Charging Systems for Electric Vehicles. Electronics. *2023; 12(20):4295. https://doi.org/10.3390/electronics12204295.*

Major thoroughfares and motorways receive more coverage. The strategic road network now has rapid charging stations every 20 miles covering about 97% of the area. Public charging stations, however, serve a much smaller portion of the rural population. To create a network that facilitates EV adoption for everyone, the public sector will need to take the lead on initiatives to make charging infrastructure both profitable and affordable in rural areas. The successful collaboration between the public and private sectors can be deduced from the implementation of broadband in rural regions.

To meet the growing demand for outside home charging of EVs and the uptake of EVs, more charging infrastructure is being built worldwide. In 2021, there were 11.8 million public chargers worldwide. China accounted for 63% of the market, followed by Europe with 23%. 10 Although most public charging networks consist of slower chargers (less than 20kW), faster chargers deliver comparatively more power. For instance, in the UK, 100kW+ chargers supplied 15% of the energy required for EVs, while making up only 3% of charging stations.

9.5.3. Future Projections of the Public Charging Infrastructure Requirements

The adoption and usage of EVs depend on the development of a future public charging infrastructure network that is reasonably priced, effective, and dependable since it will allow EV drivers to charge more quickly, go farther on their trips, and provide confidence to customers considering buying their first

EV. By 2030, there will likely be 10 million zero-emission vehicles on the road in the UK, necessitating the installation of 280,000–720,000 public charging stations. The uncertainty in future consumer preferences and charging behavior is reflected in the wide range of estimates of the amount of infrastructure required. The Climate Change Committee estimates that approximately 525,000 public charging points will be needed in the UK by 2050, while other studies, for instance, suggest that between 148,000–285,000 and 375,000–500,000 public charging points will be required by 2030.

In Europe, a comparable network density is required, with an estimated 2–4 million public charging stations needed by 2030 and an associated investment of approximately US$34 billion.

The UK Government is playing its part in delivering the future charging infrastructure network, with the ambition to ensure "the UK's charging infrastructure network is reliable, accessible, and meets the demands of all motorists." Over £1.6 billion of the £2.5 billion in government funds allocated to the EV transition since 2020 will go toward supporting the infrastructure needed for charging. Initiatives include a £950 million Rapid Charging Fund to support the rollout of powerful chargers on major A roads and motorways, as well as a £450 million Local EV Infrastructure Fund that will provide capital and resource funding to support local authorities in their efforts to collaborate with industry and change the availability of charging infrastructure for drivers without off-street parking. These programs carry on the assistance and intervention that the government of the United Kingdom has been offering the charging infrastructure sector since 2011.

Additionally, the development of charging infrastructure is being supported by other bodies. 95% of drivers should be within 50 miles of the 50 key motorway service locations, according to a plan being implemented by National Grid. Additionally, there is ongoing investment to guarantee that local grid connections can withstand the new energy charging regulations. Charging point operators (CPOs) have a big chance to take advantage of economies of scale and develop a mass-market clientele. By 2035, the installation and operation of the chargers will cost the UK approximately £16.7 billion, not including the extra money needed annually for upkeep and repairs.

A shift in consumer charging habits may be facilitated by improvements to the infrastructure for charging, such as the introduction of fast charging stations, smart charging with adjustable rates, and the use of V2G to balance out peak spikes in grid electricity demand. Although using these technologies will benefit providers as well as consumers, both have an impact on battery degradation. Degradation can result in early failure, poor performance, the need to replace or retire a product, as well as related financial and environmental costs.

9.5.4. Consumer Attitudes to UK Charging Infrastructure

One of the main factors influencing the decision to buy an EV is one's assessment of the standard, extent, and accessibility of the charging infrastructure in the UK. This decision is primarily driven by worries about the driving range of EVs.

A common complaint from customers is that there aren't enough public chargers, which deters them from buying an EV. A review of 45 surveys revealed that the main obstacles to EV adoption were almost all related to range anxiety and the inability to carry out any trip as conveniently as with a gas or diesel vehicle.

After adopting an EV, range anxiety frequently gives way to range awareness, but for those without off-street parking, concerns are especially pressing. According to a 2021 study, two-thirds of drivers without off-street parking cite the inaccessibility of EV charging stations as a major obstacle. The top three reasons given by those who said they did not plan to buy an EV in the upcoming year were cost (72%), charging station accessibility (47%), and EV dependability (18%). According to key market players interviewed, accessibility for disabled users of public chargers—particularly on-street infrastructure—is also a concern. These obstacles will be partially addressed by the creation of a new BSI standard for EV charging accessibility.

Ownership as perceived by consumers can differ from actual ownership. The issues that EV owners are more concerned with are location, availability, and upkeep of charging stations. Relatively less important factors are the distance to chargers and the charging time. Routine and preparation are cited by EV drivers as essential to EV ownership. Typical procedures include remembering to use public charging stations during slower periods of the day, topping off charges frequently, and fully charging once a week.

Concerns concerning the infrastructure for electric vehicles are not unique to Europe. In Europe, public charging infrastructure ranked higher on average (17%) among those who intended not to buy an EV than in the United Kingdom (14%). Purchasing price (18%) and driving range (20%) were the other two main concerns of European consumers.

Long car trips are relatively uncommon, Although range concerns are common. Approximately 99% of private automobile trips are shorter than 100 miles, accounting for 87% of total mileage driven. To encourage EV adoption, it will be essential to inform prospective EV owners about how EV batteries can essentially meet current usage patterns.

The fact that there are currently too many different connectors and charging methods is another issue that EV drivers frequently bring up. The Renewable

Energy Association (REA) underlined the significance of interoperability and concluded that widespread adoption requires cooperation between the charging sector and the government as well as the adoption of common protocols and standards. Additionally, the National Infrastructure Commission argued that the government ought to establish minimal requirements for a system of compatible smart charging stations. Lately, there has been a significant standardization effort. For instance, all new plug-in models now come with the Combined Charging System standard, while other connectors—like the CHAdeMO charging standard—are being phased out.

The 2018 Automated and Electric Vehicles Bill standardized what is needed for both private and public charging stations. Along with standards for payment and dependability, compatibility with public charging stations, and other factors, these should all aid in meeting consumer demands. The development of the UK charging infrastructure fit for the future is further supported by more recent regulations on smart functionality for private charging points, with protection around grid stability, cyber and data security, and interoperability (**Meyer and Wang, 2018; Anwar et al., 2022**).

9.5.5. The Role and Importance of Home Charging

A significant amount of charging is anticipated to continue using home chargers, with each home charger delivering comparatively more energy annually than non-home chargers. Approximately 78% of homeowners who drive vehicles have access to off-street parking36. Although it takes longer, charging at home is affordable and convenient. Regarding how home charging use will evolve over time, there is still some debate and uncertainty. In contrast, a DfT survey found that 78% and 52% of EV drivers said they "regularly" (i.e., once a week or more often) charged their vehicle overnight and during the day, respectively. Ofgem's consumer survey found that 59% of EV owners "usually" charge their vehicle at home at the moment. Furthermore, even though public chargers are only anticipated to make up 13% of charging stations, BNEF projected that by 2040, the public charging network could supply about 65% of the electricity needed for EVs outside of homes.

Nowadays, the majority of electric vehicle owners choose to install a 3–7kW home charger—which can be installed for approximately £750–£1000—as their dedicated home charging point. On compatible models, charging with a typical single-phase home UK 3-pin plug is possible, but it is extremely slow. For example, a Tesla Model 3 would require 24–36 hours to charge from 20% to 80% capacity when using a regular home plug, but only 10 hours when using a slow charger that is specifically designed for that purpose. It is estimated that 6.6 million homes, or more than 25% of households in Great Britain, lack access to off-street parking. This problem is especially severe in big cities like London. Among EV drivers who don't charge at home, 36% claimed it was impossible to install a home charger, 28% claimed the cost of installation was too high, and 19% said they had no idea how to do it. Local governments are increasingly offering dedicated on-street charging stations. This lessens the risks associated with individuals dragging cables from their houses across sidewalks, which can lead to accidents and legal complications. The Local EV Infrastructure Fund, which will provide capital and resource funding to support local authorities, and the On-street Residential Charge Point Scheme, which covers up to 75% of the costs of the charging point and parking bay, are just two of the grants that OZEV offers to local authorities to encourage the installation of on-street parking.

9.5.6. Energy Supply, the National Grid and Bilateral Charging

By 2050, the UK's need for electricity may have doubled due to the electrification of transportation and residential heating. Future net-zero grid operations will need to be flexible, particularly since generation will be more weather-dependent and there will be more chances for adaptable demand management.

Transportation-related demand growth could be offset, for instance, by smart charging and vehicle-to-grid (V2G) applications that both shift and lower energy consumption. According to National Grid's Future Energy Scenarios, the adoption of EVs would only raise peak energy demand by 7–16 GW by 2050 if these applications were to reduce peak energy demand by 32 GW47. With the help of V2G applications, EV batteries will be able to function as a significant independent energy source, meeting the demands of the electrical grid by providing services to the National Grid system operator during peak demand.

The switch to EVs is predicted to result in an increase in global electricity demand of 5% by 2030 and 25% by 2050. The future EV business model will include bidirectional charging capability or 14 V2G applications, which will boost the resilience of the UK energy system.

There will also need to be upgrades made to the electrical grid, especially to strengthen local distribution networks where there is expected to be a high demand for electric vehicles. Nearly £1 billion has been made available in the UK through the Rapid Charging Fund to assist with grid upgrades at key locations. In general, improvements to the electricity network for low-carbon development are being made possible by £300 million thanks to Ofgem's Green Recovery Scheme. Approximately 16 GW of daily flexible grid capacity could be produced if half of the EVs on UK roads in 2030 were V2G compatible.

The financial benefits of utilizing V2G technology for individual consumers are estimated to vary greatly depending on the situation, but they may total approximately £436 annually for each user.

However, the financial and economic benefits provided by EVs supplying energy to the grid need to be balanced against the impact on battery degradation (Meyer and Wang, 2018; Anwar et al., 2022).

9.6. POWERING THE FUTURE: BUILDING SUSTAINABLE CHARGING INFRASTRUCTURE FOR ELECTRIC VEHICLES

The adoption of electric vehicles represents a significant advancement in the quest for sustainability, as they have quickly and gradually become a part of daily life. However, as EV use increases, there will inevitably be a rise in demand for the infrastructure needed for EV charging. These charging stations must always represent environmental responsibility as we work to satisfy this growing demand while staying in line with sustainability goals. The establishment of sustainable charging infrastructure is essential to accomplishing this goal because it promotes a more sustainable and environmentally friendly future for future generations while also assisting in the widespread adoption of electric vehicles.

Conventional EV stations have primarily used steel, plastic, and concrete over the years. These materials are functional but have limits. For example, the production of concrete requires a lot of energy and produces carbon emissions. Despite its durability, steel corrodes and requires regular maintenance. Another common material that causes environmental problems is plastic because it doesn't decompose. To increase sustainability and efficiency, it is crucial to switch to environmentally friendly charging stations and incorporate intelligent EV charging options. For EV charging stations to have a positive environmental impact and be climate-friendly, they must include composite materials, solar panels, sustainable coatings, weather-resistant features, recycled materials, and sustainable design principles.

9.6.1. Use of Composite Materials

The sustainability of EV charging stations is being revolutionized by the introduction of composite materials. For example, composites, like carbon fiber and fiberglass, offer a lightweight substitute for conventional materials like steel or concrete, which lowers installation-related carbon emissions and transportation energy. Moreover, their resilience to deterioration and corrosion reduces the need for routine maintenance and repairs, which reduces the amount of resources used and waste produced during the station's lifetime. Composite materials' flexibility and modularity allow for the development of creative, adaptable charging station designs that fit a variety of settings and situations. Furthermore, designers can create aesthetically pleasing stations that blend in perfectly with their surroundings, increasing their accessibility and appeal. This is made possible by the aesthetic appeal of sustainable materials used in EV charging infrastructure, such as composites.

9.6.2. Solar Panels and Sustainable Coatings

The sustainability of EV charging stations is greatly improved by the incorporation of solar panels and environmentally friendly coatings. These stations can produce electricity from solar energy thanks to solar panels, which lessens their dependency on fossil fuels and, as a result, their greenhouse gas emissions. Furthermore, solar-powered charging stations can reduce or completely bypass the electrical grid, resulting in a more dependable and sustainable energy source. Sustainable coatings, on the other hand, prolong the life of the charging stations and lessen the need for regular maintenance and repair by shielding the infrastructure from rust, weathering, and degradation. By reducing the release of dangerous chemicals and pollutants, these coatings also improve the quality of the air both indoors and outdoors.

9.6.3. Weather-Resistant Features

The durability and dependability of EV charging stations are crucial in harsh weather and demanding climates. By protecting against damage from rain, snow, and temperature fluctuations, weather-resistant features are essential to improving the sustainability and resilience of these charging stations. This protection minimizes resource consumption and waste production by extending the lifespan of the charging infrastructure and lowering the frequency of repairs and replacements. Additionally, weather-resistant design features guarantee that charging stations will continue to operate even in inclement weather, giving EV owners peace of mind that the infrastructure will continue to be usable and accessible no matter what the outside weather does.

9.6.4. Use of Recycled Materials

Employing recycled materials in EV charging stations brings about notable sustainability advantages in numerous aspects. By lessening the requirement for materials like metals, plastics, and concrete, these stations aid in preserving natural resources and diminishing the ecological consequences linked to resource mining and fabrication. Moreover, the use of recycled materials consumes less energy compared to producing new ones, thereby lowering energy consumption and greenhouse gas discharges during construction. This twofold advantage of conserving resources and lessening energy consumption is in line with sustainability objectives and encourages environmental stewardship. Furthermore, incorporating recycled materials assists in diverting waste away from landfill sites, thereby lessening environmental pollution and fostering a circular economy where materials are reused and repurposed.

9.6.5. Adopting Sustainable Design Principles

The construction of charging stations and the integration of sustainable design principles play a critical role in the advancement of next-generation EV charging infrastructure. Using small smart chargers, for example, can improve functionality while saving space and increasing efficiency. These stylish and effective chargers reduce their environmental effect and maximize space usage. By incorporating charging stations into the current urban infrastructure, the need for new land development is reduced and efficient land use is encouraged. Cities can maximize land utilization and lessen environmental disturbance associated with new construction by repurposing existing spaces, such as parking lots or roadside areas, for EV charging infrastructure. In addition to improving the effectiveness of EV charging networks, this strategy supports more general objectives for urban sustainability (Meyer and Wang, 2018; Anwar et al., 2022).

REFERENCES

1. Alsharif, A., Tan, C. W., Ayop, R., Dobi, A., & Lau, K. Y. (2021). A comprehensive review of energy management strategy in vehicle-to-grid technology integrated with renewable energy sources. *Sustainable Energy Technologies and Assessments, 47*, 101439. https://doi.org/10.1016/j.seta.2021.101439.
2. Anwar, M. B., Muratori, M., Jadun, P., Hale, E., Bush, B., Denholm, P., & Podkaminer, K. (2022). Assessing the value of electric vehicle managed charging: A review of methodologies and results. *Energy & Environmental Science, 15*, 466–498.
3. Bonges, H., & Lusk, A. (2016). Addressing electric vehicle (EV) sales and range anxiety through parking layout, policy, and regulation. *Transportation Research Part A: Policy and Practice, 83*, 63–73.
4. Bunch, D. D. S., Chakraborty, J. H., Lee, G., & Tal, G. (2019). Demand drivers for charging infrastructure–Charging behavior of plug-in electric vehicle commuters. *Transportation Research Part D: Transport and Environment, 76*, 255–272.
5. Coffman, M., Bernstein, P., & Wee, S. (2017). Electric vehicles revisited: A review of factors that affect adoption. *Transport Reviews, 37*(1), 79–93. https://doi.org/10.1080/01441647.2016.1217282.
6. Drude, L., Pereira Junior, L. C., & Rüther, R. (2014). Photovoltaics (PV) and electric vehicle-to-grid (V2G) strategies for peak demand reduction in urban regions in Brazil in a smart grid environment. *Renewable Energy, 68*, 443–451. https://doi.org/10.1016/j.renene.2014.01.049.
7. Islam, M. S. (2018). A techno-economic feasibility analysis of hybrid renewable energy supply options for a grid-connected large office building in south-eastern part of France. *Sustainable Cities and Society, 38*, 492–508. https://doi.org/10.1016/j.scs.2018.01.022.
8. Kester, J., Noel, L., Zarazua de Rubens, G., & Sovacool, B. K. (2018). Policy mechanisms to accelerate electric vehicle adoption: A qualitative review from the Nordic region. *Renewable and Sustainable Energy Reviews, 94*, 719–731.
9. Madina, C., Zamora, I., & Zabala, E. (2016). Methodology for assessing electric vehicle charging infrastructure business models. *Energy Policy, 89*, 284–293.
10. Maestre, V. M., Ortiz, A., & Ortiz, I. (2021). Challenges and prospects of renewable hydrogen-based strategies for full decarbonization of stationary power applications. *Renewable and Sustainable Energy Reviews, 152*,

111628. https://doi.org/10.1016/j.rser.2021.111628.

11. Meyer, D., & Wang, J. (2018). Integrating ultra-fast-charging stations within the power grids of smart cities: A review. *IET Smart Grid, 1*(1), 3–10.
12. Mwasilu, F., Justo, J. J., Kim, E. K., Do, T. D., & Jung, J. W. (2014). Electric vehicles and smart grid interaction: A review on vehicle-to-grid and renewable energy sources integration. *Renewable and Sustainable Energy Reviews, 34*, 501–516. https://doi.org/10.1016/j.rser.2014.03.031.
13. Schäfer, A. W., Kempton, W., & Boies, A. (2017). Integration of renewable energy sources in electric vehicle charging stations. *Applied Energy, 188*, 149–158.
14. Shaukat, N., et al. (2018). A survey on electric vehicle transportation within smart grid system. *Renewable and Sustainable Energy Reviews, 81*, 1329–1349. https://doi.org/10.1016/j.rser.2017.05.092.
15. Sierzchula, W., Bakker, S., Maat, K., & van Wee, B. (2014). The influence of financial incentives and other socio-economic factors on electric vehicle adoption. *Energy Policy, 68*(C), 183–194.
16. Wang, S., Tarroja, B., Schell, L. S., Shaffer, B., & Samuelsen, S. (2019). Prioritizing among the end uses of excess renewable energy for cost-effective greenhouse gas emission reductions. *Applied Energy, 235*, 284–298.

Chapter 10

Policy, Regulation, and Environmental Standards

Contents

INTRODUCTION

Only a small number of the policies that can be used by policymakers at the local, regional, and federal levels of government to encourage the adoption of electric vehicles (EVs) directly involve financial assistance. The initial cost of buying an EV can be decreased with the aid of financial incentives like tax credits, rebates, and lower import duties. The cost of owning, fueling, operating, and parking an EV can be lowered with additional financial incentives like waived vehicle registration fees and customized tariff designs. Mandates for zero-emission vehicles and fuel economy requirements can also encourage the market for EVs. An environment that is supportive of EVs can also be produced by preferred parking, access to toll roads, and other non-financial policies. Encouraging the adoption of EVs and ensuring that they positively contribute to environmental and societal goals requires policy, regulation, and environmental standards.

10.1. GLOBAL POLICIES PROMOTING EV ADOPTION AND THEIR ENVIRONMENTAL IMPACTS

The global markets for electric vehicles (EVs) are highly diverse today, influenced by varying degrees of government backing, business involvement, consumer awareness and preference, driving habits, and cultural quirks. In particular, policy has been crucial in directing corporate strategy toward electrification and facilitating consumer adoption. Early adoption of EVs was frequently sped up by demand-spurring measures like car purchase incentives in today's major EV markets, which include China, Europe, and the US. In China, automakers were also given direct incentives. The EV markets are currently maturing in many of these nations and regions, particularly for cars, whose sales shares are rising quickly. Developed markets, including China and several European nations, are gradually reducing or discontinuing their electric car incentive programs in favor of concentrating their attention on other markets, like heavy transportation and charging.

Simultaneously, several governments in significant markets have raised their EV adoption goals even further and are attempting to address other aspects of EV supply chains, like by supporting policy for the production of vehicles and batteries as well as vital mineral supply chains. In recent years, many other nations outside of the major markets have also begun to enact policies to encourage the adoption of EVs, sometimes for the first time. Global government and consumer spending on electric vehicles has risen dramatically in the last several years, and by 2022, it is expected to surpass USD 400 billion.

In the early years of EV adoption, policy requirements were a major motivator for electrification for businesses. However, in order to maintain a competitive edge and gain market share, major incumbent carmakers must now offer EVs as a significant component of their portfolios due to the exponential growth in EV sales. The industry as a whole is being forced to accelerate decarbonization due to growing competition and the influx of new players, especially from China but also from other emerging markets and developing economies (EMDEs).

10.1.1. Policy to Develop EV Supply Chains

Supporting EV manufacturers and companies through direct incentives along EV supply chains to ramp up domestic production is not a new phenomenon in China, the world's largest market for electric cars. Over the last 10 years, domestic companies have consistently utilized both supply-side and demand-side incentives, in addition to joint ventures with foreign automakers. In China, local support has been especially strong, which has fueled national adoption and the growth of numerous significant EV businesses. Recently, cities and regions have also declared their manufacturing goals. For example, Chongqing wants to manufacture and sell more than 10% of China's new energy vehicles (NEVs), and Jilin wants to produce 1 million NEVs annually by 2025. These goals are supported by policies that emphasize EV manufacturing.

Notable announcements were also made in other important markets, like the European Union's Green Deal Industry Plan and the United States' IRA. The national Production Linked Incentive (PLI) scheme in India has received support from subnational governments, including Tamil Nadu, whose updated policy encourages local EV manufacturing, in order to further the goals of the scheme. Other instances of EMDEs that highlight this trend include Ethiopia, which is enticing investment by providing tax breaks for EVs that are assembled locally, and Indonesia, which has implemented legislation encouraging investment in battery production. Following the 2021 reduction of tariffs on lithium-ion cells to encourage assembly, Morocco celebrated the completion of its first domestically produced battery electric vehicle (BEV) in December 2022. In July 2022, the country announced plans to build a large EV battery factory.

10.1.1.1. United States: Inflation Reduction Act

Passed in August 2022, the Inflation Reduction Act (IRA) consists of several funding programs and tax breaks aimed at achieving the goal of creating a clean energy economy. A portion of the Act, which allots USD 369 billion for climate investments, is devoted to hastening the adoption of electric vehicles.

A new set of requirements is introduced by the Clean Vehicle Tax Credit for EV models to be eligible for incentives. These requirements state that starting in 2023, final assembly must take place in North America, and vehicles must be under 6 .35 t gross vehicle weight (GVW), have a suggested retail price of less than USD 80 000 for vans, SUVs, and pickup trucks, or USD 55 000 for other vehicles, and have a battery of at least 7 kWh (excluding low-range plug-in hybrid electric vehicles, or PHEVs).

The US Internal Revenue Service's cap on household income must be met by the EV buyer for them to be eligible for the incentive. Under these circumstances, a vehicle may be eligible for an incentive worth up to USD 7,500: an additional USD 3,750 if the battery satisfies the component requirement and an additional USD 3,750 if it satisfies the critical mineral requirement. Moreover, starting in 2025, vehicles that contain any critical minerals sourced from "foreign entities of concern" will not be qualified for the credit, and starting in 2024, vehicles that contain battery components from said entities will not be qualified either.

A list of particular models that meet the new requirements has been submitted by a number of the major US automakers, and the elimination of the manufacturer sales cap of 200,000 allows GM and Tesla to take part in the program. 33 BEVs and 4 PHEVs are among the 37 models that are listed. Even though this sometimes means lowering prices, the average listed retail price for the eligible PHEVs and BEVs is slightly less than the USD 55,000 and USD 80,000 limits for cars and SUVs, respectively. This suggests that the original equipment manufacturer (OEM) is using a corporate strategy to meet the IRA eligibility requirements. Between now and 2031, the Act also allots USD 1 billion for HDV infrastructure and incentive projects. The tax credits are up to USD 40,000 per vehicle for cars made by authorized manufacturers with batteries that have a capacity of at least 15 kWh.

The Advanced Manufacturing Production Tax Credits are included in the IRA in addition to demand-side credits for the purchase of vehicles. Under this program, the US government offers up to USD 35 per kWh in subsidies for domestic battery production, plus an additional USD 10 per kWh for module assembly. If the average price of batteries is USD 150 per kWh in 2022, then nearly one-third of the total battery price could be covered by these new US production incentives. The United States also committed to supporting the development of a productive supply chain from mining to assembly when it signed an MoU with the Democratic Republic of the Congo and Zambia in January 2023. These agreements also confirmed that countries with existing free trade agreements have preferred status as suppliers.

10.1.1.2. European Union: Green Deal Industry Plan

The Green Deal Industrial Plan, unveiled by the European Union in February 2023, is built upon four pillars that will facilitate the advancement of net zero-related projects: open trade, improved skills, financial support, and quicker permitting. The plan also calls for the creation of a Critical Raw Materials Act, whose proposal was published in March 2023 and which emphasizes recycling, extraction security, and environmental standards.

The proposed Net Zero Industry Act will formalize faster permitting for facilities, including battery production, by offering more straightforward and predictable planning approvals. The plan's financial support package aims to lower electricity demand, compensate businesses for high energy prices, facilitate faster access to essential subsidies and loans, and relax state aid regulations until 2025. The plan also intends to create Net Zero Industry Academies and retrain workers impacted by the green transition. Last but not least, the trade component emphasizes bringing in private investment, expanding trade with new partners, and strengthening the resilience of the EU's supply chains.

The European Union proposed the Net Zero Industry Act in March 2023, to have EU manufacturing capacity by 2030 to meet 40% of the EU's needs for strategic net zero technologies. The goal of the European Battery Alliance is for EU battery manufacturers to meet nearly 90% of the annual battery demand of the EU, with a combined manufacturing capacity of at least 550 GWh by 2030. These technologies specifically include battery and storage technologies. These announcements coincided with the Fit for 55 package's tightening of CO_2 standards for automobile sales over 2030–2035.

10.1.1.3. India: Production Linked Incentives(PLI)

With a budget of INR 181 billion (Indian rupees) (USD 2.2 billion), the Indian PLI on Advanced Chemistry Cell (ACC) Battery Storage was announced in late 2021 with the goal of promoting domestic battery manufacturing. With funding distributed over five years to companies based on sales of batteries made in India and contingent upon meeting a domestic value-add of at least 25% in year 1 and 60% in year 5, the government specifically aimed to reach a cumulative 50 GWh in domestic manufacturing capacity. This is especially ambitious since there isn't much domestic battery cell production going on in India right now, and the 50 GWh figure is 50% more than the domestic demand that is predicted by the IEA Stated Policies Scenario (STEPS) for 2025. The government awarded funding to three companies, Reliance, Ola, and Rajesh, for a total of at least 30 GWh of battery manufacturing capacity, with the remaining amount going to the next highest placed bidder (s). The scheme attracted 10 bids with a cumulative capacity of 128 GWh in early 2022. An extra 95 GWh are anticipated to be established by other private businesses.

The Champion OEM incentive scheme, which offers incentives for the sale of advanced automotive technology vehicles (battery electric and hydrogen fuel cell vehicles) across all vehicle segments, and the Component Champion scheme, which offers incentives for the sale of specific components for both ICE and electric vehicles, make up the Automobile and Auto Component PLI scheme. The plan has been successful in attracting investments worth INR 677 billion (USD 8.3 billion), which will be spent over a period of five years. The budgetary outlay is INR 260 billion (USD 3.2 billion) over five years. As of March 2022, 95 applications had been accepted under this program. The minimum percentage of domestic value-added required in both schemes is 50%.

10.1.1.4. Leveraging Critical Mineral Resources to Build Domestic Manufacturing

Policy action has also been sparked by the realization of the critical role that critical minerals play in the EV transition. Governments aim to establish their presence in global supply chains while concentrating on enhancing domestic capacity, while automakers and battery producers compete globally to ensure supply in response to the growing demand for electric vehicles. It will be crucial to make sure that these new critical mineral supply chains are morally and environmentally sound for the foreseeable future. New guidelines for the manufacture, recycling, and reuse of batteries are introduced in the European Union's December 2022 proposed revisions to the EU Battery Directive. In March 2023, the European Union also proposed the Critical Raw Materials Act, which aims to achieve 10% extraction capacity, 40% processing capacity, and 15% recycling capacity of the EU's annual consumption of strategic raw materials by 2030. To improve supply chain security and resilience, other goals include diversifying the origin of imported materials and enhancing the environmental sustainability of mineral-related activities.

The requirement under the IRA that half of the vehicle subsidy is contingent on meeting the critical mineral requirement has highlighted the significance of critical minerals and the preference for developing domestic capacity in the United States.

Many other nations are also implementing programs to guarantee the supply of essential minerals and boost domestic manufacturing. Australia wants to speed up the production of lithium for both export and domestic downstream applications. A consultation on the Australian Made Battery Plan, which outlines a budget proposal of AUD 100 million (Australian dollars) (USD 65 million) for domestic battery manufacturing projects in Queensland, and on a new Critical Minerals Strategy are two recent policy developments. Similar to Chile, Australia and the US have a free trade agreement, which means that their countries and the IRA have more in common than those that do not. Argentina is contemplating the implementation of a domestic market quota of 5%, with the possibility of increasing to 20%, in order to ensure domestic industry access to lithium. The country sees an opportunity to develop a battery manufacturing industry that would generate 2 500 jobs by 2030. Japan's Green Growth Strategy calls for raising domestic car battery production to 100 GWh by 2030. Japan set aside JPY 331½ billion (US$ 2.5 billion) in 2022 to develop magnet and battery materials that will lessen reliance on lithium and rare earth elements, including those used in electric vehicle applications. A decentralized public organization called Lithium for Mexico was established in August 2022 with the goal of nationalizing the value chain and recognizing lithium as a strategic mineral. By 2030, half of all vehicles made in Mexico are expected to be zero-emission vehicles. Russia wants to use its mineral resources to grow the battery industry and produce at least 10% electric vehicles by 2030.

Based on trade data, India imported USD 1 billion worth of lithium-ion cells in 2020–2021, with over 95% coming from China and Hong Kong. India released the Battery Waste Management Rules in 2022 to recycle or rehabilitate all battery types, including EV batteries, in an effort to lessen reliance on imports. The regulations also establish a target of raising EV battery recycling to 20% by 2030, encompassing imports. India reported its first 5.9 Mt of inferred lithium deposits in February 2023. If verified, this could revolutionize the cell manufacturing industry both in India and worldwide.

10.1.2. Policy Support for Electric Light-Duty Vehicles

10.1.2.1. New Standards

The Fit for 55 package's goal of reducing emissions in the European Union by at least 55% by 2030 has made significant headway toward becoming law during the past year. The European Union enacted new CO_2 regulations for cars and vans in March 2023. These regulations mandate a reduction in emissions of new cars and vans of 55% and 50% by 2030 (in comparison to 2021) and 100% for both by 2035. In accordance with the regulation, the Commission will also present a plan to permit the registration of automobiles powered solely by CO_2-neutral e-fuels following 2035.

The default utility factor for plug-in hybrid electric vehicles (PHEVs) is the percentage of total distance driven that is powered by the battery as opposed to internal combustion engines (ICE). In June 2022, the European Parliament also demanded changes to the regulations governing this factor. As a result, starting in 2025, the utility factor—which affects vehicle emissions—will be based on actual data from the 2022 ICCT-Fraunhofer study rather than estimates, significantly raising the assumed emissions. As a result, automakers should have much less motivation to employ PHEVs in order to meet European fleet targets. Since company cars have been the source of an even larger discrepancy between expected and actual performance, the updated methodology will be based on an even distribution of private and company cars as of 2027. Additionally, it is suggested that the utility factor be adjusted starting in January 2027 to take into account the most recent findings from on-board fuel consumption monitoring devices.

10.1.2.2. Tighter Regulations

The European Commission suggested new Euro 7 emissions regulations for automobiles, vans, trucks, and buses in November 2022. The goal of the proposed emission standards is to make them more representative of actual driving situations, especially in cities. They will control not only emissions from tailpipes but also emissions from brakes and tires, including electric vehicles. Additionally, the standards are meant to support the more stringent ones that were proposed in October 2022 for air quality.

Similar to the UN, the Euro 7 emissions standards would impose minimum performance requirements for LDV battery durability, stating that 80% of batteries should be durable by year five or 100,000 km, and 70% should be durable by year eight or 160,000 km. Reductions in tailpipe particle emissions of 13% for LDVs and 39% for HDVs, as well as reductions in nitrogen oxide (NOx) emissions of 35% for LDVs and 56% for HDVs, are also included. These restrictions cannot be broken before driving 200,000 km or 10 years, which is twice the previous requirement. In comparison to Euro 6/VI, they indicate an expected cost increase of EUR 90–150 (USD 95–160) for LDVs and EUR 2 600 (USD 2 750) for HDVs.

10.1.2.3. Advanced Clean Cars II

The Advanced Clean Cars II (ACC II) rule, which establishes a minimum ZEV sales share for passenger LDVs ranging from 35% in 2026–100% in 2035, was approved by the California Air Resources Board in November of 2022. Any car sold after 2035 has to be a plug-in hybrid electric vehicle or PHEV. This comes after decades of consistent support for low- and zero-emission vehicles in California, including the zero-emission vehicle credits. Other states have already followed California's lead, even though the regulation won't go into effect until the US Environmental Protection Agency grants California a waiver. With their combined adoption of ACC II, Vermont, Washington, Oregon, and New York account for nearly 20% of US passenger LDV sales. As of right now, ACC II covers nearly 25% of sales because Massachusetts, Delaware, and Colorado have either started the adoption process for the rule or have made announcements about it.

10.1.2.4. National Plans in Many Countries Pave the Way for Zero-emission Car Transport

A growing number of nations put forth policies in 2022–2023, either to strengthen current programs or to introduce support mechanisms for the first time, to hasten the adoption of electric vehicles. In Europe, the UK accelerated the deadline for selling all ICE vehicles to 2030, five years ahead of schedule, with a complete switch to selling only zero-energy vehicles by 2035. To meet the 2035 target, the government proposed ZEV sales share trajectories for cars and vans in March 2023.

Greece tightened its regulations to only permit the sale of zero-emission LDVs from that point on, abandoning its earlier goal of achieving a 30% ZEV sales share by 2030. Switzerland has exceeded its 2018 target of 15% of sales of electric cars by 2022, with a 25% share by 2022. This success is a result of strong cooperation and investment from approximately 50 leading organizations through the Electromobility Roadmap. The new goal is to have 50% of the market for electric vehicles by 2025. Similar to a Singaporean program, Italy's subsidy program has also been renewed with an emphasis on getting rid of older, more environmentally harmful ICEs. A similar emphasis on scrappage and assistance for those with lower incomes are seen in Spain. While import and annual taxes for electric vehicles have been drastically lowered in Finland, tax changes in Denmark are intended to increase the appeal of EV company cars. In 2023, Austria extended its EV subsidy at the same rates as the year before. In 2022, EV purchases were first subsidized in Cyprus and Croatia.

Canada raised its national ambition for LDV deployment in 2022. As of right now, the country wants to achieve a 20% zero-emission sales share by 2026, a 60% share by 2030, and a 100% share by 2035. British Columbia has raised its 2030 target to 90% on a provincial level.

After funding a program for charging infrastructure since 2020, Australia began offering competitive grant funding in February 2022 to support the purchase of EVs for commercial use, including both light- and heavy-duty models. The funding, which comes from the expanded AUD 500 million (USD

346 million) Future Fuels Fund, totals AUD 128 million (USD 89 million). New Zealand released its first all-inclusive plan for reducing emissions in May 2022. By 2035, it wants 30% of the LDV stock to be ZEVs. Even though New Zealand only started offering purchase subsidies in the middle of 2021, electric cars accounted for more than 10% of new sales in 2022. Given the popularity of used cars in the nation, this also includes many non-purchase-related incentives and imports of used cars.

The goal of Japan's Green Growth Strategy, which was unveiled in 2021, is to electrify all vehicles by the year 2035. The 2023 Act on the Rational Use of Energy both tracks and expedites these targets. Governments in EMDEs are also becoming more ambitious. EVs are seen as a chance to lessen dependence on imported energy, lessen the effects of climate change, and reduce air pollution. The EV policies that are being proposed typically include purchase incentives, mandates, deployment targets, and tax exemptions for EVs, accessories, and equipment.

Government vehicles in Indonesia must run on electricity as of 2022, and incentives for buying electric vehicles will be available starting in 2023. 30% of new car sales in Seychelles are expected to be electric by 2030, and the upcoming National Electric Mobility Strategy will include a goal of having all buses electric by 2050. By 2030, 40% of Panama's fleet of designated public vehicles will be electric vehicles. With a ban on the manufacture, assembly, and import of vehicles powered by fossil fuels in 2040, Viet Nam has set a net zero emissions target for the transportation sector by 2050. The goal is to employ vehicles that are powered entirely by electricity and renewable energy. The goal of Ghana's recently unveiled strategy is for EVs to account for 4%, 16%, and 32% of new sales in 2025, 2030, and 2050, respectively. Furthermore, Angola, Brazil (continuing tax exemptions for electric car imports dating back to 2015), Ecuador, Pakistan, Trinidad and Tobago, Tunisia, Uzbekistan, and Viet Nam are among the nations that have recently proposed tax-related policies.

10.1.3. China: Local Action Ramps up as National Subsidies Phase Out

After being gradually reduced since 2017, the national NEV subsidy program in China came to an end in December 2022. The program was initially scheduled to be phased out two years earlier. In the meantime, China's NEV market has outperformed its goal of a 20% sales share by 2025, with a sales share of nearly 30% in 2022. The primary national financial incentive for NEVs will continue to be the extended vehicle purchase tax exemption through the end of 2023. Furthermore, a number of guiding strategies list the electrification of road transportation as a goal. China wants to reduce air pollution by having 50% of its cars sold as NEVs in its "key air pollution control regions" by 2030. These regions together make up almost 80% of China's car market. Furthermore, a goal for the sales share of NEVs to reach approximately 40% by 2030 is outlined in China's national action plan to reach carbon peaking before 2030.

In the meantime, goals and helpful policies at the regional level are still crucial. 18 Chinese provinces currently have explicit NEV targets as part of carbon peaking policies, aiming for 40% of NEV sales share by 2030, with Tianjin at 50%, in accordance with the nation's national action plan. In certain instances, regional sectoral strategies and local 14th Five-Year Plans also establish more aggressive goals. For example, Shenzhen wants to attain a 60% NEV sales share by 2025, while Shanghai wants to reach a 50% BEV sales share target by the same year. At the start of 2023, a number of regional governments announced local incentives. For instance, Zhengzhou and Wuxi are offering vouchers for the purchase of NEVs, and Beijing and Shanghai both announced incentives for switching from ICE to NEVs.

The focus and level of incentives are changing as markets mature:

- Value-added tax (VAT) was reinstated in Norway in 2023 for EVs that cost more than 500,000 Norwegian kroner (approximately $52,000 USD). This meant that only the priciest models would see price increases. A new subsidy program was proposed by Norway to replace the VAT exemption, however not many specifics have been made public as of yet. There has also been a progressive reduction in other benefits and incentives.
- In the UK, electric vehicle subsidies came to an end in 2022 after surpassing a 20% sales share, following a gradual reduction in the amount of grants that were made available between 2016 and 2021. Electric company cars with new tax exemptions, as well as electric taxis, vans, and trucks, are still eligible for subsidies. Additionally, there is a growing emphasis on charging.
- In 2023, the levels of purchase incentives were reduced in Germany and Ireland.
- The Netherlands has been lowering premiums every year; in 2022 and 2023, there was a reduction of EUR 400 (USD 420).
- Following an increase from SEK 60,000 (USD 6,000) in early 2021, Sweden had originally intended to tighten the cap on purchase incentives for electric cars from SEK 70,000 (Swedish kronor) (USD 7,000) in 2022 to SEK 50,000 (USD 5,000) in 2023. Then, Sweden declared that the purchase incentive would end in November 2022 due to the increasing price parity between ICE and electric vehicles.

- France decreased incentives from EUR 7,000 (USD 7,400) in 2021 to EUR 6,000 (USD 6,300) in 2022 and EUR 5,000 (USD 5,300) in 2023 after raising the incentive to increase sales during the Covid-19 pandemic. To improve equity of access to EVs, however, the purchase incentive for lower-income households was raised from EUR 6 000 (USD 6,300) in 2022 to EUR 7 000 (USD 7,400) in 2023.
- The Korean government encourages the use of EVs by offering larger subsidies to cars with longer fully electric ranges. Although consumers can still take advantage of local subsidies, the maximum passenger car incentive in 2023 dropped from KRW 7 million (Korean won) (USD 5,400) to KRW 6.8 million (USD 5,300). Incentives per vehicle for electric LCVs dropped in 2023, from KRW 14 million (USD 10 800) to KRW 12 million (USD 9,300). However, the total amount of government funding for electric LDV subsidies increased from KRW 1 561 billion (USD 1.2 billion) in 2022 to KRW 1 676 billion (USD 1.3 billion) in 2023 due to the subsidization of about 30% more passenger cars and commercial vehicles. Furthermore, Korea discontinued its free light commercial vehicle (LCV) registration program in April 2022. Despite this, Korea continued to lead the way in the deployment of electric LCVs in 2022, accounting for an average of 27% of sales annually.
- Japan too doubled the support in 2022.

10.1.3.1. Stronger Policy Preference for Fully Battery Electric Options

Recent events suggest that policy support for battery electric vehicles over plug-in hybrid electric vehicles may be strengthened. For instance, during the 2019–2022 period in Europe, Belgium, Finland, the Netherlands, Portugal, and the United Kingdom supported BEVs but not PHEVs. Ireland opted to discontinue PHEV subsidies in 2022, and Germany did the same in 2023. While several Indian states have set BEV-specific targets without including PHEVs, Shanghai, China, offers purchase subsidies for BEVs but not for PHEVs. Since 2019, PHEV support has remained consistent in both the US and Canada. However, California is requiring PHEVs to have a minimum fully electric range of 40 miles (65 km). Low-range PHEVs are no longer eligible for subsidies from the IRA because of its new requirement of a minimum 7 kWh battery capacity.

10.1.4. Policy Support for Electric Heavy-Duty Vehicles

10.1.4.1. New EU CO_2 Standards

In February 2023, the European Commission published proposed changes to the HDV emissions regulation. Smaller trucks, city buses, long-distance buses, and trailers are all included in the proposed regulation's scope, which will increase sectoral emissions coverage from the current roughly 70% to 95%. The International Council on Clean Transportation (ICCT) projects that the new regulations could achieve sectoral emissions reduction of over 75% by 2050 relative to 2020, which shows a step up in ambition.

The changes would raise the goals for reducing CO_2 emissions from 2019 levels to 45% by 2030, 65% by 2035, and 90% by 2040. By 2030, all city buses ought to be zero-emission vehicles. Between 2025 and 2040, compliance will be evaluated differently for freight and passenger vehicles (bus and coach) using a system of credits and debts that are paid off at the conclusion of five-year intervals. In order to accommodate dual fuel, the definition of ZEV is also being modified, going from 1 g CO_2 per vehicle-km to 5 gCO_2 per ton-km (tkm). Trade among entities that are economically connected (e.g., between brands under one parent company) is permitted in order to assist in achieving goals. There are financial penalties for noncompliance with the 350 km minimum range requirement for long-haul trucks.

In their national plans, a number of European Union members are stepping up their support for heavy transportation with zero emissions. Germany offers truck purchase subsidies that are among the highest in Europe, covering 80% of the vehicle's additional costs and/or the infrastructure needed for charging. The revised Climate Action Plan for Ireland calls for the production of 700 low-emission heavy-duty trucks by 2025, a 30% sales share of zero-emission heavy-duty trucks by 2030, and HDV provisions. Ireland unveiled its first target for electric buses as well: 300 by 2025.

10.1.4.2. Increasing Inclusion of HDVs

70% of the funding package for charging infrastructure in the state of California will go toward charging for HDVs. The IRA in the United States provides USD 1 billion for vehicles and infrastructure specifically for HDVs, including subsidies. In the UK, even though subsidies for electric vehicles have been phased out, HDVs are still covered by programs. In a similar vein, Australia's Future Fuels Program subsidy program allocates funds expressly for HDVs and HDV charging, but not for LDVs.

There are several nations that have set goals for heavy transportation with zero or low emissions. By 2035, the goal in New Zealand is to cut emissions from freight transportation by 35%. By 2030, Norway wants to have almost zero emissions from the distribution of goods in its largest cities. While Viet Nam's goal of net zero transport by 2050 includes HDVs, Tianjin (China) aims to have 80% NEVs in public transportation, rental, logistics, and delivery vehicle sales by 2025. With a plan to electrify HDVs and taxis worth JPY 13.6 billion (USD 120 million), Japan hopes to have 5,000 electric HDVs on the road by 2030. Additionally, some nations are first offering financial assistance for the purchase of electric trucks. The Incentives for Medium- and Heavy-duty Zero-Emission Vehicles (iMHZEV) program in Canada launched in July 2022 and provides incentives for the purchase or leasing of ZEV trucks across a range of categories up to CAD 200 000 (USD 160,000).

Compared to trucks, electric buses are generally more affordable due to their higher overall model availability and better cost-competitiveness. In fact, a number of EMDEs have started to include aggressive goals for the deployment of electric buses in national plans, frequently surpassing LDV targets. For instance, Panama declared in April 2022 that it intended to shift to electric public transportation, with a goal of 10% by 2025, 20% by 2027, and 33% by 2030. Tamil Nadu, India, unveiled a goal in February 2023 to have 30% of its fleet of buses run on electricity by 2030. This goal is a part of a larger promotion of EVs and related infrastructure. In Canada, subsidies for electric buses have been introduced in advanced economies. Additionally, the United States contributes significantly to the cost of leasing or buying electric buses, covering up to 90% of the infrastructure and up to 85% of the cost of the vehicles.

10.1.5. Policy Support for EV Charging Infrastructure

10.1.5.1. Support Shift

The emphasis is shifting away from more costly private vehicle subsidies and toward supporting charging infrastructure in nations where electromobility is more advanced. For instance, the government of the United Kingdom declared in June 2022 that it would shift its attention back to charging following the termination of the electric vehicle subsidy program. An estimated GBP 11.6 billion (USD 21.1 billion) in government funds have been allocated to the EV Infrastructure Strategy, to install 300,000 public chargers by 2030. Additionally, the UK government is trying to make more on-street chargers available, especially in residential areas where there isn't any off-street parking.

Since charging has been found to be the most important factor by Chinese consumers, Guangdong, Hunan, and Shanghai, among other regions, have introduced subsidies for the installation of charging stations. By 2025, Shenzhen hopes to have 43,000 fast chargers, 790,000 slow chargers, and 240,000 chargers in Chongqing.

Germany has increased funding for fast chargers while simultaneously winding down its car subsidy program. To meet the target of 50% EV sales share, the Swiss government extended its Electromobility Roadmap in 2022 and set a goal of establishing 20,000 public charging stations by 2025. Poland started funding vehicles in 2021 and is currently funding charging infrastructure as well. Recently, Finland, Denmark, and Ireland released revised ZEV plans with a focus on public charging, and all pledged to support the roll-out of chargers more.

10.1.5.2. Additional Support for Charging

Some nations are making the decision to finance the construction of infrastructure before EVs are widely adopted. One prominent example is Australia, which started supporting electric HDVs in 2022 in addition to charging them. Before 2020, the country had grants to fund charging, and since 2020, it has had a dedicated infrastructure funding program. Building a network of charging stations is part of Vietnam's initial strategy phase, which does not prioritize vehicle subsidies. Comparably, charging is where Bulgaria's EV promotion campaign begins. In India, state-level objectives and incentives are being created to support national goals; the Tamil Nadu government, for example, plans to subsidize the installation of 750 chargers. By 2030, Japan intends to build 150,000 publicly accessible charging stations—of which 30,000 will be fast charging stations—as part of the Green Growth Strategy.

Over USD 1.5 billion has been authorized in the US under the National Electric Vehicle Infrastructure (NEVI) Formula Program to support the construction of EV chargers along roughly 75,000 miles of highways. To qualify for funding, national standards were established in 2023 to build a network of chargers that is accessible, reasonably priced, dependable, and fair across the nation. These standards aim to ensure minimum numbers and types of chargers, including fast charging, to support the achievement of the target of installing 500,000 chargers no more than 80 km apart along major routes by 2030. They will also harmonize pricing and payment information for charging. Furthermore, as of January 2023, buyers of home chargers can claim a tax credit of up to $1,000 under the IRA, and installations of vehicle charging infrastructure can receive a tax credit of up to $100,000. In November 2022, California, which is frequently at the forefront of EV policy development, announced an additional commitment of USD 1 billion to charging infrastructure. The construction of medium- and heavy-duty truck charging stations is financially supported both nationally and through California's funding program.

To extend the Zero Emission Vehicle Infrastructure Program through March 2027, Canada has increased funding available by an additional CAD 400 million (USD 310 million). The program finances strategic infrastructure projects for urban delivery and fleet applications, and it supports the installation of both public and private chargers, including those at multi-unit residential buildings and workplaces. In addition, the Canada Infrastructure Bank declared in 2022 that it would fund extensive charging infrastructure with CAD 500 million (USD 385 million). All of this is done to help the government reach its goal of having 84,500 chargers in use by 2029.

The 2014 Directive was replaced in 2021 with the Alternative Fuels Infrastructure Regulation (AFIR) by the European Commission. A provisional agreement for the AFIR between the European Council and the European Parliament is in effect as of March 2023. It specifies requirements for coverage of the trans-European network transport (TEN-T) for both light- and heavy-duty vehicles, as well as requirements for total power capacity based on the size of the light-EV fleet. By the end of 2023, over EUR 10.5 billion will be made available for alternative fuel infrastructure, including EV charging, thanks to a deal between the European Investment Bank and the European Commission. Funding for the implementation of EVSEs can also be obtained through the Connecting Europe Facility (CEF) for Transport initiative.

10.1.6. International Initiatives and Pledges

10.1.6.1. Accelerating to Zero

The Accelerating to Zero (A2Z) coalition was established at the Conference of the Parties (COP 27) with the overarching objective of having all new car and van sales worldwide be ZEVs by 2040, and in the leading markets, by no later than 2035. The coalition expands upon the Zero-Emission Vehicles Declaration, which now has 223 signatories after receiving about 130 signatures at its launch during COP26. 30 governments in advanced economies, 11 governments in EMDEs, 73 local and regional governments, 14 automakers, 47 fleet owners and operators, 15 investors holding shares in the auto industry, two financial institutions, and 30 other signatories are among them. Together, the 40 governments were responsible for nearly 25% of global LDV sales each year and nearly 20% of electric LDV sales in 2022. Certain subnational governments or non-governmental signatories have also established more precise and ambitious targets. For example, fleet owners and operators, as well as regional or city authorities, want to electrify their fleets by 2030.

10.1.6.2. Zero-Emission Government Fleet Declaration

Nine nations joined forces to create the Zero-Emission Government Fleet Declaration in 2022, realizing the leadership and demand-signaling power that national governments can wield. By 2035 at the latest, they want all government fleets to be 100% emission-free cars and vans, with the goal of also having 100% emission-free trucks and buses. Timelines differ from nation to nation. For instance, by 2025, 75% and 100% of purchases in Australia and Israel, respectively, should be electric; by 2027, 100% of purchases in the US should be electric; and by 2030, 100% of Canada's LDV stock should be emission-free.

10.1.6.3. Zero-Emission Vehicle Emerging Markets Initiative

The Zero-Emission Vehicle Emerging Markets Initiative was introduced in 2022 at COP27 by the United States, the United Kingdom, and the World Business Council for Sustainable Development (WBCSD). In order to hasten the shift to zero-emission road transportation, it seeks to improve collaboration between public and private actors in EMDEs. With the intention of facilitating and announcing agreements at COP28, the initiative specifically opened a dialogue between governments and corporations on public support and private investment to achieve ZEV deployment targets in EMDEs. It brings together businesses that have promised to invest a total of USD 50 billion and to install more than 2 million EVs in EMDEs by 2030. The series' first dialogue took place in India in February 2023, marking the beginning of a national pilot program there.

10.1.6.4. Global Memorandum of Understanding (MoU) on Zero-Emission Medium- and Heavy-Duty Vehicles

The MoU on Zero-Emission Medium- and Heavy-Duty Vehicles was introduced in 2021 by the Dutch government and the Drive to Zero program. Signatories pledge to collaborate to reach the goal of 100% ZEV bus and truck sales by 2040, with a 30% interim target by 2030. The Memorandum of Understanding was signed in 2022 by the United States, Aruba, Belgium, Croatia, Curaçao, Dominican Republic, Ireland, Liechtenstein, Lithuania, Sint Maarten, and Ukraine. With this, the total number of signatory nations and territories is now 27, representing more than 15% of global new medium- and heavy-duty vehicle sales annually. The regional government, the manufacturing sector, fleet owners and operators, and others have also endorsed the MoU.

10.1.6.5. Greening Corporate Fleets

Through the Greening Corporate Fleets initiative, the European Commission pledged to support the commercial use of zero-emission vehicles by 2023. 30 private and state-owned companies, along with industry associations, demanded that the Commission's initiative incorporate a mandatory goal that by 2030, all new corporate light-duty vehicles (LDVs) be zero-emission vehicles (ZEVs) in an open letter to the Commission shared by the campaign group Transport and Environment. The businesses comprise EV infrastructure developers as well as fleet owners and operators. The EV100 campaign, which currently has 130 member companies, also signed the letter. By 2030, EV100 wants to convert all fleet vehicles below 7.5 t to electric vehicles and set up infrastructure for staff and customers to charge their phones. Thus far, EV100 members have pledged to electrify more than 5 million leased cars in addition to approximately 725,000 cars in their own fleets.

10.1.6.6. EV100+

The Climate Group, who created the EV100 campaign targeted at LDVs, started the EV100+ campaign in 2022. The EV100+ campaign's signatories pledge to convert their fleet of over 7.5 t of cars to zero-emission cars by 2040 in China, India, and OECD markets. Several corporate fleet owners or users are among the founding members. A wider coalition of truck manufacturers, transport companies, shippers, retailers, energy providers, and infrastructure operators have recently come together to demand ambitious targets in the (then upcoming) EU HDV CO_2 regulations. These targets include a binding ZEV target for 2035 with very few exceptions, as well as interim goals of reducing emissions by 30% by 2027 and 65% by 2030. CALSTART also signed the call. A similar open letter from Transport and Environment was also signed by EV100+, indicating their agreement on the matter.

10.2. REGULATORY FRAMEWORKS FOR REDUCING THE ENVIRONMENTAL IMPACT OF EVS

Finding clear and concise information on the various policies and regulations pertaining to EVs can be challenging. Research has put together many factsheets that concisely and fully describe the regulations that are in place, their objectives, the parties who must comply, and the requirements for compliance to save decision-makers time. In addition to helping stakeholders plan for current and future regulations, we hope that this compilation of resources will help them understand when and how regulations will impact their businesses, make informed purchasing decisions, and maximize the economic and environmental benefits of electrification. The following regulations serve as strong indicators that the transition to electrified transportation is already underway. By drastically lowering transportation-related emissions, these policies will both improve air quality and health outcomes and help prevent the catastrophic consequences of a 1.5°C or greater rise in global temperature. These policies, along with others similar to them, will greatly reduce emissions and the associated harms, as the transportation sector is the largest emitter of greenhouse gases in the United States.

10.2.1. State Regulations

A few such regulations are summarized below.

10.2.1.1. Advanced Clean Cars

The 2012-passed Advanced Clean Cars (ACC) rule is expanded upon by the ACC II regulation. Through the Zero-Emission Vehicle and Low-Emission Vehicle programs, ACC II reduces emissions by boosting the sales of electric vehicles.

- **Who Must Comply:** Original equipment manufacturers (OEMs) who sell more than 4,500 light- and medium-duty (MD) vehicles a year are required to comply. Until the state's 100% ZEV mandate takes effect in 2035, small-volume OEMs that sell fewer than 4,500 light- and MD vehicles annually in California are exempt.

10.2.1.2. Advanced Clean Truck (ACT)

In order to minimize emissions, the Clean Air Act (ACT) mandates that, starting in 2024, a greater proportion of medium- and heavy-duty vehicle OEM sales must come from ZEVs or near-zero-emission vehicles (NZEVs), like plug-in electric hybrids. The rule employs a cap-and-trade mechanism, imposing annual sales percentage requirements that cap the quantity of fossil fuel vehicles sold. Manufacturers can generate compliance credits by selling ZEVs or NZEVs, or by trading compliance credits, in order to comply with the regulation.

- **Who Must Comply:** Manufacturers who sell more than 500 vehicles annually are required to disclose the total number of vehicles they sell in the state. Manufacturers who sell 500 or fewer HD trucks are free from deficit accrual and are not obligated to sell NZEVs or ZEVs. They have the option to voluntarily report ZEV and NZEV credits and to bank or trade them.

10.2.1.3. Advanced Clean Fleet(ACF)

ACF mandates that fleets use a growing proportion of ZEVs, such as hydrogen fuel cell MDHD trucks, long-range plug-in electric hybrids, and battery electric vehicles. It is intended to supplement the ACT rule. The type and purpose of the truck determines the compliance requirements.

- **Who Must Comply:** Public fleets owned by state and local government agencies that own, lease, or operate medium- and heavy-duty trucks; fleets with more than 50 trucks or those belonging to private companies that generate more than $50 million in revenue annually; and drayage truck fleets with trucks operating at ports or intermodal rail yards in California.

10.2.1.4. Clean Miles Standard (CMS)

Ride hailing companies can use CMS, an organized framework, to plan their strategies for EV adoption. It will play a crucial role in bolstering the EV ecosystem by offering low- to middle-class drivers financial support, putting strong pressure on builders to complete the necessary EV charging infrastructure, and promoting collaborations between EV charging providers and transportation network companies (TNCs) to speed up adoption. In order to lower emissions in California, CMS mandates that ride-hailing service providers purchase zero-emission vehicles. It mandates that 90% of all vehicle miles traveled (VMT) must be powered by EVs by 2030 and that there must be no emissions from passenger miles traveled.

- **Who Must Comply:** TNCs and autonomous car companies with more than 5 million VMTs annually that provide passenger ride hailing services. TNCs are exempt from complying with emissions and electric VMT targets if they have fewer than 5 million VMTs. The fleet emissions calculations will not include trips made in wheelchair-accessible vehicles.

10.2.1.5. Low Carbon Fuel Standard (LCFS)

The Low Carbon Fuel Standard (LCFS) is a regulation that uses a market-based mechanism to cap the carbon intensity (CI) of fuels in an effort to reduce greenhouse gas (GHG) emissions and air pollution from transportation.

- **Who Must Comply:** The producers in California or the importers of fossil fuels into California are usually the regulated parties when it comes to fossil fuels used in transportation. When it comes to fossil natural gas (FNG), the regulated party is the company that owns the fueling equipment or the fuel facility before the fuel is dispensed. Producers or importers of biofuels are eligible to apply for LCFS credits. Regarding electricity, the electric distribution utilities (EDUs) are subject to specific regulations regarding the use of LCFS revenue and have the option to claim LCFS credits for both residential and non-residential EV charging. The EV service provider (EVSP) or site host may choose to submit credit claims in lieu of EDUs for public charging. Additionally, owners of businesses and fleet operators have the option to apply for credits related to private access charging for workplaces and fleets, respectively.

10.2.2. Federal Legislation and Regulations

10.2.2.1. The Inflation Reduction Act (IRA)

The IRA's enactment in 2022, which offered incentives for the development of domestic battery production, charging infrastructure, and passenger and commercial vehicle use, marked a turning point in the adoption of electric vehicles. In the report *How Policy Actions Can Spur EV Adoption in the United States*, outlines the obstacles to IRA implementation as well as solutions and gives them the analysis they need to make data-driven decisions.

- **Who Must Comply:** Since the IRA is an incentive rather than a rule, nobody is required to follow it. The purchase of both new and used EVs is encouraged by the IRA. Tax credits are available to qualified individuals and business entities when buying an eligible EV through the IRA.

10.2.2.2. The Proposed Standard

The Environmental Protection Agency (EPA) is proposing a regulation that would lower CO_2, hydrocarbon, nitrogen oxide, and particulate matter emissions from passenger cars, light trucks, and MD vehicles. The proposed standards would reduce vehicle emissions by 8,000 billion metric tons of CO_2 between 2027 and 2055. Compared to 2055 levels without the proposal, the reduction in 2055 would include roughly 9,800 tons of particulate matter, 44,000 tons of nitrogen oxides, and 200,000 tons of volatile organic compounds.

Model year (MY) 2032 will see an industry-wide average target for light-duty fleet CO_2 emissions of 82 grams/mile under the proposed light-duty vehicle standards. This represents a 51% decrease in projected fleet average greenhouse gas emissions target levels from the current MY 2026 standards. The MD vehicle standards are expected to reduce CO_2 emissions by 44% from the current MY 2026 standards, with an average target of 275 grams/mile by MY 2032 when fully implemented.

- **Who Must Comply:** Light-duty and medium-duty vehicle manufacturers.

10.2.2.3. Greenhouse Gas Emissions Standards for Heavy-Duty Vehicles–Phase3 (Proposed)

With the implementation of these stricter standards, HD vehicles' greenhouse gas emissions are expected to drop starting in MY 2027. According to the EPA's website,

"The new standards would be applicable to HD vocational vehicles (such as delivery trucks, refuse haulers, public utility trucks, transit, shuttle, school buses, etc.) and tractors (such as day cabs and sleeper cabs on tractor-trailer trucks). Specifically, EPA is proposing stronger CO_2 standards for MY2027 HD vehicles that go beyond the current standards that apply under the HD Phase2 Greenhouse Gas program. The EPA is also proposing an additional set of CO_2 standards for HD vehicles that would begin to apply in MY2028, with progressively lower standard search model year through 2032. This proposed 'Phase3' greenhouse gas program maintains the flexible structure created in EPA's Phase2 greenhouse gas program, which is designed to reflect the diverse nature of the heavy-duty industry."

- **Who Must Comply:** Producers of heavy-duty vehicles (including school buses, shuttles, delivery trucks, garbage haulers, public utility trucks, etc.) and tractors (such as tractor-trailer trucks' day cabs and sleeper cabs).

10.3. ELECTRIC VEHICLE SAFETY STANDARDS AND REGULATIONS

The awareness that climate change poses a serious risk to our ability to maintain a sustainable lifestyle has led to a boom in the availability of fossil fuel-free transportation options. Electric-gas hybrid and fully EVs have become more and more common since the early 2000s. Electric vehicle sales are on the rise thanks to developments in machine learning and battery technologies, and most conventional automakers now offer electric vehicle lines. The future of transportation will include EV vehicles due to sustainability goals, increased emissions regulations, declining costs for EV technology, rising battery energy densities, and a wider availability of charging infrastructure.

10.3.1. The Need for Electric Vehicle Charging and Safety Standards

Standards must be set to guarantee the dependability of EV technologies since both established players and recent entrants are utilizing diverse technological approaches to promote electric vehicles. Failures in the car's electronics and the infrastructure supporting charging could be fatal for the occupants as well as other people involved, including rescue personnel. To ensure peace of mind and prevent accidents, it is necessary to guarantee the safe operation and dependability of batteries, controls, plug connectors, switches, and wires. Regulations that create standards for different EV component technologies and give suppliers access to a certification procedure will boost supplier compliance, consumer trust, and safety. Establishing international standards and certifications has the following main advantages such as:

- Safety of personnel, product, and charging infrastructure;
- Interoperability for utilizing common infrastructure ;
- Cost reduction for mass production;
- Increased adoption of new technologies underpinning the EV revolution.

The International Electrotechnical Commission (IEC) and the International Organization for Standardization (ISO) publish many global standards that are subsequently adopted in national and supranational contexts.

10.3.2. Overview of the Main International Electric Vehicle Safety and Charging Standards

There isn't yet a single worldwide EV standard. Several of the largest EV manufacturing hubs, such as China, Europe, North America, and Japan, are pushing opposing viewpoints in a number of domains. Regulatory certifications are a necessary step to enter the electric vehicle market, even though they typically come after technological advancements. Regulatory standards influence the advancement of technology by providing fundamental guidelines for environmental compliance and safety. The majority of regulatory and standard-setting efforts for EV vehicular technologies are focused on four key areas (Figure 10.1):

- Safety and security;
- Charging connectors;
- Charging topology;
- EV charging-related communications.

Figure 10.1. *Overview of key EV standards.*

Source: By Tomás Freres – Own work, CC BY-SA 4.0, https://commons.wikimedia.org/w/index.php?curid=99723478.

10.3.2.1. Safety and Security Standards

EVs must undergo extensive safety testing. EVs must adhere to the same safety regulations as conventional cars. The safety standard includes a great deal of specific information about installation, privacy, information management, occupant injury prevention, and insulation against electric shock. The international standard ISO 6469 covers a lot of the safety aspects of electric vehicles (Table 10.1). Three components make up this standard:

- On-board electrical energy storage, i.e., the battery;
- Functional safety means protection against failures;
- Protection of persons against electrical hazards.

Table 10.1. *IEC 61851 Standard*

	Charging Method	No. Phases	Nominal Voltage (V)	Max. Current (A)
AC	Mod 1	1	≤250	≤16
		3	≤480	
	Mod 2	1	≤250	≤32
		3	≤480	
	Mod 3	1	≤250	≤250
		3	≤480	
DC	Mod 4	-	≤1000	≤400

Source: Aksoz A, Asal B, Biçer E, Oyucu S, Gençtürk M, Golestan S. Advancing Electric Vehicle Infrastructure: A Review and Exploration of Battery-Assisted DC Fast Charging Stations. Energies. 2024; 17(13):3117. https://doi.org/10.3390/en17133117.

10.3.2.2. Electric Vehicle Charging Connectors

Different models and regions have different standards for EV charging connectors or plugs. The Combined Charging System (CCS) in North America and Europe is supported by a critical mass of automakers worldwide, despite the lack of agreement on a universal plug technology. Chinese automakers use GB/T, while Japanese automakers use CHArge de MOve (CHAdeMO), the world's largest market for electric vehicles. The goal of all the standards is to establish a common architecture for conductive charging systems for electric vehicles, along with functional and dimensional specifications for the vehicle's inlet and mating connector as well as operational requirements.

In North America, the standard for electrical connectors for electric vehicles is SAE J1772 (IEC 62196 Type 1), popularly referred to as a J plug. The general physical, electrical, communication protocol, and performance requirements for the electric vehicle conductive charge system and coupler are covered by this document, which is maintained by SAE International and is formally titled "SAE Surface Vehicle Recommended Practice J1772, SAE Electric Vehicle Conductive Charge Coupler (Figure 10.2)."

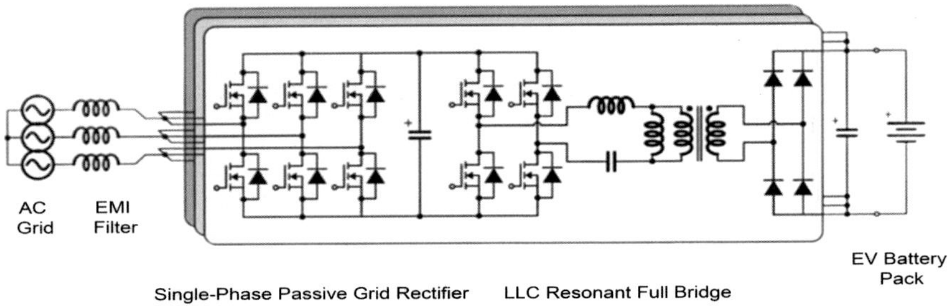

Figure 10.2. *Circuit topology providing bidirectional power transfer.*

Source: Aksoz A, Asal B, Biçer E, Oyucu S, Gençtürk M, Golestan S. Advancing Electric Vehicle Infrastructure: A Review and Exploration of Battery-Assisted DC Fast Charging Stations. Energies. *2024; 17(13):3117. https://doi.org/10.3390/en17133117.*

10.3.2.3. Charging Station and EV Communication

Nowadays, very few cars allow for V2G (vehicle-to-grid) connectivity, and even fewer charging stations—both public and at home—are smart grid-enabled. But as EV adoption rises, there will probably be a greater demand for uniform standards for the infrastructure supporting charging as well as interoperability among distribution networks, EVs, and charging stations. In order to prevent vendor lock-in in the charging infrastructure and to enable affordable connectivity of EVs with a variety of charging infrastructure and metering, interoperability is essential.

A V2G communication interface for bi-directional charging and discharging of electric vehicles is defined by ISO15118, an international standard for bi-directional digital communications between electric vehicles and the charging station. The "plug and charge" feature of electric vehicle charging is made possible in large part by ISO15118, which permits drivers to simply plug in the charger, wait for it to charge, and then take off. The vehicle's digital certificate, which enables communication between it and the charging point management system (CPMS), facilitates this process. This eliminates the need for an RFID card, an app, or PIN memorization and allows for a smooth end-to-end charging process with automated authentication and billing. The following is a list of common EV battery standards (Table 10.2).

Table 10.2. *Advantages and Disadvantages of Cell Types*

	Pouch Cell	Cylindrical Cell	Prismatic Cell
Advantages	The outer structure of the cell has minimal impact since the packaging material has a low mass, allowing for high energy density in the battery.	The process exhibits high maturity levels, featuring a well-developed shell structure, and it incurs low manufacturing costs.	The cell offers high protection, facilitating efficient internal heat conduction through reduced single-cell thickness, leading to significantly improved safety performance. Additionally, individual cells boast high capacity, resulting in the need for fewer cells in the battery pack.
Disadvantages	Large-capacity batteries pose challenges in the sealing process, being less reliable and more complex. The mechanical strength of the aluminum-plastic composite encapsulation film used is relatively low, limiting the battery's service life based on the film's lifespan.	The collector may experience uneven current density distribution, leading to inconsistent responses from internal components. This can hinder the quick release of heat generated within the battery cell, potentially creating a safety hazard due to heat accumulation.	The internal structure is complex, requiring high production process standards. Initially, the cells had low energy density, but there have been significant advancements in recent years, leading to major breakthroughs in this area.

Source: Aksoz A, Asal B, Biçer E, Oyucu S, Gençtürk M, Golestan S. Advancing Electric Vehicle Infrastructure: A Review and Exploration of Battery-Assisted DC Fast Charging Stations. Energies. 2024; 17(13):3117. https://doi.org/10.3390/en17133117.

10.3.2.4. Electric Vehicle Charging Standards

For conductive charging systems for electric vehicles, there is a standard called IEC 61851. Four charging modes are described in the standard. Mode 4 bypasses the on-board charger and sends DC current straight to the battery, whereas the other three modes supply AC current to the EV on-board charger. To ensure public safety, Mode 3 uses several control and protection features (Figure 10.3).

Table 10.3. *Analysis on Several Aspects of Fuel Cell Electric Vehicles*

Fuel Cell Electric Vehicles (FCEV)	
Powertrain	Uses a fuel cell or hydrogen fuel to generate electricity for the motor.
Performance	Offers excellent performance, with good acceleration and top speed, but is limited by the availability of hydrogen fuel.
Cost	Higher cost than BEVs due to the more complex fuel cell technology.
Range	Longer range than BEVs due to the fuel cell technology, but limited by the availability of hydrogen fuel.
Environmental Impact	Low environmental impact due to the reduced emissions from the fuel cell and hydrogen fuel.
Important Issue	Fuel cell price. ✓Reliability and lifecycle. ✓ Facilities for hydrogen refueling and conditioning. ✓

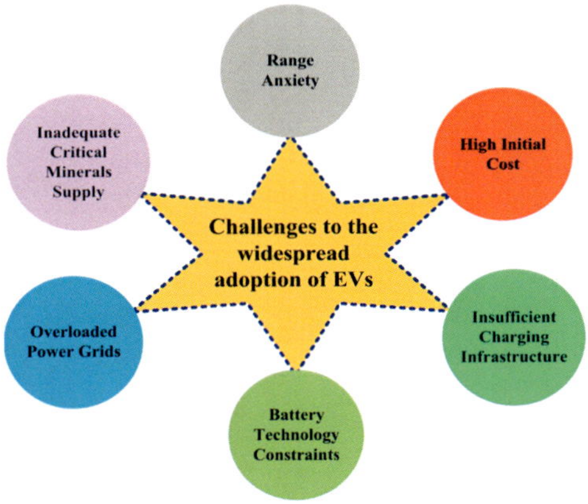

Figure 10.3. *EV challenges.*

Source: Kumar M, Panda KP, Naayagi RT, Thakur R, Panda G. Comprehensive Review of Electric Vehicle Technology and Its Impacts: Detailed Investigation of Charging Infrastructure, Power Management, and Control Techniques. Applied Sciences. *2023; 13(15):8919. https://doi.org/10.3390/app13158919.*

10.3.3. Future of Electric Vehicle Standards and Regulations

The globalization of electric vehicles will occur as environmental sustainability and climate change become more pressing issues. Significant advancements in EV technology are being driven by many suppliers, cost reduction, and technological advancements. International standards for electric vehicles have the potential to accelerate the uptake of EV technology and improve EV safety compliance. Similar to conventional cars, the main goals of the numerous international standards in use are dependability and safety. The primary topics covered by today's standards are connectors, charging topology, vehicle and charging infrastructure safety and security, and EV communications. Although disparate standards are more common, there is enough demand in different regions to propel the global adoption of EV standards in unison (Karmaker et al., 2021).

REFERENCES

1. Amir, R., Gama, A., & Werner, K. (2018). On environmental regulation of oligopoly markets: Emission versus performance standards. *Environmental and Resource Economics, 70*(1), 147–167.
2. Azarafshar, R., & Vermeulen, W. N. (2020). Electric vehicle incentive policies in Canadian provinces. *Energy Economics, 91*, 104902.
3. Cato, S. (2010). Emission taxes and optimal refunding schemes with endogenous market structure. *Environmental and Resource Economics, 46*(3), 275–280.
4. Conrad, K., & Wang, J. (1993). The effect of emission taxes and abatement subsidies on market structure. *International Journal of Industrial Organization, 11*(4), 499–518.
5. Fageda, X., Flores-Fillol, R., & Theilen, B. (2022). Price versus quantity measures to deal with pollution and congestion in urban areas: A political economy approach. *Journal of Environmental Economics and Management, 115*, 102719.
6. Gersbach, H., & Requate, T. (2004). Emission taxes and optimal refunding schemes. *Journal of Public Economics, 88*(3, 4), 713–725.
7. Karmaker, S. C., Hosan, S., Chapman, A. J., & Saha, B. B. (2021). The role of environmental taxes on technological innovation. *Energy, 232*, 121052.

INDEX

D

E

F

O

P

U

V

W

Z